AUTOMATED SYSTEMS BASED ON HUMAN SKILL

(Joint Design of Technology and Organisation)

A Proceedings volume from the 5th IFAC Symposium,
Berlin, Germany, 26 - 28 September 1995

Edited by

D. BRANDT
University of Technology RWTH, Aachen, Germany

and

T. MARTIN
Forschungszentrum Karlsruhe, Karlsruhe, Germany

Published for the

INTERNATIONAL FEDERATION OF AUTOMATIC CONTROL

by

PERGAMON
An Imprint of Elsevier Science

UK   Elsevier Science Ltd, The Boulevard, Langford Lane, Kidlington, Oxford, OX5 1GB, UK

USA   Elsevier Science Inc., 660 White Plains Road, Tarrytown, New York 10591-5153, USA

JAPAN   Elsevier Science Japan, Tsunashima Building Annex, 3-20-12 Yushima, Bunkyo-ku, Tokyo 113, Japan

First edition 1996

Library of Congress Cataloging in Publication Data

A catalogue record for this book is available from the Library of Congress

British Library Cataloguing in Publication Data

A catalogue record for this book is available from the British Library

ISBN 0-08-042379 5

This volume was reproduced by means of the photo-offset process using the manuscripts supplied by the authors of the different papers. The manuscripts have been typed using different typewriters and typefaces. The lay-out, figures and tables of some papers did not agree completely with the standard requirements: consequently the reproduction does not display complete uniformity. To ensure rapid publication this discrepancy could not be changed: nor could the English be checked completely. Therefore, the readers are asked to excuse any deficiencies of this publication which may be due to the above mentioned reasons.

The Editors

Printed in Great Britain

Transferred to digital printing 2008

5th IFAC SYMPOSIUM ON AUTOMATED SYSTEMS BASED ON HUMAN SKILL

Sponsored by
International Federation of Automatic Control (IFAC)
Technical Committee on Social Impact of Automation

Co-sponsored by
IFAC Technical Committees on
- Advanced Manufacturing Technology
- Man-Machine Systems
- Cultural Aspects of Automation
- Chemical Process Control
- Power Plants and Power Systems
- Air Traffic Control Automation
- Transportation Systems
ESPRIT-Project HERMES
MONITOR/FAST-Program of the Commission of the European Union

Organized by
VDI/VDE-Gesellschaft Mess- und Automatisierungstechnik (GMA), Dusseldorf
in cooperation with
FhG-IPK, University of Technology, Berlin, Germany
HDZ/IMA, University of Technology, Aachen, Germany

International Programme Committee (IPC)
D. Brandt (D)   (Chairman)
H. Bolk (NL)
P. Brödner (D)
F. Butera (I)
C. Cernetic (SLO)
J.P. Durand (F)
F. Emspak (USA)
T. Hancke (D)
K. Henning (D)
C. Imamichi (J)
K. Kawai (J)
P. Kopacek (A)
V. Kucera (CZ)

L. Martensson (S)
T. Martin (D)
A. Niemi (SF)
J.P. Perrin (F)
G. Rzevski (UK)
Th. Schael (I)
T. Sheridan (USA)
H. Stassen (NL)
A. Straszak (PL)
R. v.d. Vorst (UK)
E. Welfonder (D)
W. Wobbe (B)

National Organizing Committee (NOC)
D. Brandt   (Chairman)
H. Helbing
N. Lange
B. Nickolay
H. Rosenzweig
J. Springer
H. Wiefels

PREFACE

As the previous four symposia of the same name, this symposium brings together researchers, developers and users of complex Human-Machine systems. The areas of discussions are manufacturing, process control, aircraft and air traffic control, and administrative processes. Emphasis will be on how to design such systems integrating developers and users into the design process. It means joint engineering of production processes, information technology and work organization. It may lead to re-defining the roles of human operators in process automation. These are the main issues addressed.

I would like to thank all those who have contributed to making this symposium a successful event: our friends and colleagues of the VDI/VDE-Society of Automatic Control (VDI/VDE-GMA) the International Federation of Automatic Control (IFAC), and its Technical Committee on Social Impact of Automation, the Department of Informatics in Mechanical Engineering (HDZ/IMA), University of Technology, Aachen, the Institute for Production Systems and Design Technology (IPK), Berlin, the Institute of Psychology, University of Technology, Berlin, and all contributors.

Dietrich Brandt

CONTENTS

PLENARY PAPERS

PROCESS CONTROL: HUMAN PROCESS COMMUNICATION

SHOPFLOOR CONTROL AND MANAGEMENT SYSTEMS

ADMINISTRATIVE PROCESSES I

PRODUCTION PLANNING SUPPORT IN PROCESS INDUSTRIES

IMPROVING QUALITY OF WORKING LIFE

ADMINISTRATIVE PROCESSES II

MACHINES LEARNING TO SEE AND TO ACT

ENVIRONMENTAL ENGINEERING

HUMAN-MACHINE INTERFACE

SOCIOTECHNICAL DESIGN OF PRODUCTION

AIRCRAFT AND AIR-TRAFFIC CONTROL

TECHNOLOGY - ART - CRAFTSMANSHIP - AND PEOPLE

PARTICIPATIVE ASPECTS OF COMMUNITY R & D PROJECTS

Ronald Mackay

*European Commision, DG III, Industry
Integration in Manufacturing (IiM)
Brussels, Belgium*

Industrial organisations are currently undergoing far-reaching changes in their ways of doing business. Lean production, virtual enterprises and concurrent engineering are some of the concepts which are reshaping the organisation. The quality of working life will be determined by the way in which these concepts are implemented and also by the design of the supporting information technology infrastructure.

Participative management requires the consultation with employees and their representative organisations on delegation of responsibilities, decision making and working conditions. Throughout the European Union there are many different frameworks for participation and different legal situations which affect the joint design of work and technology.

Within the European Information Technology research programme ESPRIT a number of approaches to participation have been employed. There has been work on human-centred systems, aimed at concentration on job-design rather than IT-system design and emphasising the scope for decision-making as opposed to execution of residual tasks which could not be automated. Other projects closely involve the end-users, whose working practices will change as a result of the developments, in developing scenarios at the outset of IT development projects to support concurrent engineering. In addition, work is under way on methods and tools to analyse the impact of the introduction of new information technologies in the workplace. Consultancy tools and training packages are being developed which will be used to analyse the perceived effects of changes in the work environment.

1

WORK PROCESSES, ORGANIZATIONAL STRUCTURES AND COOPERATION SUPPORTS: MANAGING COMPLEXITY

Giorgio De Michelis

University of Milano and RSO, Milano, Italy

Abstract: In this paper the service paradigm is used to analyze and characterize work processes and their complexity. A work process, from this point of view, is characterized by the communicative relations binding its participants and embedding their performances. The complexity of work processes is then related to organizational design issues, to the empowerment of professional skills and to the computer support systems capable of helping people to manage the complexity of their work effectively. The approach to change management that underlies this paper renovates the Socio-technical System Design Methods, allowing them to fully exploit the potential of existing Information and Communication Technology.

Keywords: Business process engineering, Co-operation, Group work, Organizational factors, Socio-technical system design.

1. INTRODUCTION

For many years service organizations (in both the private and public sectors) have tried to model their behaviour in accordance with the production process model, seeking improved efficiency, performance standardization, hierarchical control structures, etc.

Today the situation is radically different: some peculiar features of service processes have emerged, distinguishing them from production processes (Bowen and Schneider, 1988; Heskett et al., 1991; Fountain, 1993), meanwhile, products tend to embody value-added services (Takeuchi and Nonaka, 1986) and/or to transform themselves into services (Van Gorder, 1990). Also within organizations service relations are becoming the usual way to manage inter-departmental and/or inter-functional relations (Keen, 1991).

The emergence of service relations is due to the increasing request for personalized performances on the part of the customers; moreover, it reflects and/or sustains the growing complexity of organizational behaviours (Keen, 1991). Service relations can therefore be considered paradigmatic of today organizations and of their evolution. The analysis of work processes in the context of service relations can offer a deep insight into the main features of some widely diffused organizational phenomena, along with some consistent and innovative guidelines for change management. Moreover, it can constitute a common, rich and innovative framework to all those studies aiming to define new grounds for the different disciplines studying the organization: from economy to social sciences, from management sciences to organizational theories, from psychology to computer supported cooperative work and information systems (De Michelis, 1995).

In this paper the service paradigm is used to analyze and characterize work processes and their complexity. A work process, from this point of view, is characterized by the communicative relations binding its participants and embedding their performances.
The complexity of a work process is then related both to its value (the external complexity) and to its cost (the internal complexity). It is shown that in order to face great external complexity with small internal complexity it is necessary to enhance the knowledge creation process.

Knowledge creation processes are then related to organizational design issues, to the empowerment of professional skills and to the computer support

3

systems capable of helping people to manage the complexity of their work effectively.

2. WORK PROCESSES, COMMUNITIES OF PRACTICES AND SOCIAL COMPLEXITY

A service is generally characterized in the literature as being something that is not "a good" (Fountain, 1993). Between pure services and pure goods lie all those processes where goods are enriched by something that is not a good and where non-goods embed some goods: a clear dichotomy between them is therefore an oversimplification, and service can be considered as a viewpoint for observing any work process.

Despite the variety of categories into which they can be classified, all services have three fundamental characteristics: intangibility, customer co-production, and production/consumption simultaneity (Bowen and Schneider, 1988). Service intangibility conveys the idea that a service relation results in the experience having taken place rather than the object having been transferred. The value of a service is highly subjective, since it depends on its customers. Service co-production conveys the idea that the customer is not passive in the service relation. The unique temporal proximity of production, delivery and consumption in a service relation implies that quality control must take place in real time and that front office agents are crucial to assure the quality of the delivered services.

The three above characteristics can be considered as the main aspects of the service paradigm and they can be used to observe the complexity of work processes. as social processes.

From the service point of view work appears as a process relating its performer (the singular form is an abstraction that can be accepted at this point: generally a work process has more than one single performer; this point will be discussed in the following pages) to the customer (idem) who recognizes its value. While performing a work process the performer consumes resources and generates a value for the customer. The customer is the legitimate beneficiary of the performance of the work process (no distinction between production, delivery and consumption is meaningful from this point of view; see service characterization above) and the value she attributes to it has a social recognition.

The value of a work process is therefore intangible and subjective (see service characterization above). The social recognition of the value of a work process can. in fact, be given by means of the price the customer pays for it, but it can also be given by the resources she consumes (within a work process the performer and the customer perform together, they co-produce: see service characterization above), or finally by her declaration of satisfaction. All the three above behavioural patterns exhibit in some sense the appreciation of the customer for the performed work: it is up to the performer to determine if the socially

recognized value of her work is sufficient with respect to her cost (with respect to the resources she has consumed).

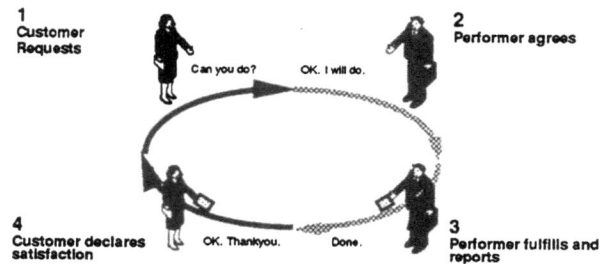

Fig. 1. The Action Workflow Model - ATI™.

No matter the means through which the value of a work process is recognized, it is clear that the value recognition is carried on through the communicative interactions between the performer and the customer occurring during the work process. Through communicative interactions the performer and the customer reach, in fact, an agreement on the actions to be performed and share the evaluation of their execution. The actions performed within a work process are embedded into the communicative interactions between its customer and its performer.

To work is therefore not simply to act with a purpose: what a person does can be considered work if she performs something in order to satisfy a request made by another person[1]. Work processes - see Fig. 1 - embed actions within a relationship, whose essential nature is communicative and pragmatical[2].

The relationship between a performer and her customer is basic within a work process, but customer - performer relationships are not the only ones relevant within it. A work process involving only one customer and one performer is in fact a simple limit case (many observers call it individual work - e.g. (Schmidt and Bannon, 1992) - as there is only one performer). Generally work processes are more complex: they involve several customers and several performers, belonging to different organizations (Fig. 2), and they can be decomposed into sub-processes within which some actions are performed necessary to the completion of the main work process (Fig. 3).

The relationships between the actors of a work process can be characterized from the point of view of the positions they occupy within it. Beside the basic relationship between the customers and the performers of a work process, the relationships between its customers as well as the relationships between its performers have also to be taken into account, and the

---

[1] The above definition of work is due to a personal rethinking (De Michelis, 1994, 1995) of the language/action perspective as it has been proposed by Fernando Flores, Terry Winograd and co-workers (Flores, 1982; Winograd and Flores, 1986; Medina Mora et al.. 1992).

[2] The customer - performer relationship has been inspired by the Action-Workflow™ (Medina Mora et al., 1992).

4

relationships between actors of different sub-processes of the same work process, since any positional relationship type within a work process is

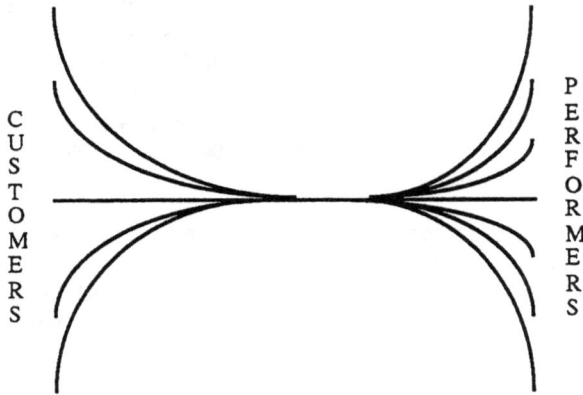

Fig. 2. A work process as a relation involving several customers and several performers.

characterized by a specific form of cooperation between its participants. Let us present them briefly here below[3].

Collaboration is the cooperation form occurring between the customers and the performers of a work process. It is the basic form of cooperation within work processes, since value is created through it. Collaborating persons interac doing something together. Successful collaboration creates a common language and a common understanding between the customers and the performers through their mutual listening.

Co-decision with equal roles is the cooperation form occurring between the customers of a work process. It logically precedes the action: if it is carried out successfully, it fixes the common expectations with respect to it. Co-decision with equal roles is a decision process where all the participants share full responsibility for the decision to be taken.

Coordination is the cooperation form occurring between persons being actors within different sub-processes of a work process. It allows the integration of the different sub-tasks that are parts of a more general common task. Successful coordination depends on the degree of awareness of their organizational context of its participants.

Co-decision with distinct roles is the cooperation form occurring between the performers of a work process. Through successful co-decision with distinct roles the participants in a work process recognize and maintain the mutual interfaces between their respective performances. It is a decision process, where each one of the participants takes a decision on the issue on which she has responsibility within the constraints defined by the decisions of the other participants.

The participants in a work process share an experience constituting them into a whole, into a community. This community has all the features that characterize the communities of practices, studied by work

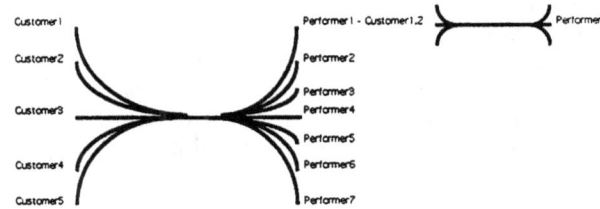

Fig. 3. A work process with one of its sub-processes.

anthropologists[4], since its members share a space (physical and/or virtual; this point will be deepened later), some artifacts (tools and resources), a language, the knowledge of the world they live in and of the possibilities it opens, the history of the work process to which they participate and the value they create within it.

As a social phenomenon a work process is complex: the community of practices involved in it is a network of social relationships that cannot be reduced to any functional model. Despite any attempt to plan its evolution with respect to its expected outcome, its history is unpredictable. The members of a community of practices during a work process in fact change their understandings, their requests, their ways of performing, their mutual agreements in a history of successes and failures, in a common experience of learning and knowledge creation, in the building of their community bonds.

Managing a work process in a normative way constraining its participants to some predefined plans is, therefore, ineffective: as the service literature points out, effective management cannot be disjoint from change management (Heskett *et al.*, 1990). Within a work process the unexpected events, the breakdowns, play a central role: on the one hand, increasing costs, menacing the possibility of satisfying customers; on the other, activating the learning process, updating ineffective plans, exploiting new opportunities.

Within a community of practices various breakdowns may occur, either generated by the changing conditions of satisfaction of the customers, or by the difficulties the performers meet in order to satisfy them. In both cases, breakdowns can generate new breakdowns, either when their solution goes beyond the capabilities of the members of the community of practices, or when they fail to cooperate in accordance with the positional relationships they occupy in the work process. The breakdowns of a work process depend on its complexity and reflect it: if a work process is simple it does not create surprises.

---

[3]For a detailed discussion of cooperation forms see: (De Michelis, 1994).

[4]See the contribution of Susan Stucky to: (Bagnara *et al.*, 1995).

Complexity is a rather fuzzy concept, characterizing in some cases the difficulties an observer encounters facing a phenomenon, in other cases, a quality of the phenomenon itself. The second case implies the first one, while the converse is not true. In this paper complexity is therefore characterized as the quality making a phenomenon not reducible to any explicative model. The complexity of a work process can be considered, from this point of view, as the combination of two factors: its multiplicity and its autonomy.

The multiplicity of a work process characterizes its complication: it depends on the number of its actors (customers, performers) and of the relationships binding them, on its duration, etc. The autonomy of a work process characterizes its unpredictability: it depends on the quality of the professional skills of its actors, on the difficulty of their tasks, on the degree of parallelism of their perfomances, on the frequency of breakdowns occurring to them, etc.

Pure multiplicity can be controlled through well structured plans; pure autonomy can be controlled through close observation of everyday behaviour (single persons autonomy can be approximatively anticipated); but the combination of the two is not controllable, since autonomy imposes frequent plan changes, and multiplicity hinders close observation (any person sets her own pace, has her aims, her image of the world, the images of any relationship are multiplied by its observers, by the different timings of its observation, etc.).

The community of practices performing a complex work process is therefore constituted by several actors with high competence, facing non routinary problems and interacting in a rather creative way. The complexity of a work process affects both its value (with low complexity only small values can be created) and its cost (high complexity consumes time and other types of resources). Managing a complex work process requires containing its cost without affecting its value. The cost of a complex work process is not predictable through the traditional accounting methods of traditional management. It is constituted more by transaction costs (Williamson and Winter, 1993), i.e. by the cost of the interactions between the actors, than by production costs (the cost of the actions performed by the actors).

If external complexity, generated by the customers and their requests (qualifying the value of the work process), is distinguished from internal complexity, generated by the performers and the requests they do to other persons in order to perform their tasks (determining the amount of time and other resources necessary to perform it), the quality of a work process can be characterized by its capacity to face great external complexity with small internal complexity.

It seems a paradoxical objective, since it is a widely diffused opinion that there is a direct correlation between external and internal complexity[5]. It is well known, on the contrary, that different communities of practices exhibit different effectiveness facing a problem: where one community exhibits all the needed competences ready at hand, a smooth way of interacting, a prompt mutual understanding and performances fitting the requests, another one is continuously fighting against misunderstandings, missing persons, lack of competences, wrongly focused performances.

Despite the direct correlation between the internal and external complexity of a work process, different communities of practices are characterized by a different correlation parameter. The quality of a work process depends on it.

What is the factor determining a low value of this correlation parameter? The factor allowing to face a growing external complexity without letting the internal complexity grow consequently? Taking into account the dynamic nature and the impredictability of a complex work process, this factor has to do with the transparency and/or visibility of the community of practices performing it, with the richness and flexibility of its communication network, with the quality of the competences of its members and, finally, with its plasticity with respect to the external complexity of the work process.

The analysis of work processes proposed so far indicates a good candidate for playing this role: it is the knowledge created and shared by the community of practices performing a work process through its individual and collective learning practice (about the work process history and expected outcomes, about the competences needed to perform it, about the community of practices performing it, its language, the space it occupies, the artifacts available in it, its members, their competences, their mutual relations, etc.).

Knowledge creation processes within organizations have been characterized by I. Nonaka (1988, 1991, 1993) in terms of tacit and explicit knowledge transformation: he considers the knowledge transformation processes the means through which, in its evolution, the individual knowledge of any member becomes knowledge shared by the whole community and, conversely, the knowledge of the whole community projects itself into the individual knowledge of the members. Knowledge transformation processes are Organizational Learning processes performed by autonomous and responsible persons: they can be associated to the above presented cooperation forms, since any type of knowledge transformation is performed through a specific type of interaction, i.e. through a specific form of cooperation, between the members of a community of practices (De Michelis, 1995b).

---

[5]Cybernetics has stated the requisite variety law on this issue. See for example: (Ashby, 1956).

Good cooperation is the principal means through which communities of practices create knowledge, through which they learn. Managing complex work processes is a radically different matter with respect to traditional accounting management: its problem is to enhance and direct the learning capability as well as the learning practice of the community of practices performing a work process towards customer satisfaction[6].

Knowledge creation processes are strengthened by any change reducing the obstacles to effective cooperation (e.g.: simplifying the process plan, making control easy, designing process-oriented organizational structures) and enhancing the learning capability of the community of practices (e. g.: creating responsible roles, designing adequate work spaces, providing adequate tools).

The simplification of the work plan as well as the design of easy control systems are actions that are consistent with the Business Process Reengineering (BPR) approach (Hammer, 1990; Hammer and Champy, 1993; Davenport, 1993), since they cancel all the non necessary intricacies making the flow of actions difficult and therefore creating a confused background to the learning process.

Professional empowerment (see Section 3) as well as organizational design (see Section 4) issues integrate the BPR approach with the Organizational Learning one proposing a revisited Socio-technical System Design approach (Butera, 1994), taking into account not only the need for improved process performances but also the need for preserving the value generated by previous experiences.

The failures of many BPR projects (for example, Gartner Group claims that 50 to 70% of BPR projects fail to meet their original objectives) have been attributed to their lack of paying attention to human and organizational problems (Davenport, 1994; Hall *et al.*, 1993): a revisited Socio-technical System Design approach allows to overcome those limits through a more comprehensive and effective change management.

Within this context, also the role of computer support systems for work processes has to be considered again, since they can play a relevant role in reducing the internal complexity of work processes (see below, in particular Section 5).

3. PROFESSIONAL SKILLS WITHIN COMMUNITIES OF PRACTICES

The members of a community of practices are persons sharing a meaningful common experience: their effectivenes within the work process they perform together depends on their capability of playing a

responsible role in it. A responsible role requires not only mastering the domain knowledge characterizing the tasks assigned to it, but also the capability of interacting with the customers and with the other members of the community of practices, creating value with them. The members of a community of practices, therefore, are not only specialized technicians but also organization professionals (Butera, 1994) capable of performing for customer satisfaction. The growing complexity of the work processes to which they participate requires, among other, that they are able to manage the breakdowns which may occur without creating further complexity.

Any person, as anyone has probably experienced, is capable of managing a certain degree of complexity, beyond which her perfomances lose effectiveness: she is overwhelmed by her duties, she is continuously reordering her agenda, modifying the commitments she has with other people, she fails to do what the other expect from her. Let us call this behavioural limit of any person her threshold of sustainable complexity (De Michelis, 1994).

A person that has surpassed her threshold of sustainable complexity is provoking a growth of the internal complexity of any process to which she participates, since she is continuously generating new breakdowns, involving the other actors of the work process with whom she cooperates without contributing to solve any of them. The threshold of sustainable complexity of any person is not fixed once for all, it evolves, on the basis of her experiences and it depends on various interconnected factors.

As first, it is affected by the eventual asymmetries between the tasks she has to accomplish and her competences: if a person has not the professional competence to perform her tasks, if she is not capable to act in a responsible manner, then her threshold of sustainable complexity decreases, since the difficulties she encounters to accomplish her duties are stressing her behaviour so that she loses effectiveness and is paralyzed by the eventual breakdowns.

As second, it depends on her awareness of the work process she is participating and of the community of practices performing it: if she does not know what has happened and what is going on, if she does not know the positional relationships binding her with the other members of the community, if she does not know their professional competences, if she does not where are the resources, the tools, the information, the persons she may needs, then her threshold of sustainable complexity decreases, since she has not ready at hand the context within which she is acting and inter-acting.

As third, it depends on the nature of the space she is shares with the community of practices whose she is member of and on its tool equipment. The locations where the members of a community of practices work (no matter if they are physically adjacent or distant) are integrated by the media they can use to communicate each other: the electronic media

---

[6]This observation rephrases Organizational Learning imperatives; see: (Argyris and Schoen, 1978; Senge, 1991).

interconnect the physical locations creating the unique virtual space of the community of practices. If the space where a community of practices works, offers a rich set of communication possibilities to its inhabitants, if it is capable to bring forth the context of any interaction they may perform each other, if it provides efficient and highly usable tools ready at hand when necessary, then it increases their threshold of sustainable complexity, since it reduces the effort they need for managing the complexity of the work processes they are performing. Computer support systems are today the main components of the space of a community of practices and of its tool equipment, but they are not alone: their design must exploit their individual functionalities as well as their integration in the physical space.

The three above factors are mutually inter-dependent, since they are mutually enhancing each other: the professional skill and the awareness of the work process to which a person is participating are in fact increased by the quality of the space where she is acting and of its tool equipment, while poor professional skill and poor awareness do not allow a full exploitation of the services offered by the space and its tools.

The threshold of sustainable complexity characterizes the individual factors affecting the learning capability of a community of practices: increasing it means empowering its members from the point of view of their capability to create organizational knowledge.

4. ORGANIZATIONAL STRUCTURES FOR PROCESS-ORIENTED ORGANIZATIONS

The work process point of view presented in Section 2 offers an original account on how organizations emerge within complex work processes.

The knowledge created within a work process can be analyzed in two different contexts. On the one hand, it allows to reduce the internal complexity the community of practices performing it will create in any new work process in which it engages itself; on the other, it allows to reduce the internal complexity its performers (the persons participating to it in the performer position) will create in any new work process of the same type in which they engage with other customers. While the creation of knowledge shared by the whole community of practices characterizes the growing partnership relations binding each other its members, the creation of knowledge shared by its performers (or by some of them) characterizes their increasing efficiency and effectiveness in performing a particular type of work process, i.e. their increasing capability to perform consuming less resources and creating greater value.

In order to reduce the resources (the cost) of their performances, the performers enforce their being performers, separating themselves from the customers and trasforming their experience into (practical) knowledge usable within other work processes (with other customers): they create an organization on the performer side.

Organizations emerge as means for trasforming the experience some persons do performing a work process into a structure enabling them to increase their efficiency and/or effectiveness. An organization gives a collective identity to its members: defining membership conditions; characterizing their roles (both in terms of responsibility and competence) and the products/services they deliver; providing them with work spaces, with material, technical and/or financial resources, with behavioural patterns (procedure templates, forms, ...), with professional information and, last but not least, with a language through which they can share all those things and give sense to what they do.

The necessity of restoring efficiency and effectiveness is frequently reappearing in organization today, since the turbulence of the environment continuously improves the external complexity of work processes. While working within work processes the organization continuously evolves in order to try to restore effectiveness. The process of working and the process of maintaining the organization are (mutually related) autonomous processes whose coupling can be granted only in dynamic terms: the latter in fact involves an interpretation of what is going on within the former as well as of the organization itself and implies the creation of a shared understanding. Since language allows the members of an organization to give sense to their behaviour within various organizational metaphors (Morgan, 1976), sharing is more a matter of practice than of semantics: the organization is performed (Bowers, 1993) by its members.

It cannot be taken for granted that an organization is performed in such a way that its effectiveness is restored: the history is and will be full of organizations which fail to do this: maintaining an organization, in fact, is not a deterministic process; its results can be different from its aims. The changes an organization introduces while its members perform it in fact mediate between its current structure and its new needs, between its current language and knowledge and the improved performances required by the environment: performing an organization has the aim of reducing its costs but it is exposed to the risk of increasing its internal complexity, and consequently to fail its objective.

The work process point of view sketched in the second Section suggests, in order to avoid the risk of increasing through organizational change the disalignment between the value and the cost of the performances of an organization, to increase its learning capabilities, to improve its knowledge creation processes.

The organizational units (offices, departments, etc.) are, from this point of view, the repositories of the knowledge created by an organization: its units of an organization give duration and stability to the knowledge creation process and, moreover, make that knowledge available for other members and other units of the same orgnaization. Where pure adhocracy has been implemented, creating and sharing knowledge becomes difficult and unsafe. Organizational units are not the teams performing a work process: while the latter ones have the objective to improve their performances to meet customer needs, the former have the objective to maintain and develop the knowledge that is created and used within work processes.

Within an organization performing complex work processes three different types of managerial roles can be distinguished: (1) the work process owner, who is responsible of the performances of a class of work processes: her focus is on the customer needs, she is, in some sense, an internal customer of the professional (and material) resources of the organization; (2) the responsible of a group of professional (and/or material) resources: she is responsible of maintaining and increasing their value, i.e. of exploiting their learning capabilities, of enhancing their knowledge creation processes; (3) the executive who designs the business of the organization, balancing the quality of its performances, the value it offers to its customers, with the exploitation of its resources, with its internal value.

Since the success of an organization depends on the cooperation between the three above sketched managerial roles, on their ability to maintain their own knowledge creation process, it is necessary that their mutual relationships are defined in terms of mutual services, and not in terms of pure hierarchy. Learning organizations are always more similar to network organizations (Butera 1993). Computer support systems are very important to let a network organization perform effectively, since, creating and equipping the virtual space where its managers cooperate (the electronic network is in some sense a model of the organizational one), it helps them to sustain its internal complexity.

5. COMPUTER SUPPORT SYSTEMS FOR PROCESS-ORIENTED ORGANIZATIONS

Information and Communication Technology (ICT) has been widely applied from the mid of the sixties as a mean for reducing the costs of organizations: as the Socio-technical school has pointed out from those early years, its introduction within an organization has to be considered as part of its process of organizational change. The Information Systems that were developed up to the late eighties were mainly oriented to increasing the efficiency of the organizations and not to reduce their internal complexity. It is a popular complaint today that Information Systems failed. The above observation contributes to explain why: efficiency does not suffice to meet customer needs and to improve performances.

In the late eighties a new family of organizational computer-based system and a new way to conceive the use of ICT within organizations appeared, orienting their application in a rather different direction with respect to traditional Information Systems: ICT is proposed as a means of supporting the cooperation of the persons involved within work processes and not as a means of substituting them. Computer Supported Cooperative Work (CSCW) is the multidisciplinary research area facing the problem of analyzing, designing and evaluating cooperative work processes as well as their support systems (AA.VV., 1986, 1988, 1990, 1991, 1992, 1993, 1994; Bowers and Benford, 1991).

CSCW systems (called also Groupware (Ellis et al., 1991) or Workgroup Computing Systems) are the systems aiming to support people cooperating within work processes and therefore our attention can now focus on them. Among the systems and prototypes developed within the CSCW field there are Workflow Management Systems, Knowledge Sharing Environments, Conversation Handlers, Group Decision Support Systems and various multimedia communication supports (videoteleconferencing systems, computer supported meeting rooms, ...). The variety of the proposed systems is very rich, but CSCW systems are far from having reached a steady state, where offered services, underlying architectures and human interfaces can be characterized in accordance with some standard categories. Moreover, the relationships between CSCW systems and the traditional Information Systems have still to be investigated and fully understood. It is very difficult therefore to define ways to choose and/or design the support system needed by an organization on the basis of the work processes it performs and the qualities it needs while performing them.

The conceptual framework developed so far offers some hints in this direction, that are briefly sketched below. From the work process point of view, computer support systems are, on the one hand, artifacts equipping the space where a community of practices works, on the other, means through which that space is enhanced to a plastic dynamic virtual space. Moreover, computer support systems can contribute to reduce the internal complexity of the work processes incresing the threshold of sustainable complexity of its actors and creating the electronic network giving shape to the organization performing it.

a) Workgroup computing systems are not tools to process information, but enabling artifacts. Information processing is the way through which computer based systems enable their users, not the service they offer to them. Through information processing, in fact, they can do many things that are relevant to help them to sustain the complexity of the work processes where they are acting: they can make available the knowledge created and/or shared within a work process; they can automatically maintain the multimedia information basis of the process; they can offer a knowledge basis modelling the organizational environment. In all these sample cases their utility depends on the transparency and/or visibility they offer to their users about the organizational structures, the work processes, the communities of practices in whose context they perform. It is therefore very important that they are ready at hand when necessary: designers should pay attention to the way they are physically designed and located.

b) Workgroup computing systems transform the physical space into a virtual space supporting multimedia communication. Their quality depends on the multiplicity of communication media they support: if they support both synchronous and asynchronous communication; if they link together sequences of interrelated communication events; if they link effectively communication to action, then the virtual space they create allows its inhabitants to cooperate effectively independently of the time and space conditions constraining them.

c) Workgroup computing systems absorb routinary tasks and dissolve pure hierarchical control supporting the workflow. If they create automatically distributed to-do lists, if they integrate any type of productivity tools, if they keep track of the work done, if they are able to support exception handling, then they create a work setting where the actors can exploit their capabilities and responsibilities.

d) Workgroup computing systems support the awareness of their users, as much as they make visible and/or transparent the knowledge created within work processes, as they reflect their changes. Thank to the success of Internet workgroup computing systems can create an open virtual space where changes of membership, of roles, of relationships are adequately reflected.

At the Cooperation Technologies Laboratory of the Department of Infomation Sciences of the University of Milano a prototype of a system supporting cooperation within work processes is currently under development, that is inspired by the above remarks (De Michelis and Grasso, 1994; Agostini *et al.*, 1995). Its aim is to show the feasibility of a system with the above characteristics.

6. CONCLUSION

This paper proposes a conceptual framework based on the analysis of work processes and their complexity to analyze and design organizational behaviour. It offers some first hints on how to characterize organizational behaviour from the point of view of the distinction between internal and external complexity and of knowledge creation processes (Nonaka, 1993). It sketches some observations about the professional skills, the organizational structures and the computer support systems well suited for process oriented organizations. But any issue raised by it requires further investigation and discussion.

Its main claim is that the analysis of work processes and of their complexity, on the one hand, can be the basis for updating the Socio-technical System Design

approach with respect to new popular change management methods as Business Process Reengineering, Organizational Learning, Professional Empowerment and Workgroup Computing Systems, avoiding the limits of each one of them, on the other, can offer a renewed unified ground for a paradigm shift of the various disciplines studying the organizations.

With respect to both these issues this paper is only posing the problem and indicating a research direction.

ACKNOWLEDGMENTS

This paper presents a research that has been conducted with the financial support of the EC within the ESPRIT- BRA Project 6225, COMIC and within the IMPACT project of the COST-14 Action. Alessandra Agostini and Maria Antonietta Grasso of the Cooperation Technologies Laboratory of the Department of Information Sciences of the University of Milano deserve a particular credit for their contribution to the development of the ideas proposed in the paper and for their careful reading of a preliminary release of it. The professional collaboration of the author with his colleagues at RSO, mainly Federico Butera, Thomas Schael and Buni Zeller, has also influenced him deeply.

REFERENCES

AA.VV. (1986). *Proceedings of the Computer Supported Cooperative Work Conference 1986.* MCC, Austin.

AA.VV. (1988). *Proceedings of the 2nd Computer Supported Cooperative Work Conference 1988.* ACM, New York.

AA.VV. (1990). *Proceedings of the 3rd Computer Supported Cooperative Work Conference 1990.* ACM, New York.

AA.VV. (1991). *Proceedings of ECSCW'91.* Kluver, Dordrecht.

AA.VV. (1992). *Proceedings of the 4th Computer Supported Cooperative Work Conference 1992.* ACM, New York.

AA.VV. (1993). *Proceedings of ECSCW'93.* Kluver, Dordrecht.

AA.VV. (1994). *Proceedings of the 5th Computer Supported Cooperative Work Conference 1994.* ACM, New York.

Agostini A., G. De Michelis, M. A. Grasso, W. Prinz and A. Syri (1995). Contexts, Work Processes, Work spaces, *Computer Supported Cooperative Work. An International Journal,* (to appear).

Argyris C. and D. A. Schoen (1978). *Organizational Learning.* Addison-Wesley, Reading.

Ashby R. (1956) *An Introduction to Cybernetics.* Chapman & Hall, London.

Bagnara S., S. Stucky and C. Zucchermaglio, Eds. (1995). *Organizational Learning and Technological Change.* Springer Verlag, Berlin.

Bowen D. E. and B. Schneider (1988). Service Marketing and Management. In: *Research in Organizational Behaviour*, Vol. 10, JAI Press.

Bowers J. (1993). Understanding Organization Performatively. In: *Issues of Supporting Organizational Context in CSCW Systems* (L. Bannon, and K. Schmidt, Eds.), COMIC Deliverable 1.1, pp.49-72 (Available via anonimous ftp from: ftp.comp.lancs.ac.uk).

Bowers J. and S. Benford, Eds. (1991), *Studies in Computer Supported Cooperative Work*. North Holland. Amsterdam.

Butera F. (1994). Network of Enterprises or Network Enterprise? In: *Proc. of CEFRIO Conference*, University of Quebec Montreal.

Davenport T. H. (1993). *Process Innovation: Reengineering Work Through Information Technology*. Harvard Business School Press, Boston.

Davenport T. H. (1994). Saving IT's Soul: Human-Centered Information Management. *Harvard Business Review*, **72**.2.

De Michelis G. (1994). From the analysis of cooperation within work-processes to the design of CSCW Systems. In: *Proceedings of the 15th Interdisciplinary Workshop on Informatics and Psychology: Interdisciplinary approaches to system analysis and design*, Schaerding, May 24 - 26.

De Michelis G. (1995). *Computer Support for Cooperative Work: Computers between Users and Social Complexity*. In: (Bagnara *et al.*, 1995).

De Michelis G. (1995b). *Work, Knowledge and Computer Support Systems*. unpublished notes.

De Michelis G and M. A. Grasso (1994). *Situating conversations within the language/action perspective: the Milan Conversation Model*. In: (AA VV 1994), 89-100.

Ellis C. E., S. J. Gibbs and G. L. Rein (1991). Groupware: some issues and experiences. *Comm. ACM*, **34**.1, 39-58.

Flores F. (1982). *Management and Communication in the Office of the Future*. Hermenet, San Francisco.

Fountain J. E. (1993). A Customer Service Literature Review. In: *Customer Service Excellence*, JFK School of Government - Harvard University, Cambridge.

Hall G., J. Rosenthal and J. Wade (1993). How to Make Reengineering Really Work. *Harvard Business Review* **71**.6.

Hammer M. (1990). Reengineering Work: Don't Automate, Obliterate. *Harvard Business Review*, **68**.4, 104-111.

Hammer M. and J. Champy (1993). *Reengineering the Corporation*. Harper Business, New York.

Heskett J. L., W. E. Sasser Jr. and C. W. L. Hart (1990). *Service Breakthrough: Changing the Rules of the Game*. Free Press, New York.

Keen P. G. W. (1991). *Shaping the Future: Business Design through Information Technology*. Harvard Business School Press, Boston.

Medina-Mora R., T. Winograd, F. Flores and R. Flores (1992). *The Action Workflow Approach toWorkflow Management Technology* In: (AA VV, 1992), 281-288.

Morgan G. (1986). *Images of Organizations*. Sage Publ., Newbury Park.

Nonaka I. (1988). Toward Middle-Up-Down Management: Accelerating Information Creation. *Sloan Management Review*, **29**.3.

Nonaka I. (1991). Managing the Firms as Information Creation Process. In: *Advances in Information Processing in Organizations* (J. Maindl, Ed.), Vol. 4, JAI Press.

Nonaka I. (1993). *On a knowledge creating organization*. Paper presented at the XII AIF Conference, Parma, Italy, October.

Schmidt K. and L. Bannon (1992). Taking CSCW Seriously: Supporting Articulation Work. *Computer Supported Cooperative Work. An International Journal*, **1**.1/2, 7-40.

Senge P.(1991). *The Fifth Discipline: the art and practice of the learning organization*. Doubleday Currency, New York.

Suchman L. A. (1987). *Plans and Situated Actions*. Cambridge University Press, New York.

Takeuchi H. and I. Nonaka (1986). The New New Product Development Game. *Harvard Business Review*, **64**.1.

Van Gorder B. E. (1990). Total Customer Service - The Real Competitive Edge. *Credit*, **16**.6.

Williamson O. E. and S. G. Winter, Eds. (1993). *The nature of the firm*. Oxford University Press, New York.

Winograd T. and F. Flores (1986). *Understanding Computers and Cognition*. Ablex, Norwood.

APPLYING HUMAN-PROCESS-COMMUNICATION TO PROCESS INDUSTRY

M. Heim, N. Ingendahl, M. Polke

RWTH Aachen, Lehrstuhl für Prozeßleittechnik, 52056 Aachen, Germany

Abstract: Classical monitoring and control of production processes may be characterized by its plant and individual-signal oriented view. It's an important task of process control engineering to overcome the resulting shortcomings by a systematic approach to process-oriented human-process communication. The aim of human process communication is to make the production process itself transparent to the operating personnel and to support the effective use of its working experience. For this, structuring and analysis of process information are of basic importance. Flexibility of modern information technology and media are used to make process knowledge accessible to the operating personnel and present information in a task- and situation oriented way.

Key Words: Human-process-communication; process control engineering; information model; human factors

1. INTRODUCTION

Nowadays, technical processes are controlled by the operating personnel in control rooms. The operating personnel takes the responsibility for:

- controlling the operation according to set rules;
- detect critical development of the process;
- find potential improvements of process or process control

The process itself may only be perceived by way of technical monitoring and control devices. The collected and presented information has to enable the operating personnel to fulfill the mentioned tasks.

In this context, typical plant-oriented presentation reaches to its limits. Insight into the process itself requires transparent presentation of the multiple relations between process properties. A systematic approach to this problem is titled "human-process-communication", refering to the headline of (Färber *et al*, 1985).

Methods and concepts of process control engineering deliver the basis for human-process-communication. Process control engineering integrates subsystems for information acquisition, information processing and distribution, information presentation as well as intervention into the process to an application oriented overall system, cf. Fig. 1.

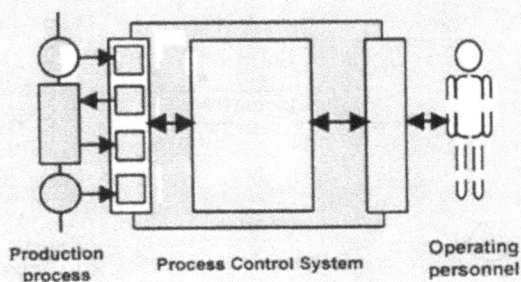

Fig. 1: Process Control Engineering

In this article, aspects of actual work on human-process-communication are demonstrated in applications for waste-water treatment. For this, the chosen process is described and problems in typical operational situations are explained. Hereafter, the preliminaries of human-process-communication are introduced. In its main part the article shows up solutions for human-process-communication. The use of methodic selection of presentation forms is demonstrated. Finally, the importance of integrating operation experience of the technical staff into process control is underlined. For that, ways of effecive support for the operational collection of process knowledge are shown up. The solutions shown in this article have been realized in common process control systems.

2. CHARACTERISTICS OF OPERATIONAL SITUATIONS

Waste-water is treated by sequentially applying different physical, chemical or biological processes. The operating personnel may influence the individual processes especially by varying the amount of mass-flows. For example, biological decomposition is controled by the mass-flow of recirculation water, return mud and oxygen. Moreover, nutrients, chemical additives and special feeds may be added.

The complexity of operating waste-water plants results from variations in boundary conditions of process control and the connected necessity of process variations. For example, the effects of alternation of the inflow (quantity and concentration of substances) on purification processes are often known only qualitatively. The operating personnel finds itself confronted with the task to fix the probable process developement and to prevent any deterioration of bacteria population by intervening with appropriate steering actions. Competing goals (minimal deposits vs. economic and energetic restraints) and dependent relations of product and

implementation of single-loop controlers, all relations between individual signals must be extraced by the operator. The augmented power of graphic presentation systems is used for presenting three-dimensional apparatus instead of three-dimensional relations between relevant properties! Thus, the operational staff has a hard job in reacting to critical situations as loadsurge, mud floatation or lack of nutrients. Important indicators and characteristic factors often are not presented at all.

3. PRECONDITIONS OF HUMAN-PROCESS COMMUNICATION

Profound knowledge of the production process is needed, to come to solutions for human-process communication. Therefore, relevant information have to be defined by analysing the demands faced by the production process and by structuring these in an application oriented way. For this, methods are adequate which suit to the description of static and dynamic properties of production processes. Applying adequate methods results in an information model of the process. Process control engineering has to care about these information models during the whole life-cycle of the production process

The phase-model of production is a method suitable to the outlined necessities (Buchner *et al.*, 1994)(cf. Fig. 2). It supports a hierarchical structuring of production processes, the systematic analysis of

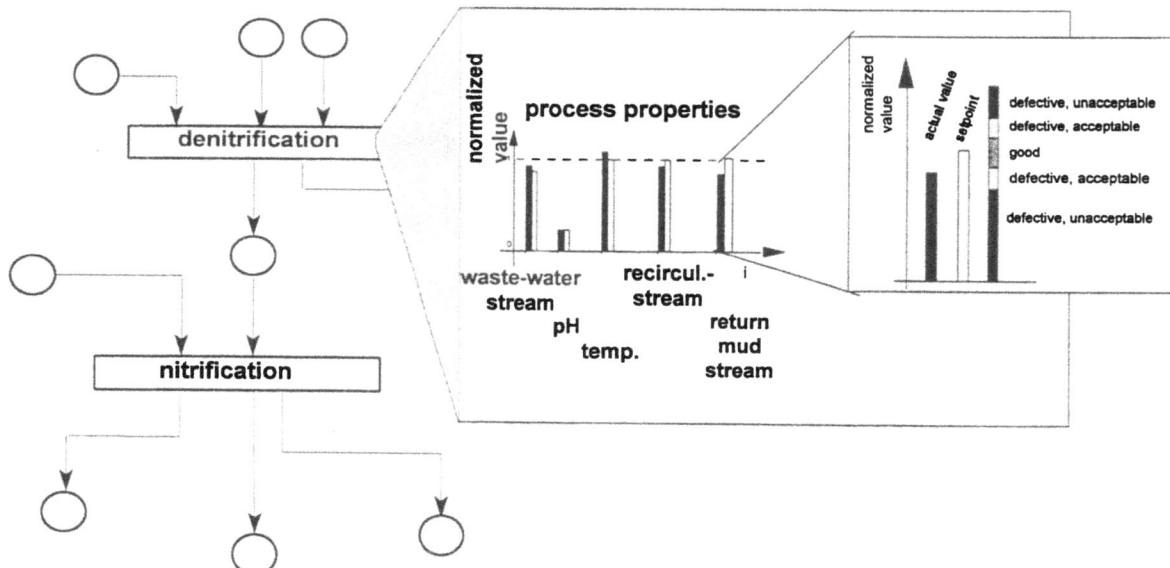

Fig. 2: Excerpt of phase model of waste water treatment

process properties call for compromise.

In sight of this complexity, todays's design of monitoring and control shows obvious deficits. For example, in typical P&I-diagrams, which follow the conventional structure of control room

production demands (quality, safety, economy, etc.) by definition of information about product and process properties and its dependency on time and space. Moreover, relations between the individual properties must be known for the design of human-process-communication. These relations form the basis to explicit presentation of process behaviour. As far as these relations can be described analytically, they may be used for designing process control functions. However, in many cases there is a lack of analytical knowledge about relevant relations; often, process knowledge is implicitly given by experience of operating personnel or in stored data of production.

In the following sections, solutions for human-process-communication based on the outlined

products, intermediate products, recycling and final products as well as by the respective process elements. In this way the structure of the process, which doesn't necessarily correspond to the apparatus structure, is easy to percieve. In biological clarification processes, for example, denitrification and nitrification are conducted in different basin sections. Therefore, if an apparatus-oriented representation is used, important separations of the the individual processes cannot be carried out and essential intermediate products, e.g. nitrate, are

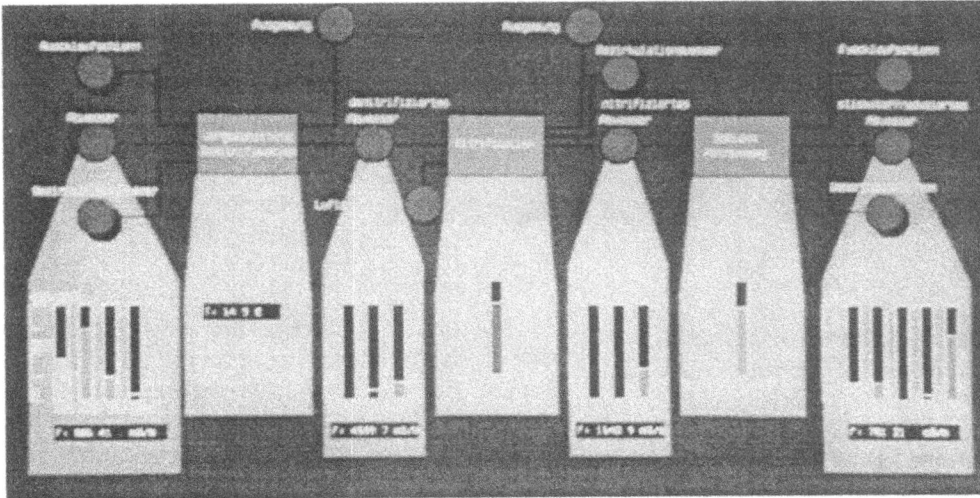

Fig. 3: Process-oriented Overview

information modelling are presented.

4. DESIGN OF HUMAN-PROCESS COMMUNICATION

To ensure smooth operation, the personnel needs first of all a general overview over the process. In Figure 3 such an overview for the "biological unit" of a waste water treatment plant is shown. The process structure is represented by essential input

displayed at unsuitable places.

The simple elements are described by significant attributes in the general view. They increase the interpretability of an overview representation considerably and can be either "direct" measurement data, such as BSB (required oxygen for biological decomposition), CSB (required oxygen for chemical decomposition), TOC (total organic carbon) or "indirect" (e.g. calculated) characteristic values such as 'period of dwell' or 'mud age'. It should be noticed

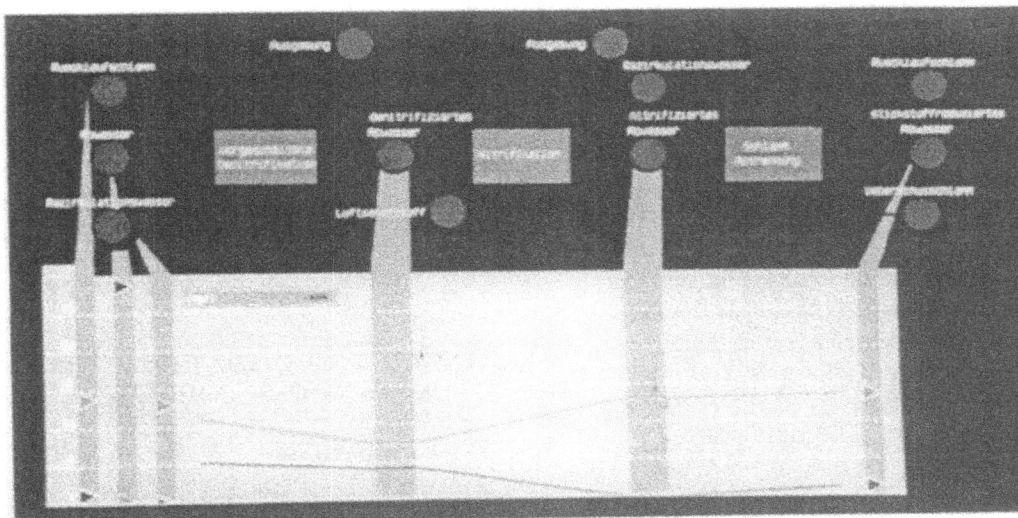

Fig. 4: Process Structure and Nitrogen Balance

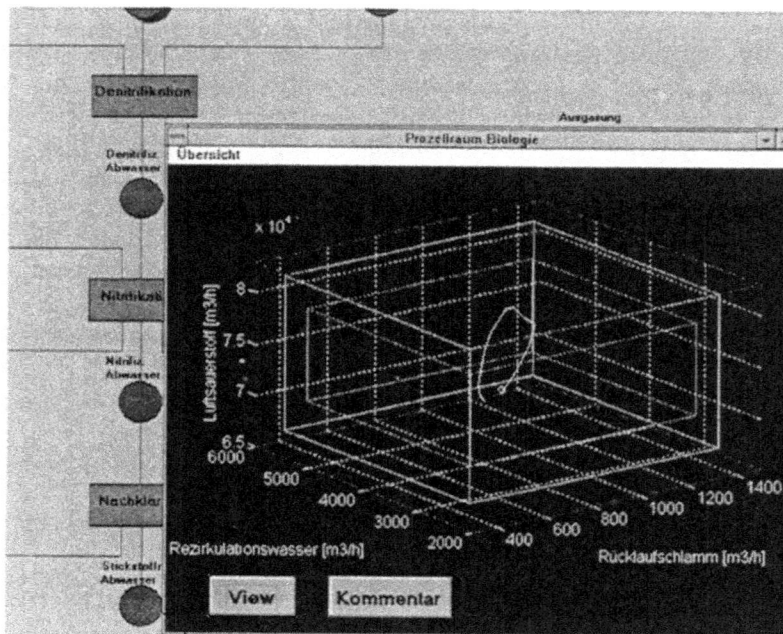

Fig. 5: Process Space

In particular when boundary conditions of process change, the personnel needs not only an easily surveyable "over-view" but also detailed and situation dependent views on the happenings in the process; this can be done appropriately with process and product spaces. Figure 5 shows a process space for regulated quantities which can be changed. In the case of a biological unit these are input air volume, returning mud ratio and recirculation ratio. By comparison of the actual value changes of set points to their tolerances and maximum values an important hint can be given on ressource of performance. Herewith, the premature detection of errors is supported.

that in order to achieve a presentation, which can be easily surveyed, the essential attributes for each element of the process structure must be presented.

Furthermore, during the operation, it is likewise important that actual values are put opposite to setpoint values. Thereby, a comparison between actual and desired values can be carried out and an evaluation of the actually running process with regard to the product specifications, as for example the fulfillment of legal requirements, is facilitated. By choosing certain forms of graphical presentation, the personnel's capability to recognize process conditions by typical patterns is used and enhanced.

For the illustration of the process it is important to present the charceristics dependency on time and space. This is shown in Figure 4 for the nitrogen balance. The ammonium and nitrate concentrations of vital products are presented according to their relative quantity of the inflows along the locally distributed processes. With this the process, here the ammonium decomposition and the resulting formation of nitrate, is rendered transparent. The ascerts of the correlation lines, which result from the values of the stationary concentrations, are a distinct indicator for the rate of reaction. Moreover, the change of the ascents provides the operating personnel with a clear and understandable indicator for the dynamics of the process. The described presentation is especially suited as far as the process structure is concerned and allows to prematurely detect critical developments in the process, for example problems with "sensible" nitrificants.

In connection with further product or process spaces important conclusions (which often cannot be modelled) can be drawn. For the realization of the described process oriented presentations of information, a large variety of powerful multimedia tools exists. However, it is important to note that the projecting engineer must clarify some vital questions in order to relate the information to appropriate forms of presentations.

From a process oriented point of view, these questions concern the characteristics of the information such as scale, type, time and space dependency etc. On the other hand, characteristics of forms of presentation, which have to be chosen according to the application, must be kept in mind.

In overviews (general views) significant graphical forms of presentation supporting psychological association are highly advantageous, whereas in the domain of analytical problem solving, information must be evaluated in sequence. In the latter case, an alphanumerical form of presentation provides precise results. To support the projecting engineer, he should only be offered such forms of presentation which are suitable because of application oriented criteria. As a basis for future tools classification schemes (taxonomies) have been developed. They render a preselection possible.

5. UTILIZATION OF OPERATING EXPERIENCE IN HUMAN-PROCESS COMMUNICATION

While running process facilities the knowledge about the process grows continuously and gets more

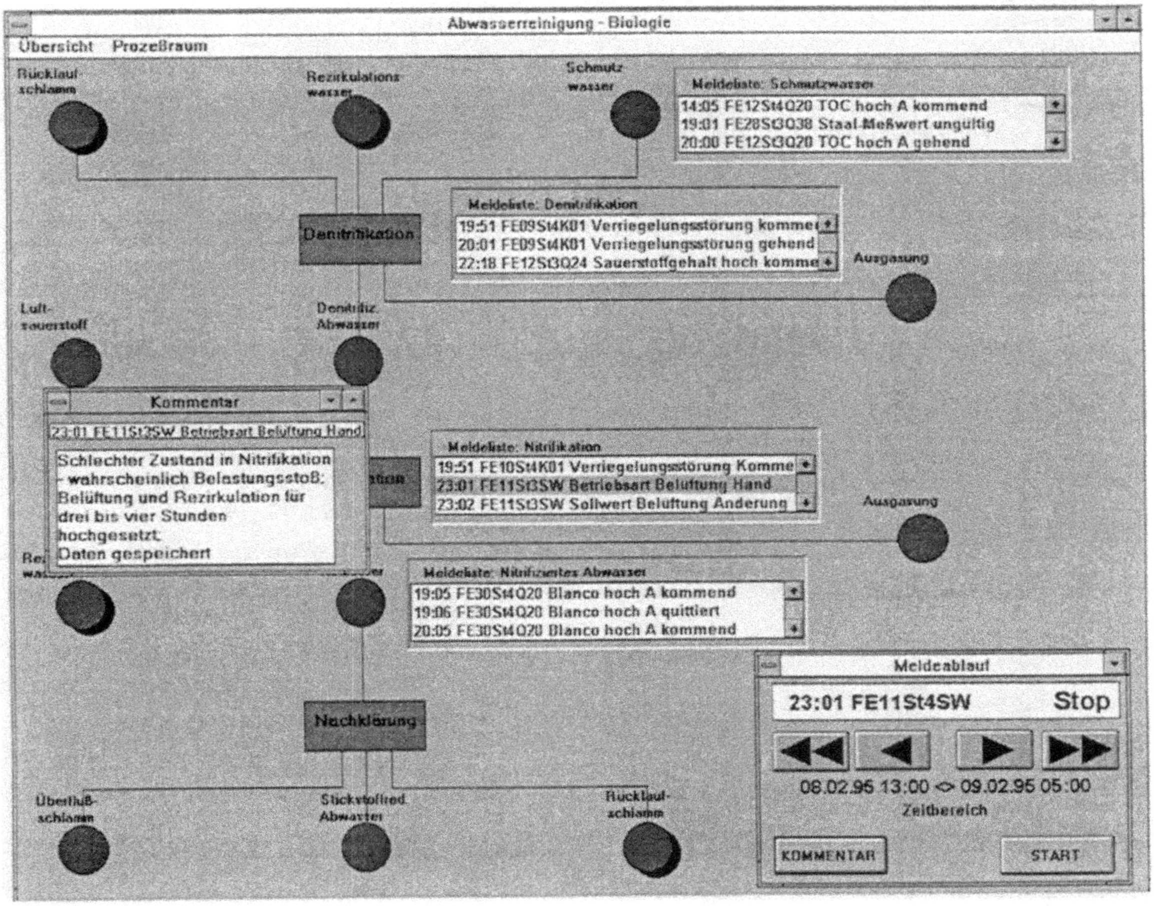

Fig. 6: Process oriented event browser

detailed. Peculiar experiences can be e. g. disturbances during the operation. The operating personnel can gain important knowledge about typical operating conditions and relevant relations between process properties by examining the list of alarms and events. Process oriented handling is supported if alarms concerning specific process elements or products in defined time intervals become accessible and can be browsed chronologically, cf. Fig. 6.

In case of doubts about appropriate intervention, e. g. when having to cope with load surge, information about similar situations can be called for in a process-oriented way. Thus, single events can be condensed to process behaviour.

Using operating experience of the technical staff for refining process knowledge helps to intensify optimization of processes and process control. However, due to that permanent enhancement, adaption of presentation and interaction becomes necessary. This requires on-line configuration, which again has to be supported by taxonomies to help chosing adequate image structures and forms of presentation.

Single events, comments and observations by the operating personnel have to be gathered systematically. Focusing this information to process structure and its properties is an adequate way to ensure common and methodic acquisition, notice, verification and interpretation. This is supported e. g. by on-line addition of comments as shown in Figures 5 and 6.

Finally, statistical tools adapted to the operator needs and usable without special programming knowledge serve for deepening the understanding of relations betweeen process variables. The use of on-line simulation starting with actual or assumed operating conditions allow to widen process knowledge and to foresee probable process developments.

Human-process communication described above requires flexible communication in a distributed control system. Typical needs are integration of on-line simulation and tools for statistic functions. Actual research and the development of object-oriented communication systems will help to overcome actual deficits.

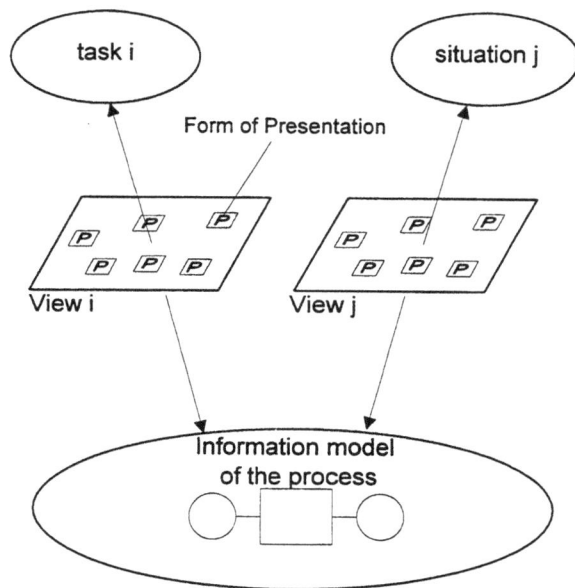

Fig. 7: Human-Process Communication

6. SYSTEM AND DESIGN REQUIREMENTS OF HUMAN-PROCESS-COMMUNICATION

The different tasks of controlling a process can be hierarchically structured according to their degree of abstraction. According to certain situations and the system's state the operators may change between these different levels. Due to the different responsibilities, the technical staff has to be enabled to mesh into the process at each level and at any time. Consequently process-oriented functional units (see Polke, 1994 Chap. 4.5) need an interface which allows to switch between manual control and automatic control mode. The implementation of this concept (hardware, software) depends, among other things, on requirements of process reliability and system availability. Further design aspects can be taken from software ergonomics and studies on socio-technical systems (Hartmann *et al.*, 1993).

7. SUMMARY

Aspects of actual work on human-process communication have been demonstrated in applications for waste-water treatment. The use of its systematic approach for the control of plant operation, the detection of critical developments of the process and the improvement of process control have been made obvious. Human-process communication requires a profound description and analysis of the production process. Here, modern information structuring techniques prove to be helpful. Flexibility of modern information and media technology are used to make process knowledge

accessible to the operating personnel and present information in a task and situation oriented way (cf. Fig. 7).

REFERENCES

Färber, G.; Steusloff, H.; Polke, M. (1985). Mensch-Prozeß-Kommunikation. *Chem.-Ing.-Technik* **57** (1985), 307-317.

Buchner, H.; Lauber, J. and Polke, M. (1994). Das Informationsmodell: Basis für die interdisziplinäre Prozeßbeschreibung. *Automatisierungstechnik at,* **42** (1994), 5 - 10.

Polke, M. (1994). *Process Control Engineering* (Polke, M., Ed.). Verlag Chemie, Weinheim, 1. Edition, 1994.

Hartmann, E.A. and Fuchs-Frohnhofen, P. (1993). Von Menschen und Handrädern. *Technische Rundschau*, **85**, 33 - 38.

IMPROVED PRODUCTION AND PRODUCT INFORMATION FOR CHEMICAL PULP MILL OPERATORS

Sven-Olof Lundqvist and Egils Kubulnieks

STFI (Swedish Pulp and Paper Research Institute)
Box 5604, S-114 86 Stockholm, Sweden

Abstract: Industrial production is often performed in a series of processes or operations. Long and varying time delays often create major problems. The operator is deprived of crusial feedback information and cannot perceive important relationships. To handle these problems, STFI and partners have developed a new information concept for the fiber line of the pulp mill. Models are used to calculate all time delays and product data without time delays are created. Through new task-oriented information functions, FLIPs, the operator is provided with high quality information on process conditions, product quality and production costs from a production line and product perspective. The goal is improved competence of the operator and the quality and efficiency of production. An off-line system is now running on real data sampled in a pulp mill. The next step is to install and evaluate the system on-line. The concept is applicable also in other industries.

Keywords: Flow heterogenity, information systems, models, operators, paper industry, process identification, quality control , time delay estimation, training, work organization.

1. INTRODUCTION

Industrial production is often performed in a series of processes or operations. There are often strong interrelations between the conditions in different processes and the overall product quality and production cost. For high efficiency and quality, the operation must be understood and optimized from a *production line and product perspective*. This is especially difficult in a line of continuous flow processes with non-uniform flow patterns and mixing in process and storage vessels, as well as in lines with both batch and continuous processes.

Because of varying raw material properties, disturbances, quality and production rate changes, the production line is normally in a dynamic situation. The operators are continuously performing coordinated adjustments to optimize the overall result for each produced unit. This requires a very

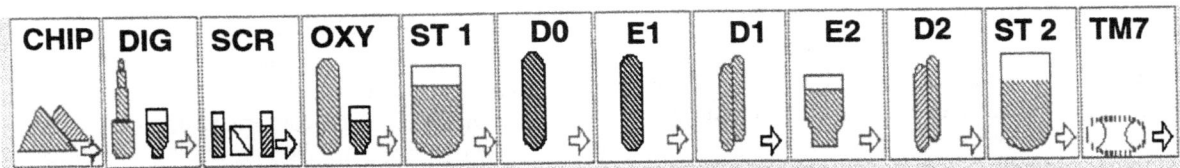

Figure 1: Pulp is produced in a chain of processes and pulp storages, the fiber line. The figure illustrates the reference fiber line of the STFI project: chip storages, a digester, screening, oxygen delignification, unbleached pulp storage, bleaching in five stages with different chemicals, bleached pulp storage and finally a pulp dryer (TM7). 8 intermediate pulp washers are not shown. (This fiber line is now modified, not least in the bleach plant. This is not, however, covered in this paper.)

good understanding of the production line as a whole, good knowledge on the relationships between process conditions, qualities, costs, environmental effects, etc and efficient on-line information on the overall production and its results. In the future, such competence and perspectives will be even more crucial, because of increased *competition* and *customer orientation*, more *specialty products* of higher and more uniform *quality*, more *complex production systems* and *fewer operators* with *wider resposibilities*.

To develop such competence and to run the mill efficiently, the operator needs *improved information* from the general perspective and information on *the full production history of each product unit*. The information systems of today do not provide such total production line information. There is little information on *the successive creation of the product*. The operator gets a *very poor feedback* on the consequences of his actions. *The long and varying time delays in the fiber line are the core of the problem*.

STFI and partners have developed a new information concept to match these needs in such a production line, the fiber line of the pulp mill, *to improve the efficiency and quality of the production and the competence of the operator*. The STFI concept will give the operator high quality information on process conditions, product quality and production costs from a production line and product perspective. An off-line system is now running on real data sampled in a mill. The next step is to install and evaluate the system on-line.

The concept is generally applicable. Interest has been shown from the steel and food industries. The concept may be further developed to handle also mixtures of continuous and batch processes.

Pulp is produced in a chain of processes with complex interactions. *Figure 1* represents the fiber line of the reference mill three years ago, before changes in the bleach plant.

A pulp mill operator normally works for 8 hours. The total production time in the process chain is 20-30 hours. Thus, the operator will normally not see the final result, nor experience the full perspective on the production. The long time delays deprive the operator of this crusial feedback. When examining the process data, he loses track of the pulp and can not perceive the relationships between process conditions, qualities and costs in different processes and in the fiber line as a whole, not even the full results of his own actions. Thus, tools to handle the varying time delays are strongly needed and also more adequate information presentations. Therefore, STFI has developed a new concept to match these needs.

2. THE STFI CONCEPT

> **The STFI Concept**:
>
> **1.** Fiber tracking
>   **Pulp Data** without time delays
>
> **2.** Task-oriented information functions
>   **Fiber Line Information Presentations (FLIPs)**

The STFI concept involves firstly *the tracking of the fibers* through the fiber line, the calculation of the time delays and the creation of *Pulp Data without time delays*. In the project a new approach to this has been developed, including models for the flow patterns of the different storage and process vessels, algorithms for parameter identification (Andersson and Pucar, 1994), the removal of time delays through "resampling" of time series, etc. Similar problems have been treated by others earlier. The project is, however, unique in its approach of addressing so many of the crusial practical and theoretical problems connected to continuous flow production in a unified manner and in the way the information functions have been developed. It is believed that this new approach renders more reliable data than the methods used before and will improve the result of fiber tracking. A final on-line evaluation has not yet been performed. Some other evaluations have, however, been carried out on the the quality of the fiber tracking and Pulp Data with very positive results. This will be presented in another paper.

Secondly the STFI concept involves *task-oriented information functions*. They are called *Fiber Line Information Presentations*, or in short *FLIPs*. This paper will focus on Pulp Data and their use in the FLIPs, especially in FLIPs for the operators. In the paper, several FLIPs will be presented, giving new types of process, quality ocd cost information from the production line and product perspective, both for feedback and forcasts.

3. TIME DATA VS PULP DATA

Figure 2 shows regular *time data* for the state of chemical treatment of the pulp after different processes during 2 days. The upper graphs are the kappa numbers (lignin concentrations) of the pulp after the digester, after oxygen delignification, before prebleaching and before final bleaching. The lower graphs show lab test brightnesses from the pulp after three bleaching stages and from the final product at the dryer during the same time.

At 0.4 days a severe disturbance occurs, caused by overcooking in the digester. Two hours later it

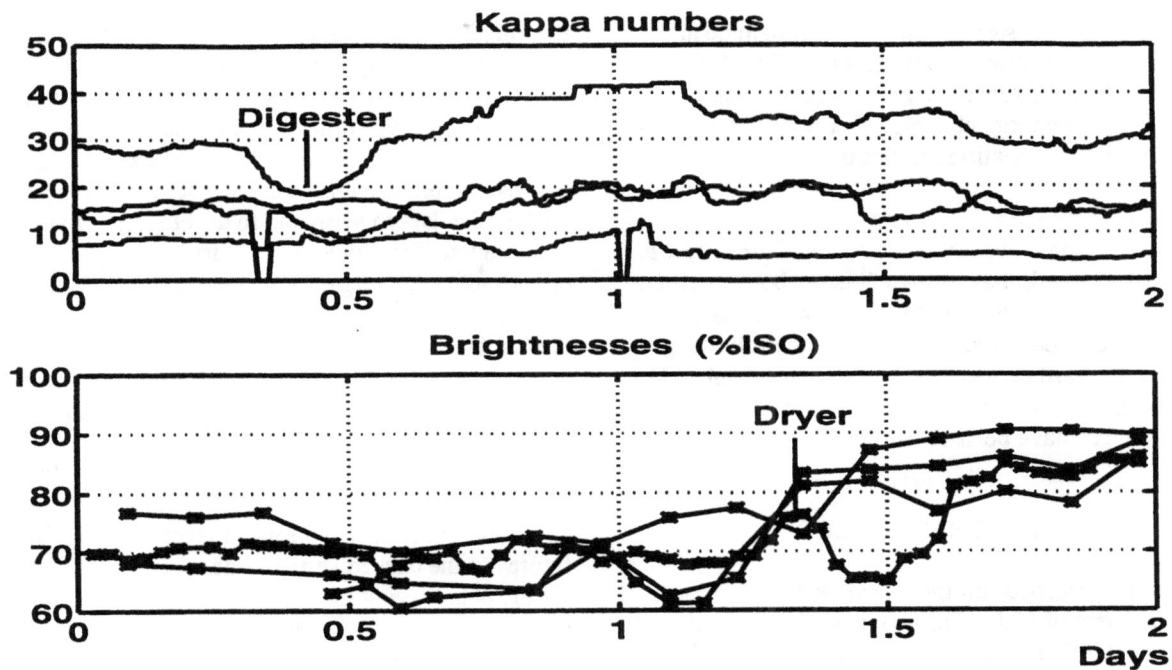

Figure 2: Time data for the kappa numbers and brightnesses all along the fiber line during 2 days. The residence time between the digester and the dryer is about 1 day.

appears after oxygen delignification. Later it enters the bleachplant, etc. Finally, at 1.3 days, about 1 day later, the consequences of the disturbance appears in the final product on the dryer. The final brightness and the production cost are unnecessarily high and probably there are also other quality consequences. Even at such an evident disturbance, it is not easy to track the disturbance and to perceive the quantitative relationships between different variables in time data.

The time series data are transformed into *Pulp Data*.

The new models and algoriths are used to calculate and remove the time delays. For each bulk of pulp, data have been collected from all along the fiber line into a vector.

Figure 3 shows the Pulp Data vectors for the same kappa number and brightness data seen in figure 2. The Pulp Data vectors are presented by the *Pulp Data Information FLIP* of the off-line system. Here you can easily see the relationships. You can develop your knowledge and find more successful ways to handle disturbances, perform grade changes, etc. In

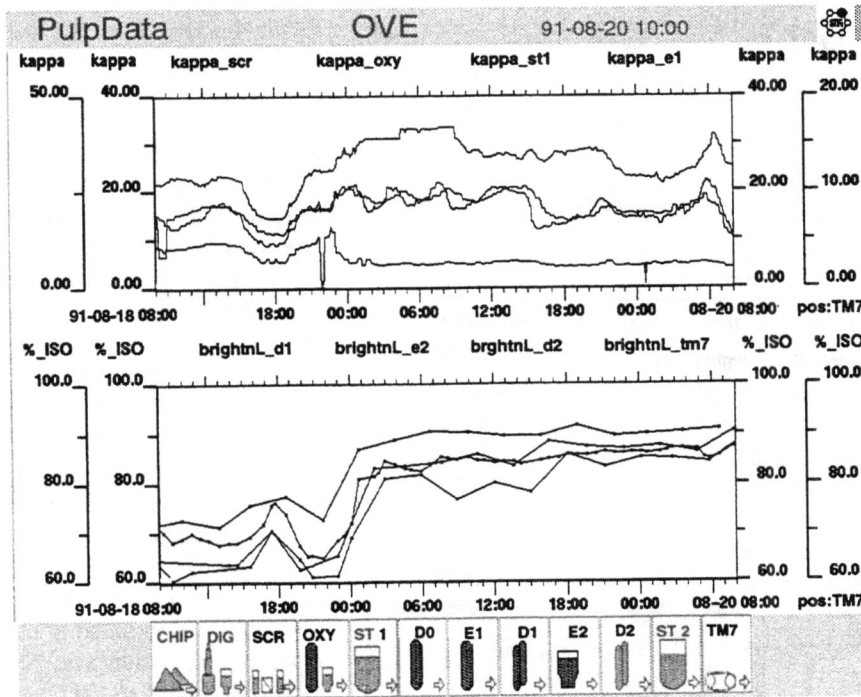

Figure 3: The standard layout to present Pulp Data Information. Kappa numbers and brightnesses from all the fiber line are shown with reference to the time when the pulp leaves the dryer (TM7). Each variable has an index indicating the measurement position. For legend, see the fiber line scheme at the bottom. Digester disturbances affect the pulps ready at about 18:00 on 910818 (the same disturbance as in figure 2) and at about 06:00 on 910820. A grade change from semi-bleached to fully bleached pulp arrives after the dryer at about 00:00 on 910819. It takes about 4 hours to reach full brightness quality.

a ready on-line system, Pulp Data will include all important information from the whole fiber line: Wood and chip properties, process conditions, chemical and energy charges, intermediate and final pulp properties, qualities and costs.

What you now see are not time series data, but information on *a sequence of produced tons of pulp*. Each bulk of pulp has its own "history book" written in a Pulp Data vector and the information has been fetched from these vectors. This and the following figures are hard copies from the system's display. For demonstration, two days with unnormally big disturbances have been chosen.

4. OPERATOR FLIPS

As the project focus on the *operator*, the development of FLIPs has so far been limited to a set of *Operator FLIPs*, especially matching the tasks and needs of the operator. More detailed information on these FLIPs has been presented earlier (Lundqvist and Kubulnieks 1994). The paper will comment on the FLIPs called:

Operator FLIPs:
+   Pulp Data Information
+   Last Shift Survey
+   Forecast
+   Passage Times
+   Costs and Quality
+   Processing Profile

The first FLIP has already been shown in figure 3. The *Last Shift Survey FLIP* is a special version of

the Pulp Data Information FLIP, giving the operator the complete history and result of *all the pulp he processed during his latest shift*. This is a new and very important type of feedback to the operator. He can see the full perspective of what he earlier only in part has experienced. He can use enclosed tools to analyse the data to learn and to develop better ways of operation. This could provide fruitful opportunities for "organised reflexion", which is today rare in the mill operation situation, but is important for learning and improvement.

Figure 4 shows the *Forecast FLIP*, with which you look into the future. In this example the future passage times off the dryer are forcasted for all bulks of pulp still in the fiber line. These data have been added to Pulp Data as vectors "projecting" into the future. At the dashed vertical lines you see Pulp Data for the pulp leaving the dryer *just now*. And to the right of these lines you see data on pulp which will arrive to the dryer *in the future*. In this way the operator can get a feeling for *what* will happen in the future and *when* it will happen. From the figure you can also see in what vessels the pulp of a disturbance is now physically located or how far in the fiber line a grade change has proceeded.

It is believed that the operator will be happy to use this Forecast FLIP when he arrives on a new shift, to foresee what will happen during his shift and to plan his activities accordingly. The forecast will also support the short term quality and production planning, to match temporary quality variations with current orders of different quality specifications.

If the operator wants to see the forecasted passage

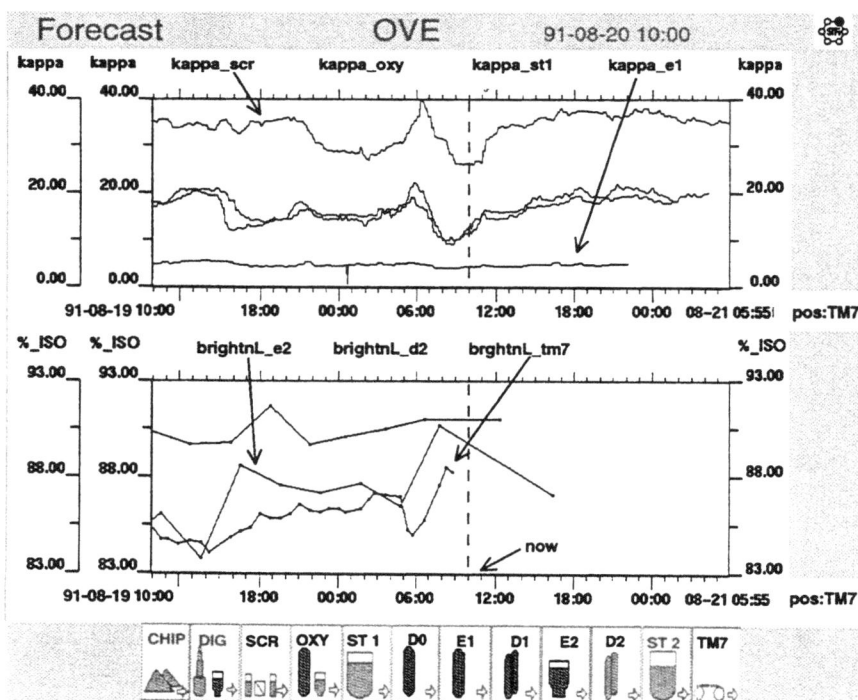

Figure 4: In this Forecast FLIP, the Pulp Data for the pulp now leaving the dryer are seen at the dashed vertical lines. The graphs are "protruding into the future", telling when pulp still residing in the process and storage vessels will leave the dryer. This creates a forecast of when disturbances, grade changes, etc will occur in the final product. The pulp now leaving the E1 bleaching stage is forecasted to leave the dryer at about 22:00 and the pulp now leaving screening at about 06:00 in the next morning. Other passage positions may of course be selected. (Note the poor information value of the low frequency brightness lab tests after the E2 and D2 bleaching stages.)

times more clearly, for instance to advise his fellow operators furher down the line, he can use the *"Passage Times" FLIP*. Here the residence times and the passage times in the different storage and process vessels are presented numerically and in graphics. With the *Costs and Quality FLIP* of *figure 5*, a summary of key variables, costs and qualities from all along the production line is provided for a *specified amount of pulp*. In the example of the figure, the amount of pulp which entered the bleach plant during one specific hour have been selected. The choice is very free. You might select the pulp passing any process position during 2 hours, a shift, a grade campaign, or whatever might interest you. Here the costs are given as relative costs, compared to the costs of the reference treatment for the current pulp grade. As an alternative, absolute costs as well as relative or absolute charges may be presented.

Please observe that the *data quality* here is *incomparably higher* than that of time based averages, for instance a *traditional shift average report*. Here information is given on *the same pulp* all along the fiber line, rather than on *different* bulks of pulp in different processes.

Cost information has traditionally been used in a quite defensive manner. With this Cost and Quality FLIP, the cost information may be used *offensively* as a tool for continuous improvments.

It might be hard to quickly detect deviations, etc, from all the information condensed in the Cost and Quality table. The *Processing Profile FLIP* is a *"snap shot"* version, presenting the most important parts of the table information in graphics, where different profile forms will correspond with different

pulp quality deficiencies. The idea is that the operator will be able to judge the pulp just from a glance at the form of the processing profile.

The system will also include tools for *Process Analysis* to support the operator in operation, training, fault detection, development, etc. One very special facility for fault detection is the *Production Calibration Support* function, which continuously checks the data quality of the pulp flow and consistency measurements. This calibration support is very important for successful fiber tracking.

5. OPERATOR PARTICIPATION AND PERCEPTUAL NEEDS

A group of operators from the reference mill was deeply envolved in the work of selecting, specifying and developing these Operator FLIPs. Operator engagement is of ultimate importance in the development of this kind of information systems. The needs in relation to process characteristics, production demands and a new work organisation were discussed with mill operators and staff. The operators were interviewed by a work psychologist. Based on this and knowledge on new *information technology, perception and ergonomy*, new presentations have been *jointly designed and engineered*.

One example of this perceptual analysis is the presentation of the Pulp Data information, depending on what task to be supported. For *customer service* tasks, you might refer to a specific amount of pulp by its accumulated produced ton number or by bale or roll numbers. To support the

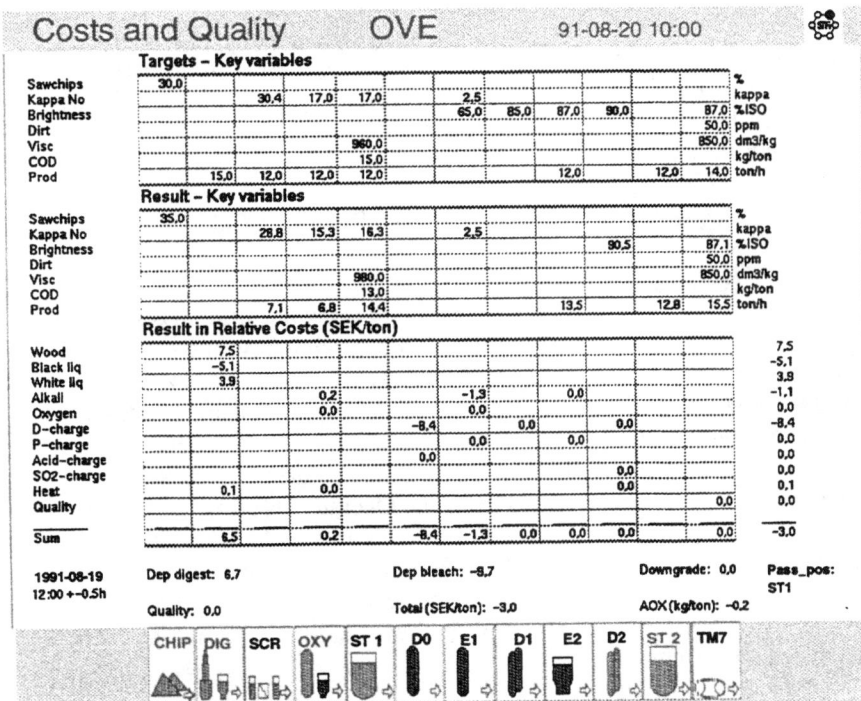

Figure 5: Costs and Quality information for a specified bulk of pulp, in this example the pulp entering the bleach plant during one hour. Raw materials, charges, intermediate results, qualities and costs all along the fiber line are compiled and presented. At the top, the preferred processing of the current grade is shown as a set of targets for key variables, followed by the actual results. The costs are given in relation to this preferred reference processing of the pulp grade. (Note that a corresponding traditional shift average report is based on different bulks of pulp in different processes.)

operator, who lives with time and act on time, you get the best reference to a sequence of product units by telling at what period of time it passed a certain process positions. So in this figure, you do not see time graphs. You see the history of chemical treatment for each bulk of pulp produced during 2 days and these bulks are tagged with their passage times off the dryer. It is also possible to tag them with their passage times in any other process position.

6. THE GENERAL CONCEPT

These are the major FLIPs developed in the project, which totally focus on the operator. *The general concept* includes not only FLIPs for operation, training and process analysis. For quality and process development there should, for instance, be functions to separately accumulate data for different pulp grades, to search in a historic relational database, to make calculations based on the Costs and Quality information and for more powerful process analysis. There might be other functions to support customer service, traceability for ISO 9000, maybe maintenance, etc. The general concept, thus, brings furher possibilities through a *broader application* of the concept.

```
The general concept includes FLIPs for:
+ operation
+ training
+ process analysis
+ quality development
+ customer service
+ traceability (ISO 9000)
+ maintenance
```

The *general concept* is also applicable on the integrated pulp and paper mill as well as other types of production in chains of processes, maybe even prolonged towards the customers. This has been confirmed in discussions with the iron and steel industry.

7. A COMMON LANGUAGE FOR IMPROVEMENTS

The installation of this kind of system make data on process conditions, qualities and costs available to many employees and much easier to understand. The *production history of each produced unit* will be available and form a *common language within the mill* and a *tool for improvements*. This is in many cases also a necessity for future competitiveness of the mill.

The project is based on ideas and experiences from STFI. The project partners are the reference mill, the

Department of Psychology at Uppsala University and the Division of Automatic Control at the University of Linköping. The project has been financed by the reference mill and by NUTEK, the Swedish National Board for Industrial and Technical Development, within its DUP-program.

Cooperation with vendors and mills is now discussed for the development of a product based on the concept and for installing and evaluating this on-line in a mill. There is also a big interest in developing other system functions and for other industries. Hopefully, there will on later occasions be opportunities to inform you on the:
+ on-line installation and evaluation of the *fiber tracking algorithms and the operator* FLIPs,
+ results from pulp and paper mill applications of the *general concept*.
+ applications of the concept in *other industries*

8. SUMMARY

- For efficient operation and high quality, a production line and product perspective is necessary.
- The operator needs improved information and competence.
- STFI and partners have developed a new concept for this.
- It involves fiber tracking and Pulp Data...
- ..and task-oriented information functions, FLIPs.
- The objective is to enhance human skill and improve information, to allow better overall operation of the production line
- An off-line version is now running on sampled real data. The next step will be to install and evaluate the system on-line.
- The concept has a wider applicability for product tracking and the creation of Product Data, not only in the pulp and paper mill as a whole but also in other industries.

REFERENCES

Andersson, T.,Pusar, P. (1994).
  Estimation of residence time in continuous flow systems with varying flow and volume. *Proceedings, SYSID '94* Session CP11, 10th IFAC Symposium on Systems Identification, Copenhagen, Denmark, July 1994.
Lundqvist, S.-O., Kubulnieks, E. (1994).
  Fiber Line Operator Information for Improved Efficiency, Quality and Competence. *Proceedings, Control Systems '94.* Conference on control systems in the pulp and paper industry, Stockholm, Sweden, May 1994.

LOGIC OF AUTOMATIC SYSTEM AND LOGIC OF OPERATOR

A.E. Kiv, V.G. Orishchenko, I.A. Polozovskaya

South Ukrainian Pedagogical University,
26, Staroportofrankovskaya 270020 Odessa, Ukraine

Abstract: These investigations show that an interaction of human-operator with complex automatic systems depend on the operator thinking style. New computer testing methods are described that allow to determine operator's thinking structure and to select operators for given creative task. The consideration is applied to operators who interact with automatic systems of nuclear objects.

Keywords: Automatic systems, computer diagnosis, error probability, nuclear plants, personnel qualifications, tests

1. OPERATOR'S CREATIVE THINKING TESTING

Operator behavior in modern industry is a subject matter of numerous investigations. The effects of modern process control systems on the operator's supervisory control task was described in (Hoonhout and Zwaga, 1993). In similar cases only technical aspects of a problem are considered (Zwaga, 1993). But it is clear that psychological characteristics of operator play an important role in his interaction with complex systems. This paper is devoted to analysis of psychological aspects in the problem of human interaction with nuclear objects. In previous works (Kiv *et al.*, 1994a, 1994b) computer methods of creative thinking testing were described. They are based on a new mathematical model of creative thinking. Computer testing games were worked out and each of these games corresponds to concrete schemes of creative thinking. It was shown that one can obtain a set of new psychological parameters for creative thinking by using mentioned computer methods. These parameters characterize creative thinking structure of a person in detail (for instance: intuition (I), logics processing (LP), volume of thinking space, volume of long time memory (VM) etc). New testing methods are suitable in cases when we are interested to check person's abilities

to fulfil a certain creative task. By and large we always can choose special computer testing problems that include corresponding logic schemes. We can also work out computer testing games that are based on these logic schemes. They reflect peculiarities of creative thinking of a given person. In the case of nuclear objects operators have dealings with non-determined processes and testing problems that we must use are based on non-classical logics. It must note that our mathematical description of creative thinking model in (Kiv *et al.*, 1994b) corresponds to classical logic. This means a twice-valued character of any statement. Each step in thinking process is true or false. Here we formulate a set of equations that correspond to many - valued logic with resources (Kiv *et al.*, 1995). The typical equations are following:

$$\frac{dN_i}{dt} = I\left(R\right) + \alpha N_i + \alpha' N_m^k - \beta N_j^p N_p^s \quad (1)$$

Here N_i is a number of steps that are proportional to a probability to reach the solution. Functions N_m and N_n describe intermediate steps that form N_i steps in processes of complexes formation. Steps N_j and N_p form complexes that prevent to N_i steps accumulation. Parameter of intuition $I(R)$ depends on resource R that may change. α and

25

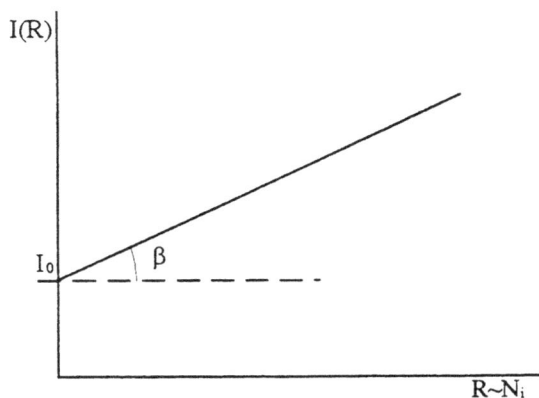

Fig. 1 Illustration of the I(R) function shape

α' are coefficients of mastering accumulated information, β is coefficient of critical potential (see also (Kiv et al., 1994a). Thus the main peculiarities of (1) are following:

1. There are no completely wrong steps.

2. During the accumulation of thinking steps and corresponding moving of a person to a solution the resources are changing.

3. The number of equations depends in each case on creative thinking processes that are described. It is important to choose testing problems that have logical structure in accordance with equations structure.

Computer testing programs that were developed for non-classical logic problems allow to obtain in addition to above mentioned psychological parameters (Kiv et al., 1994a, 1994b) an important function I(R) that shows how the intuition coefficient depends on the different information accumulation. It is obvious that I is proportional to N_i. For some people a strong dependence between I and N_i takes place. As it is shown on Fig. 1 two parameters are taken into account: the initial value of intuition parameter I_0 and the angle of inclination β.

2. BEHAVIOR OF OPERATORS DURING CHERNOBYL ACCIDENT

Electrical engineers assume control of the reactor to test a generator's capacity to power emergency systems as it coasts after steam is shut off. As a result of control rods were lowered into the core the thermal-energy level drops from normal 3,200 MW to 1,600 MW. *The emergency core-cooling system, which would draw power and affect test results, was shut off. Monitoring systems were adjusted to low power levels, but the operator failed to reprogram the computer to maintain power at*

700 to 1,000 MW. Power falled to the dangerously low level of 30 MW. The majority of control rods were withdrawn to increase power, but xenon has built up in the fuel rods. This by - product absorbed neutrons and "poisoned" the reaction. *In this situation virtually all control rods were withdrawn. Power climbs and stabilized briefly at 200 MV. To ensure adequate cooling after the test, all light pumps were activated.* The combination of low power and high flow necessitated many manual adjustments. The operators turned off emergency shutdown signals. The computer indicated excess reactivity. *But operators reserved the possibility of rerunning the test by blocking the only remaining trip signal just as it is about to shut down the reactor.* The test has began and power started to rise. At this dangerously low power level any small increase in power triggers an even larger increase. Water expanded to steam and absorbed neutrons. The power began rise faster. Facing catastrophe, operators began insertion of all control rods. The rods, however, have five meters of graphite at their ends. The reaction speed was strongly displaced. In the next four seconds power surged to 100 times the reactor's capacity. The uranium fuel disintegrated, bursted through its cladding, and came into contact with cooling water. An enormous steam explosion shears 1,600 water pipes, flings the reactor's cap aside, blows through the concrete walls of the reactor hall, and throws burning blocks of graphite and fuel into the compound. Radioactive dust rised high into the atmosphere.

3. ANALYSIS OF OPERATOR ACTIVITIES AT NUCLEAR REACTOR

Operator of nuclear reactor works in special conditions. Nuclear reactor is a new type of industrial plant. Processes in this physical object have a probability character. These processes are not localized in space and are not uniformly distributed. So operator must analyse the local situation in nuclear reactor and do prognosis for the whole plant. Operator cannot completely rely upon computer data. In some cases computer needs about several minutes to give an account of reactor's state. Operator cannot wait for computer data and must act in no time. In these conditions it is very important for operator to have not only a high professional level but also adequate psychological characteristics. Let us anylise typical difficulties of operators in the case of Chernobyl accident.

We cannot believe that operators had not enough preknowledge in physical processes that take place in reactor. Their mistakes in given situation may be explained by structure of their creative thinking. What did operators' mistakes show?

- They were not able to comprehend the situation at a glance.

- They did not foresee results of each stage of experiment by using previous information.

- Their interaction with computers was faint.

- They could not to compare many parameters of nuclear processes at the same time.

Proceed from comparison of general characteristics of operator's creative thinking (see paragraph 1) and real activities of operators during Chernobyl events we concluded that creative thinking structure of operators in this concrete case was not suitable for their tasks. It goes without saying that their function $I(R)$ was wrong. We may assume that their parameters of thinking space volume (VTS) and critical potential (CP) were small. Earlier (Kiv et $al.$, 1994b) we have shown by using of our computer tests that for operators who carry out computer modelling of radiation defects creation in solids it is important to check such parameters as I_0 and mutual influence of different domains of thinking space.

The following scheme of computer testing methods application is proposed (see Fig. 2). A set of computer testing games $(G_1, G_2,..., G_n)$ is used. Each game G_i allows to measure several psychological parameters $K_{1j}, K_{2j},....$ For each game one of these parameters K_{ij}^* is the most important. So if operator shows for given game G_i a result $K_{ij} < K_{ij}^*$ he is not allowed pass to following game. By this way we can obtain an effective approach for corresponding operators selection. In (Chislov et $al.$, 1995) problems of interface improvement for described testing methods development are discussed.

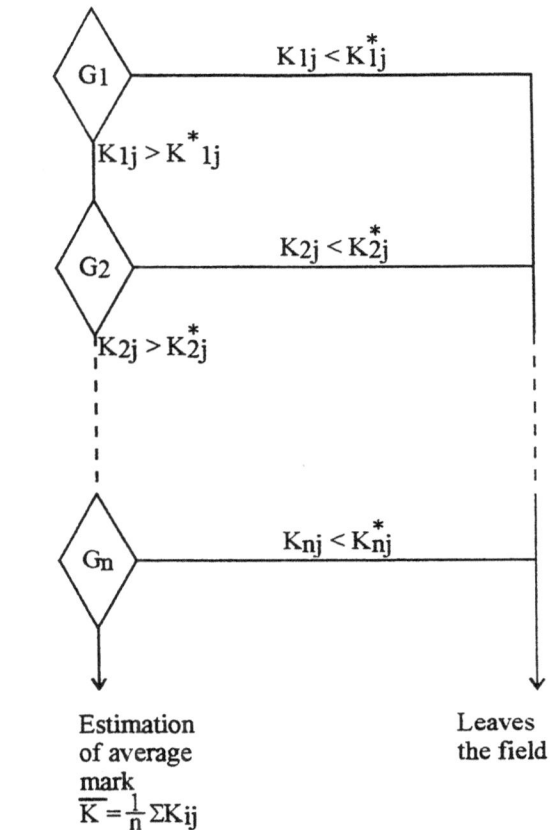

Fig. 2 Scheme of operator's creative thinking testing

REFERENCES

Chislov V.V., V.L. Maloryan, I.A. Polozovskaya, G.V. Shtakser, A.I. Uyemov, I.G. Zakharchenko and M. Athoussaki (1995). The interface improvement for the creative thinking computer testing. In: *Proceedings of 6th International conference on Human-Computer Interaction (HCI International'95).* (Y. Anzai and K. Ogawa (Eds.)), Elsevier, Amsterdam.

Hoonhout H.C.M. and H.J.G. Zwaga (1993). Operator behavior and supervisory control systems in the chemical process industry. In: *Human-Computer Interaction: Applications and case Studies.* (M.J. Smith and G. Salvendy (Eds.)), pp. 109-114. Elsevier, Amsterdam.

Kiv A.E., V.G. Orishchenko, I.A. Polozovskaya and I.G. Zakharchenko (1994a). Computer modelling of the learning organization. In: *Advances in Agile Manufacturing.* (P.T. Kidd and W. Karwowski (Eds.)), pp. 553-556. IOS Press, Amsterdam.

Kiv A.E., V.G. Orishchenko, I.A. Polozovskaya, I.G. Zakharchenko, V.V. Chislov and V.L. Maloryan (1994b). Creative thinking process simulation and computer testing. In: *Proceedings of the Symposium on Human Interaction with Complex Systems* pp. 234-237. North Carolina A&T Univercity, Greensboro, USA.

Kiv A.E., V.A. Molyako, V.L. Malorayn, I.A. Polozovskaya and Z.I. Iskanderova (1995). The creative thinking testing by using of testing problems based on different logical schemes. In: *Proceedings of 6th International conference on Human-Computer Interaction (HCI International'95).* (Y. Anzai and K. Ogawa (Eds.)), Elsevier, Amsterdam.

Zwaga H.J.G. (1993). Developing process control systems: procedural requirements in desing. In: *Human-Computer Interaction: Applications and case Studies.* (M.J. Smith and G. Salvendy (Eds.)), pp. 127-132. Elsevier, Amsterdam.

A HYPERMEDIA INFORMATION SYSTEM FOR QUALITY MONITORING AND DIAGNOSIS ON A GALVANISING LINE

E. Gnaedinger*, S. Roth*, G. Bloch* and P. Fatrez**

* Centre de Recherche en Automatique de Nancy - CNRS URA 821
ESSTIN, Rue Jean Lamour, F-54500 Vandoeuvre France
Phone: +33 83 50 33 33 ; Fax: +33 83 54 21 73 ; E-mail: gerard.bloch@cran.esstin.u-nancy.fr
** SOLLAC Ligne de Galvanisation Sainte Agathe
17, Avenue des Tilleuls, F-57191 Florange Cedex France
Phone: +33 82 51 54 78 ; Fax: +33 82 51 60 11

Abstract: This paper is based on a research project between CRAN and a flat steel company. The purpose is to define an Hypermedia Technical Information System (H.T.I.S.) for quality monitoring and diagnosis on a new galvanising line. A methodological approach is proposed to generate the Hyper-document structure with an hypergraph. The H.T.I.S. is carried out using HTML language in order to provide a useful documentation tool.

Keywords: information technology, multimedia, distributed database, supervision, galvanising line, HTML language.

1. INTRODUCTION

Manufacturing industry needs to maintain a constant and high quality production to meet customer demands. Any unexpected variations in operating conditions or a breakdown in any part of the system necessarily leads to a deterioration in the quality of production. It is therefore worthwhile to provide an on-line supervision of the plant in order to diagnose any significant deterioration in parameters characterising the quality of production.

The monitoring and diagnosis tasks make use of various information of different types: measurements, knowledge or empirical models, databases, control operators and engineers' experience, structural knowledge of the physical process. The processing is therefore hybrid and termed 'knowledge based'. Nevertheless, with the growing complexity of the plants and their control systems, with the progress of automation and the related decrease in the number of operators, the access to relevant information tends to be crucial.

In this paper, an hyperbase structuring of the information required for quality monitoring and condition diagnosis on a hot dipped galvanising line is presented. This plant of the French flat steel company Sollac is located in Florange (France).

For this hyperbase structuring, the large number of variables describing the process, the various types of information available at different places and in different forms (electronic, paper) must be handled. Thus, a classical Data Base Management System is not sufficient to deal with so much different, scattered data. It is important to build a tool that allows access to any piece of information from anywhere and which copes with the data evolution.

New information technologies providing multimedia and hypertext capabilities are now available. WWW tools based on HTML protocol allow the generation of hyperbases with easy access for the end-user. HTML browsers are commonly used in the Internet community for the consultation of distributed information.

Sollac has already put in place network architecture in accordance with the CIM (Computer Integrated Manufacturing) pyramid. Communication services are based on standards, particularly the 'Internet' TCP/IP protocol. Therefore the Technical Information System of the galvanising line presented here is developed using WWW tools (World Wide Web).

A brief description of the line and of the different knowledge for monitoring purposes is given in section 2. Section 3 summarises the state of the art in the field of hyperbases for information systems. The last section gives the architecture of the Hypermedia Technical Information System (H.T.I.S.) dedicated to the galvanising line.

Mostly dedicated to help monitoring and condition diagnosis, this system could be used for learning how to control the production line, and getting the 'savoir-faire'. The most important factor for an operator in front of such a system is to reach quickly the information needed.

2. THE HOT DIPPED GALVANISING LINE

2.1 Plant presentation

The aim of galvanising is to protect a steel sheet from corrosion, in order to increase its life time. This sheet is mostly dedicated for the car industry. Figure 1 shows the process which is divided into three main sections separated by strip accumulators.

The first stage consists in welding a new strip to the previous one to allow continuous production. The steel sheet is then cleaned, before being heated in a long furnace (6 sections) at about 800 °C, in order to obtain particular metallurgical characteristics. The main process is the zinc coating deposition. Both sides of the steel strip are covered with zinc. The cooling tower and the quenching process are then used for annealing and/or cooling the strip. Two different gauges are used respectively to measure the zinc thickness and the percentage of iron for the zinc alloy coating. The skin-pass allows the obtaining of the flatness characteristics, the roughness and the elasticity parameters of the strip. At the end of the line, the sheet is inspected, prepared and coiled.

2.2 Knowledge sources for diagnosis

A galvanising line is actually a very complex multivariable system. For such huge lines information is scattered among various representations and sources: measurements, logical and analogical variables, more or less structured events, human reports, elements of faults databases, historical recordings, faulty sequences of the Programmable Logic Controllers and control models. Their collection leads to a waste of time for users and engineers.

The process structure and behaviour knowledge could be described by the so called 'deep knowledge'

(structural and functional decomposition of the system) and 'shallow knowledge' (association of failure symptoms with system malfunctions).

The gathering of the deep knowledge is achieved in two axes, namely spatial decomposition and functional decomposition. The sections along the line are: the unrollers 1 and 2, the welding section, the cleaning section, the entry accumulator, the furnace, the zinc pot and blowing section, the cooling tower, the zinc gauges, the quenching section, the skin-pass, the iron-zinc gauges, the chromatation, the exit accumulator, the cutting section, the checkpoint, the printer, the oiling section and then the rollers 1 and 2. Variables related to the different parts, subprocesses and subfunctions are classified: tractions, strip and roller speeds, pressures, flow rates, pyrometer and thermocouple temperatures, opening ratios of valves, levels in both accumulators and logical signals.

The shallow knowledge approach deals with various sources, which provide mainly off-line information concerning the behaviour of the process. First, the product faults which are collected at the exit of the line: description, type of product, involved (sub)processes, detection means, importance of the fault, troubleshooting solutions proposed in the different company plants. Another interesting source of information can be the Statistical Process Control (SPC), in which several important parameters are particularly monitored because they have a great influence upon the final quality of the product. Finally, product FMECA (Failures Modes, Effects and Criticality Analysis) also gives a hierarchical classification of failures, but is more oriented toward the final user.

A preliminary task, the inventory of relevant documents, was made. This rather tedious work has been undertaken by S. Roth for her thesis. The main problem, as so often, is that information available is generally on paper format (technical drawing, many reports, schemes, diagrams) (Drouin, 1993). This 'paper' information must be so digitalized in order to be integrated in the technical information system. An important task is to collect, to choose, to touch up, to organise and to index all these documents in order to integrate them in a W3 system.

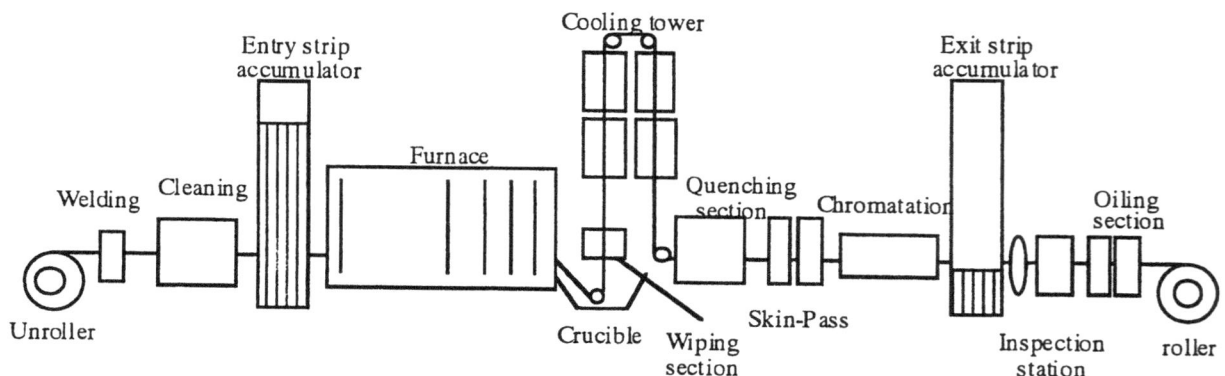

Fig. 1. The hot dipped galvanising line.

3. THE HYPERBASE INFORMATION SYSTEMS

3.1 HTML and related standards

Works about technical information systems are issued from many research programs such as CALS (Computer aided Acquisition on Logistic Support) developed by Department Of Defence and STEP (Standard for Exchange of Product data) (Mason, 1993). The established standards are built on the SGML (Standard General Mark up Language) for electronic documents management (ISO, 1986) and EXPRESS for information modelling and interchangeable format (Spiby, 1992). The electronic documentation of the galvanising line proposed here is directly in HTML format (Hyper Text Mark up Language) (Berners-Lee, 1994). The HTML language is an application conforming to the SGML standard. Most of recent documentation services on Internet use HTML language and HTTP communication protocol (Hyper Text Transfer Protocol).

Theses tools are defined for hypertext documents including multimedia information in order to:
- present heterogeneous information such us text, flowcharts, images, sound, photos and video sequences,
- improve automated indexing of documents, linking of related documents, and cross-referencing between text, images and sound,
- access from any station on a local area network,
- use an interactive mode for knowledge acquisition,
- integrate multimedia end-user systems.

The complexity of a H.T.I.S. lies in the management of data distribution through the network. Data can be produced and consumed by different workstations or industrial computers. The HTML language allows to reach data on any station on the network using URL reference (Universal Resource Locator). Gateways can be generated to link this technical documentation system with other existing applications like quality monitoring and diagnosis system.

3.2 Hyper-document meta-model

Nevertheless, it is necessary to define a methodology for the organisation and indexation of an electronic documentation system (Comparot-Poussier and Christment, 1994). This methodology is based on the representation in object classes in order to provide a conceptual graph for electronic guidance. The first step is to build a meta-model describing the documentation hierarchical structure. The second step deals with the management of the contents storage (heterogeneous information classification). To achieve these tasks, a software engineering environment including authoring tools and scenario management facilities can be used.

The architecture of the Technical Information System (called Hyper-Document) uses a documentary database approach. A structuring model applicable to different kinds of documents and the management mechanism

to allow a useful navigation are now described. Even more difficult is the challenge of defining the interactivity of various elements inside a document page.

To build an hyper-document, the definition of the general hierarchical structure (map) must first be defined, then each document content, i.e. components associated to a specific information page, must be detailed. The links, named 'Anchors' in HTML language, can point toward an URL address, to any local page, to text on any local page, to movie file/image file/audio file and so on.

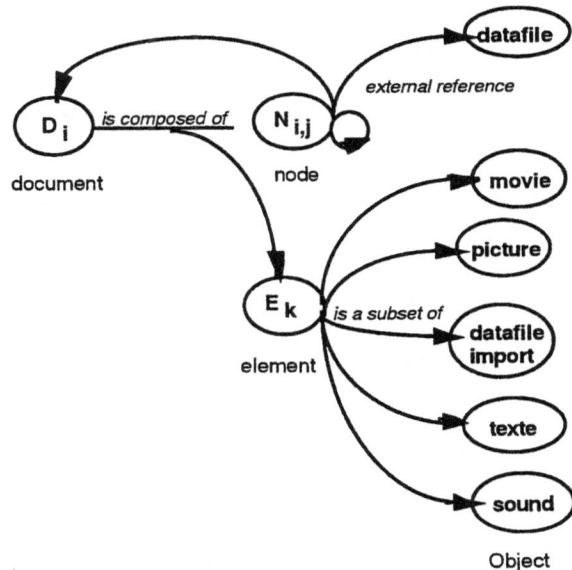

Fig. 2. Meta-model of document structure.

As presented in figure 2, a document in WWW Form is composed of a list of nodes ('Anchors') and a list of elements. Nodes allow links between documents and between documents and external files. They avoid the redundancy of information. This meta-structure preserves maintainability and coherence.

3.3 Hyper-document structure analysis

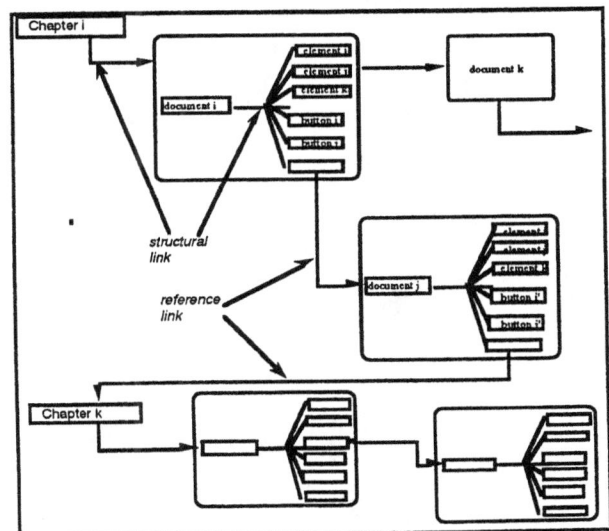

Fig. 3. Hypergraph representation.

Four levels of decomposition, presented in figure 3, have been defined: hyper-document, chapter, document, information component (element).

A database, named hyper-document dictionary, must be created in order to manage the various information elements.

Table 1 Hyper-document hierarchical structure

chapter attribute
- name
- code
- description
- list of document pages: {...}
- link with other chapter
- version
document page attribute
- page name
- code
- description
- domain
- list of components:
{(element i, internal), (element j, external), }
- version number
- number of screens
- version
element attribute
- file name
- label
- code / internal reference
- nature (text, image, sound, movie, tag)
- format (jpeg, gif, quicktime, mpeg)
- storage_volume
- storage_region (directory, local/remote,...)
- node_reference (link, tag)
- last_update_date
- version

Table 1 represents the structure of the hyper-document dictionary. At each level, entities must be defined using an object representation. This description must be completed before realisation of the Webpages (WWW documents). The dictionary will be useful for hyper-document maintenance. Cross-references are included into the objects dictionary.

4. AN HYPERBASE FOR QUALITY MONITORING AND TECHNICAL INFORMATION MANAGEMENT

The complexity of the steel plant, which incorporates the latest technologies from a variety of disciplines, is paralleled by the problems associated with accessing various information sources. An information retrieval system can greatly improve maintainability and accessibility of the information for diagnosis. The system allows the user to acquire, gather and analyse knowledge from multimedia sources and provides the user with background learning resources (Leung, *et al.*, 1995).

Building such a system involves several phases:

- information system modelling according to many points of view (equipment decomposition, process functionality, quality monitoring, maintenance...),
- communication system modelling based on a graph representation of object classes,
- documentation meta-model structuring: hierarchical structure of database, document format description, navigation storyboard,
- implementation using HTML language, automatic indexation using HTML translator (Raggett, 1994).

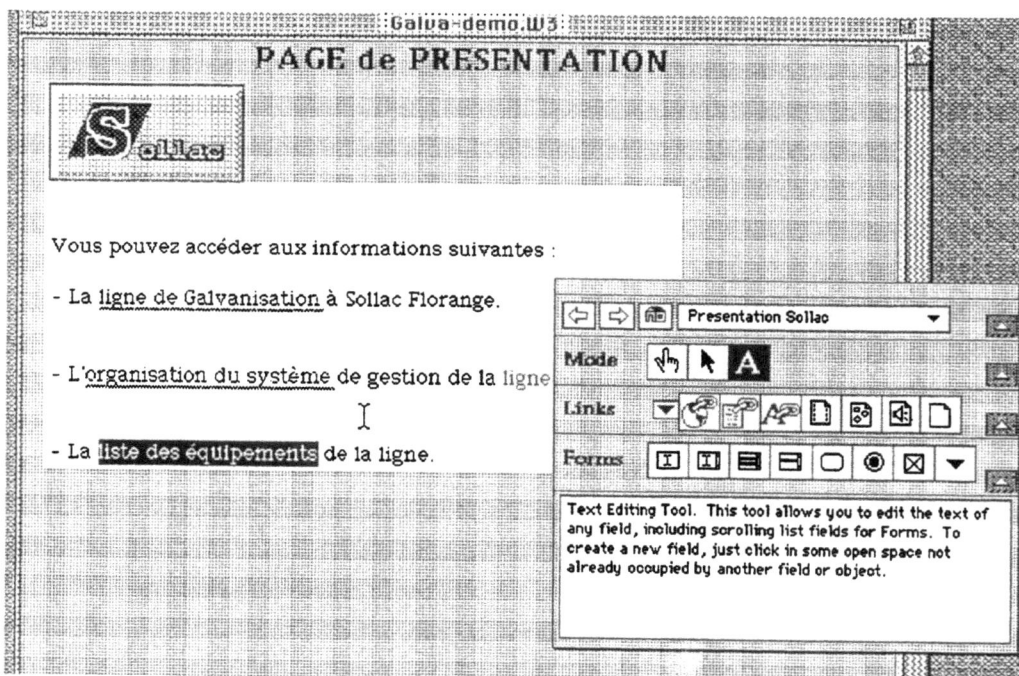

Fig. 4. Webpage design tool.

A prototype of the H.T.I.S. is still being developed for the galvanising line. The following application domains, which have been selected, are: line decomposition analysis (hierarchical equipment list), network organisation (information system architecture), functional analysis and supervision (measurement databases, faults trees, knowledge sources, etc.). This domain list must be applied to the Chapter breakdown of the Hyper-document. Graphic layouts for Webpages design have been used. The Arachnid software is a new WYSIWYG Authoring class tool, for building Webpages in the Macintosh environment (Arachnid, 1995).

All elements incorporated are visual features:
- importing function
- display or playback of audio (AIFF), video (QT) and graphics (GIF, PICT, JPEG) files
- placement of objects
- cloning of in-line GIF image files
- scaling and aligning of Header Rules
- drawing tools
- interactive objects such as editable text fields, scrolling list, radio and command buttons, Popup Menu buttons.

The first Webpages of the Hyper-document are presented below. The Meta-structure of this Hyper-document is based on a specific dictionary (database) of all information components collected. An example of the data dictionary record is detailed for page DL1.2.1.

The line domain has been broken down hierarchically into 5 levels: (cf. Hypergraph structure)
n1: galvanising line
n2: entry section, centre, exit section
n3: furnace, zinc crucible, cooling tower,, quenching section, skin-pass, chromatation
n4: furnace entrance, fire-chamber, temperature maintenance, slow cooling chamber, fast cooling chamber, equalisation chamber
n5: burner.

The dictionary database takes up the following information for each entity:
- attribute of a document (Webpage):
    name: Furnace_Z.html
    code: DL1.2.1
    description: furnace breakdown
    domain: LINE
    list of component: {(Doc_furnace.txt, int.), (sch_furnace.gif, ext.), (Anchor_C1, ext.)...}
    version number: 2
    number of screens: 1

Figure 5 shows some HTML pages generated from our H.T.I.S. using the Netscape Navigator Browser. This figure displays an example of hypermedia capabilities: a scheme ('structure de la section centrale') inside the Webpage DL1.2 and another picture ('MiniFleurage') called by external reference link (Anchor inside text document).

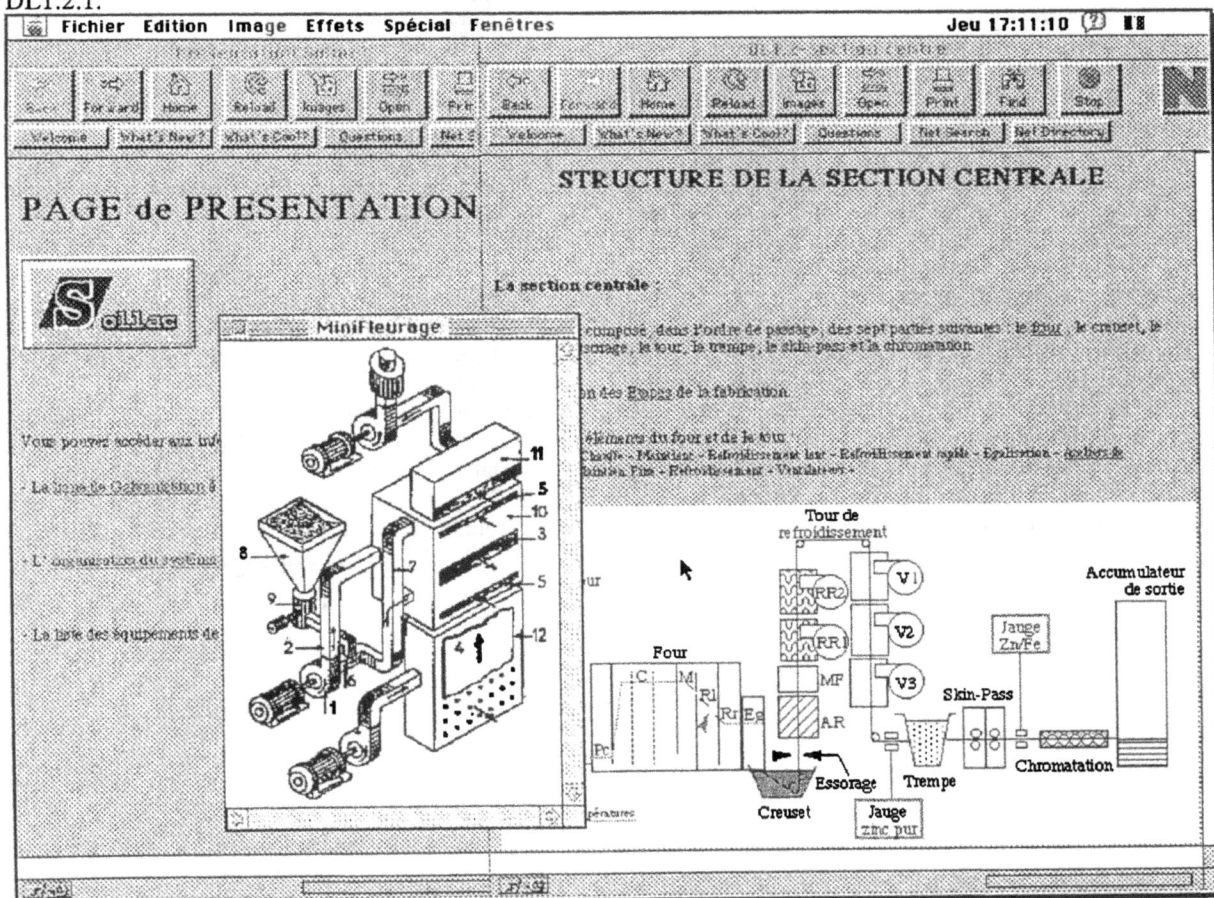

Fig. 5. Netscape WebBrowser example.

5. CONCLUSION

The contribution of this paper can be summarised as an illustration of multi-level H.T.I.S. modelling for quality monitoring and maintenance of complex manufacturing systems. A hyper-document based on HTML language provides a useful tool for making the vast amount of associated documentation readily accessible to a wide range of users.

The technical aspect of an Hyper-document development is less important than the methodological point of view. The main difficulty consists in translating the paper information into available numerical element for building Webpages. The adopted methodological approach is based on several steps. The first one consists in defining a meta-model of document structure. The second step is to realise the hypergraph representation fitted to the different information domains. The last step is to draw the script definition allowing an interactive navigation (reference link).

Technical Information Systems will be integrated in future on such Webpages design tools. All HTML browser applications are currently found on commonly used platforms (PC/Windows, Macintosh, Xwindows system, ...). New technology integration will bring more interactive multimedia features to change the character of W3 from static pages to dynamic presentation. Users use directly Webpages on-line hypermedia documentation. They can also use the Webpages interface to write and to update information directly into H.T.I.S. databases. In this case, users will appropriate the Hypermedia Documentation system.

Combining the features of on-line hypermedia documentation systems with supervision systems (Roth, *et al.*, 1994) can be useful in situations where maintaining and disseminating documentation is regarded as an important problem.

The first version of the developing hypermedia system is not yet set up at Sollac Florange. The aim of the final version will be to provide operators with a useful documentation tool and to define the gateways with different applications of line supervision. W3-tools interface are very easy to learn, so operator training will be facilitated.

Moreover, as the structure of the galvanising line is the same as many other production lines at Sollac, it would be interesting to transpose the W3 tools H.T.I.S. to the other production lines.

REFERENCES

Arachnid beta 1.5.4 (1995) Second Look Computing, University of Iowa, *http://sec-look.uiowa.edu/*

Berners-Lee, T., D. Connolly and K. Muldrow (1994). HyperText Markup Language (HTML) version 2.0. - *Request for Comments Internet (ftp://www.ics.uci.edu/pub/ietf/html/htmlspec2. ps.gz)*

Comparot-Poussier, C. and C. Christment (1994). Hyperbase for electronic technical document management, *Engineering of Information Systems (Ingénierie des systèmes d'information)*, **2**, (5), pp. 533-570, Hermès.

Drouin, B. (1993). CALS: An essential Strategy for paperless interchange of technical information for manufacturing - current and future standards, *Proc. of Computers in Design, Manufacturing and production*, 7th Annual European Computer Conf., p.73-78, May, Paris, France.

ISO, Standard General Markup Language, *ISO-8879*, 1986.

Leung Ruth, F., C. Leung Horris and F. Hill John (1995). Distributed quality manual via hypermedia, *AIAI Conference '95*, June 7-9, Nancy, France.

Mason, H. (1993). Product data representation and exchange - STEP part 1: overview and fundamental principles, *ISO / TC184 / SC4 Status IS*, November 20.

Raggett, D. (1994). A review of the HTML+ document format. *First International World-Wide-Web Conference*, May.

Roth, S., A. Filali, G. Bloch and T Cecchin (1994). Towards an object approach for quality monitoring and diagnosis on a galvanizing line, *Proc. of the Third IEEE Conf. on Control Applications CCA '94*, **2**, pp. 1139-1144, Glasgow.

Spiby, P., and Schenck, D. (1992). Industrial automation systems and integration - product data representation and exchange - part 11: Description methods: The express language reference manual. *Tech. Rep. ISO DIS 10303-11 Draft International Standard*, ISO.

BUILDING A NETWORKING CELL: THE CASE OF A FINNISH ENGINEERING WORKSHOP

T. Alasoini[*], R. Hyötyläinen[**], S. Klemola[***], P. Seppälä[***], K. Toikka[**]

[*]Ministry of Labour, Research Unit, P:O. Box 524, FIN-00130 Helsinki, Finland
[**]Technical Research Centre of Finland (VTT), Automation, P.O. Box 1301, FIN-02044
VTT, Finland
[***]Finnish Institute of Occupational Health, Topeliuksenkatu 41 a A, FIN-00250
Helsinki, Finland

Abstract: In order to improve their competitiveness, manufacturing companies changed their mode of production from functional to cellular manufacturing in the 1970s and 1980s. In the past few years, a lean and flexible mode of operation, which emphasizes cooperation and networks, has been advocated as the result of increasingly tough competition on the world market. This paper presents a model and a construction method for a production cell capable of network-like cooperation and continuous improvement.

Keywords: continuous improvement, flexible manufacturing, lean manufacturing, machining, networking cell

1. INTRODUCTION

1.1 Background

Work groups, teams and production cells have been a much-debated issue in manufacturing industry recently. Companies are interested in new forms of work organization for a number of reasons. They hope these will make production flow more smoothly, further customer orientation, promote automation, and diversify job contents.

This paper presents a model and a construction method for a production cell capable of network-like cooperation and continuous improvement. The model, called a *networking cell*, was built in the course of a multidisciplinary research programme called 'Work, Culture and Technology' (1991-1994), which comprised development projects in four engineering workshops. The programme aimed to use *experimental development research* to study and promote ways of applying the *lean and flexible mode of operation* in Finnish industry. The aim of the

research was to achieve changes in the mode of operation along four development dimensions (Alasoini et al., 1994): 1) group and network relations, 2) tools and procedures for development work, 3) professional skills, and 4) work and enterprise culture.

1.2 From functional layout to cellular manufacturing

When demands for flexibility began to grow in the 1970s, the weaknesses of functional production, typical of manufacturing companies at that time, became increasingly clear. Companies started to seek solutions to the problems of long lead times, large work-in-process (WIP) inventories, complicated material flow and 'robust' production management by adopting *group technology* and *cellular manufacturing* (Tidd, 1991).

In the 1970s and 1980s, companies changing over to cellular manufacturing managed to achieve

35

substantial improvements in their production. These mainly concerned lead and delivery times, productivity, flexibility, material flows, number of handlings and size of WIP inventories (Alford, 1994).

In the past 10-15 years, however, the business environment has changed in many ways. Demands concerning quality and delivery times are tighter, individual customer tailoring has come to be a much more important means of competing, and the speed of product renewals faster. It is no longer enough for many companies to have a good production flow and hence *operative flexibility*, i.e. the ability to respond rapidly to customer needs. Today, a company must adapt to constant change while also sharpening its competitive edge by constantly improving its products and processes. To achieve this longer-term *tactical and strategic flexibility* a company must learn how to exploit the full potential offered by new technologies and the improved competence and changed values of its workforce. Promoting multi-skilling of staff, building closer collaboration between different operations, and creating a corporate culture based on mutual trust are all key elements here.

Paying too little attention to network relations of the cell to other operations and to development work inside the cell can be considered shortcomings typical of cellular models in the 1970s and 1980s from the point of view of new demands for competitiveness.

1.3 Cells in a lean and flexible mode of operation

In the past few years, essentially Japanese production techniques have revolutionized views about how production should be organized in the West as well. Current debate on the *lean and flexible mode of operation* (Adler & Cole, 1993; Womack et al., 1990) or *world class manufacturing* (Harmon, 1992; Schonberger, 1986) derives from the use of these techniques in modern industrial production. They have also provided one important impetus driving companies to seek a new perspective in cellular manufacturing.

The cell takes on new significance in this new approach. Unlike the functional layout or the conventional cellular model, planners, foremen and the cell form a close, mutually interactive network in manufacturing management.

In production that flows fast, and practically without safety nets, support functions must be located as near as possible to the manufacturing process. The cell bears much greater responsibility both for quality and for preventive maintenance. The new idea of a cell also calls for much greater insistence on including support functions in the cell's responsibility than was the case with the conventional cell model. At the same time, it also calls for closer cooperation between the cell and its outside support staff.

A customer-oriented viewpoint demands constant improvements aimed at meeting customer needs even better. Continuous improvement, however, is not possible if it is done solely 'from the top down', solely by engineers and planners. On the other hand, effective development work is not possible merely 'from the bottom up' either, i.e. solely by the cells' own resources. These are often inadequate, if only for the reason that their main problems usually impinge on the 'territory' of many different operations. Constant improvement calls for (i) close cooperation between the cell and its support staff and (ii) systematic tools and methods to back up problem-solving, decision-making and learning by the cell and its support staff.

Networking cooperation and continuous improvement are not possible without common tools and methods which can be used to create the common frame of reference needed for communication and efficient problem-solving by the various parties involved (Cole et al., 1993). The main thing is to use these tools and methods systematically and to ensure that their use is 'democratized', i.e. extended from the engineers and planners, by means of training, to those working in cells (Adler & Cole, 1993).

2. THE NETWORKING CELL

2.1 Features of the networking cell

The present research created a new kind of cell model, called a networking cell. The crucial element in the model is that it sees the cell as part of a company's internal and external cooperation network. This demands new properties from the cell, such as multi-skilling, group work and job rotation, and also involvement in planning and support functions and development. On the other hand, it is only in cooperation with the network that the cell can achieve and sustain these properties (Figure 1).
A networking cell can be said to have the following features:
- In a cell, employee skills are expanded through training and job rotation to take in new work operations. Flexible shifts from one job to another also increase manufacturing flexibility and diversify the work.
- Each cell has its own administration and a spokesman in charge of it. This spokesman is involved in, for instance, planning detailed weekly

work scheduling in the cell and the necessary work arrangements. The cell holds meetings at regular intervals at which topical issues from the point of view of the cell are dealt with together with the production manager, foremen and support staff.

- Each cell is intimately linked with the workshop organization and is responsible for certain support functions, partly independently but mainly jointly with the necessary support staff. The scheduling of the cell's work is discussed and agreed at weekly meetings which are attended by the production manager and foremen as well as spokesmen of the other cells.

- Each cell is involved in developing products, processes and the work environment. This calls for a systematic approach in which production events and cell problems are monitored and taken up for joint discussion.

2.2 Constructing a networking cell

The planning stage is a 'laboratory' in which the first germs of the new mode of operation are created. That is why the planning and adoption of a networking cell differ from those of a conventional cell. Production targets set solely by the management are usually the starting point for planning in a conventional cell model. The cell model is presented to employees as a ready solution, and they can only be involved in 'fine-tuning' the model as far as

details of their workplaces, jobs and work environment are concerned. Planning proceeds in a linear and hierarchical manner, in successive stages and from the top down, within the organizational hierarchy.

There are two main problems with this kind of planning approach. Firstly, the aims of the change remain external and alien to the employees, which tends to prompt resistance to change. Secondly, employee expertise is not utilized, or utilized only to a limited extent. It is not possible to change the mode of operation; indeed, this may not have been the aim.

The construction of a networking cell can be described as a staged development cycle and the principles behind it can be crystallized as follows:

- *Network-like cooperation*: All the employees and necessary support staff of the future cell are involved in constructing it. This broad participation is the precondition both for sufficient expertise and for commitment. Determining the goals for the new mode of operation and the way they will be achieved calls for all these various viewpoints to be taken into account and jointly discussed. A single new realistic model can only emerge as a result of such discussion.

- *Linking the goal to problems, and potential and tools for solving them*: The aim of the change is based on a detailed analysis of the present state of the company, its problems and the potential for their solution. This prevents the goal from being kept at too general a level. The networking cell model is honed at the various stages of the development cycle. At the same time, views clarify concerning the preconditions, means and tools for achieving it.

- *Systematic approach and tools*: Constructing a new mode of operation presupposes systematic and phased progress forward, from analysis of the present mode of operation and its problems, to planning, experimenting and establishing a new one. It is impossible to deal with a large amount of information jointly without creating and using written tools such as forms, models, reports, etc. and without disciplined project work. To qualify as such joint tools, all items must be clear and simple.

- *Shifting the tools and procedures into continuous use*: The tools and procedures used to construct a networking cell remain in the continuous use of the cell and its network. The construction stage is the 'laboratory' in which solutions concerning the networking cell can be planned and tried out in special conditions, without the time pressures of the production process, and also drawing on extensive expertise. Only after this does the new mode of operation really get put to the test.

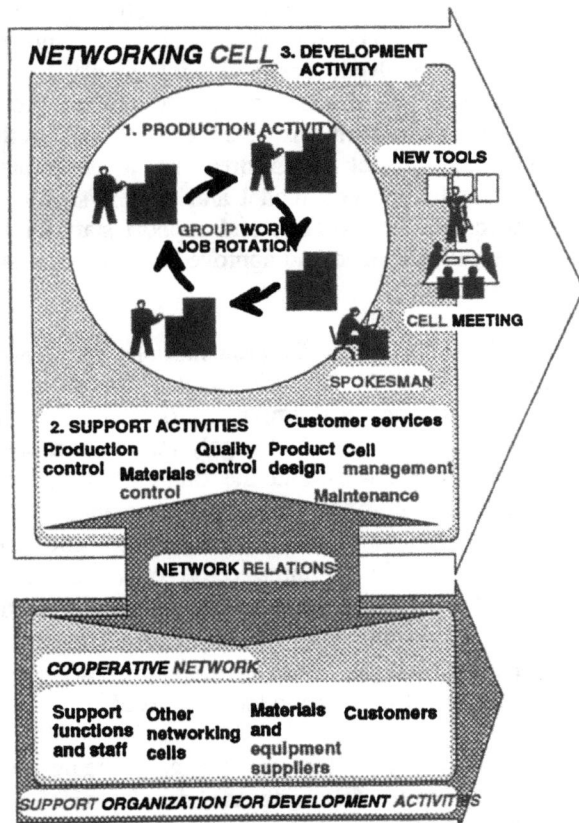

Figure 1. The networking cell.

3. A DEVELOPMENT PROJECT IN A WORKSHOP

3.1 Background to the project

The research took place in a 600-employee engineering company belonging to a large Finnish group making machinery for the chemical wood-processing industry. Total group turnover is FIM 8,300 million and the full staff figure 12,100. 64% of the turnover derives from paper and board machines, where the company has a 26% share of the world market.

At the moment, production in the company is divided into five workshops. The researchers took part in a project aimed at constructing a networking cell at the 100-employee roll workshop.

Roll workshop manufacturing comprised mostly machining. Workers usually did piece work on a single machine and there was no regular job rotation, group work or cells. Nearly all the workers were skilled machinists, however, and many of the machine tools were similar lathes. This meant that, if necessary, many employees could shift from one machine to another. Workers were also responsible for machine setup, quality assurance, preventive maintenance and to some extent the programming of

CNC machine tools. The scheduling of work was completely the responsibility of the foreman. The roll workshop had no quality circles or other permanent development groups involving all the employees.

The decision on the project was made in autumn 1992, and in spring 1993 tube roll shell machining, which involves eight employees and five machines, was selected as the pilot cell.

3.2 Carrying out the Development Project

Changeover to the lean and flexible mode of operation can be viewed as a development cycle with a particular basic structure. In practice, the boundaries between the various stages of the cycle are moveable, of course, and their intensity and duration vary from case to case. The pilot cell was constructed as a development cycle in three main stages, each one covered by the research project.

I. Basic Analysis (May-September 1993).After the choice of the pilot cell, a *project group* was set up at the roll workshop to make a detailed proposal for the construction of the cell. The group was chaired by the production manager, and also included a foreman, two workers of the future cell and the researchers.

An analysis was made of the existing mode of operation for tube roll shell machining and the problems involved, based on interviews by the researchers and discussions in the project group; solutions to the problems were then outlined. On this basis, the project group drew up a preliminary proposal for the cell model and how it should be constructed. The workers and support staff for the future cell discussed and approved the proposal at a joint meeting.

II. Planning and Experimentation (September 1993-June 1994). The plan called for an initial charting of the most important operations of the future cell. This stage had largely formed part of the project group's work and the final selection made was: manufacturing, maintenance, manufacturing management, materials management and cell management (in all three cells constructed later, product design was also included). Thereafter, a start was made on changing the way each of these was done, as well as altering the job descriptions, cooperation structures and tools involved.

According to the plan, the change is being made stage by stage, following the development cycle shown in Figure 2: first, an analysis of the operation's development history and present state (stage 1) and its problems (stage 2), followed by

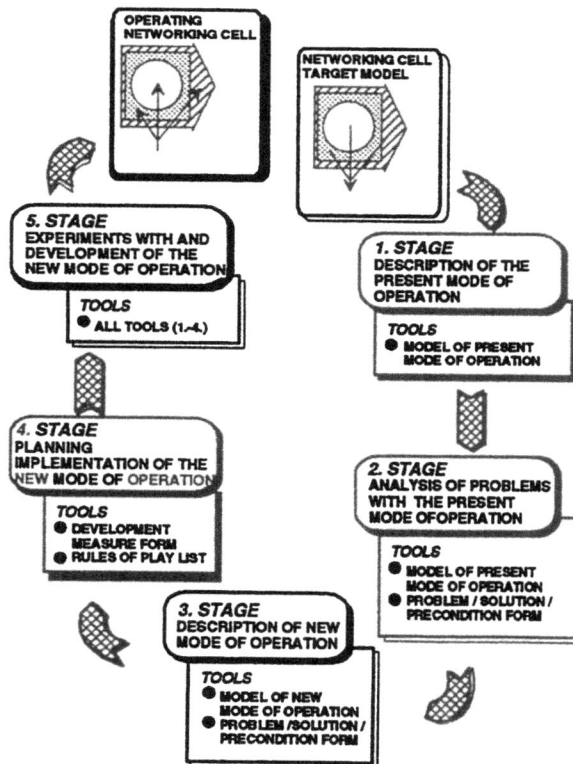

Fig. 2. Planning and experimentation of networking cell operations.

outlining of a new mode of operation (stage 3). This entity, together with the plan for a new mode of operation (stage 4), is discussed at one 4-8 hour cell meeting, preceded by preparatory work by the project group and researchers. When possible, an experimentation of the new mode of operation (stage 5) can also be launched at the cell meeting, e.g. in the form of a group work. However, the experimentation would normally only take place afterwards, as part of the cell's day-to-day operation. The various operations were discussed in the pilot cell at seven meetings phased according to the above-mentioned cycle stages. Altogether 15 people from the company took part: 8 workers, the production manager, 2 manufacturing foremen, the maintenance foreman, the purchasing manager, the product designer and the pre-production planner. Meetings were chaired by the production manager. The amount of time, responsibilities, procedures (lecture, discussion, group work, etc.) and the tools (models of operations, forms, lists, documents, diagrams, etc.) were planned carefully in advance by the project group. The researchers took the main responsibility for preparing and running the meetings.

The models illustrating how the various operations were performed took on particularly great importance in the workings of the meetings. These showed the parties involved in the operation concerned, their responsibilities and the interaction between them. These operational models made it possible (i) to form a homogeneous view of the present object of planning, (ii) to systematically analyze the present mode of operation, and (iii) to hold a rational discussion of the problems encountered in the present mode of operation.

At the cell meetings, most of the time was spent on planning manufacturing management (10 h). This included also an extensive group work, in which the workers tried out scheduling of work within the cell in different situations. The company personnel spent roughly 300-350 h on the seven cell meetings and the necessary preparation for them.

Multiskilling. One essential goal in building the networking cell was to change the 'one man - one machine' mode of working more flexible.

The change process was started by analyzing the problems of the current division of labor and potential benefits of the new situation in which each worker would manage all of the manufacturing tasks within the cell. To find out the current skills and training needs a skills matrix was constructed (including also the workers' training preferences by task). Furthermore, the training time needed to attain the acceptable skill level for each task was assessed in the cell meeting. In addition, the goal

was set on how many workers should be trained during the following two months.

Networking connections to maintenance. To ensure smooth running of production, a new division of labor and responsibilities had to be planned between the production cell and the maintenance department. The planning was carried out in several sessions in which current organization and mode of operation, as well as problems for maintaining and repairing machines were analyzed and discussed thoroughly. The maintenance foreman was consulted in these sessions. An important tool for the analysis was a chart or model, describing all the parties involved in the various operations, such as preventive maintenance, annual maintenance and minor and major repairs.

Several problems and needs for developing new practicies and tools for cooperation as well as for increasing the responsibility of the cell crew for preventive maintenance and minor repairs were identified in the planning sessions. These dealt with, e.g., intstruction sheets for preventive maitenance, hand tools, lubrication oils, etc.

Establishment (June 1994-). After the planning and experimentation period the researchers have been involved in the follow-up and assessment of the start-up of the cell. Their principal method has been 'participative observation', i.e. monthly meetings in which the members of the cell discuss production-related issues with the production manager and foremen. Separate meetings have been held to discuss the progress made in the cell's development measures. The cell has also been trained in business economics and labour protection.

3.3 Results of the Development Project

A change in the mode of operation of a company is a complex and gradual process which can take years. The change involves several development cycles of different levels and lengths. In addition, a change in the mode of operation in only one limited part of the production system, such as a single cell, does not usually remain permanent for a longer period, unless it is extended to the whole organization. At the roll workshop, the development project is being continued by changing the whole mode of operation of the factory and placing its entire production process on a cellular basis. The factory-of-the-future model at the workshop is called a 'networking factory', i.e. the key features of the networking cell are extended to the level of the entire workshop (Hyötyläinen & Simons, 1995).

There are always particular problems involved in implementing pilot projects. In tube roll shell machining, there were three. Firstly, it was not yet possible to change the foreman's job description while setting up the cell, though the foremen did have some new tasks to perform. A project aimed at changing the job description of all the roll workshop foremen was launched in spring 1995 as part of the construction of the 'networking factory' model. Secondly, the pilot cell pay scheme could not be changed at the construction stage. However, the aim is for a changeover to a workshop-wide bonus scheme in summer 1995, which will include both the workers and the white-collar employees. Thirdly, this was a *research* project which largely merely created and tested the new tools and methods needed to construct a networking cell.

So far, the main results of the tube roll shell machining cell project are:
- Worker know-how about the cell's manufacturing operations has grown. The employees still work mainly on a given machine as before, but they move as needed, and on their own initiative (before only when ordered by the foreman), from one job to another. The employees did not want systematic job rotation.
- The system of preventive maintenance has been improved and the 'rules of play' clarified. When the cell was constructed, 12 development measures were made on four machines in the cell.
- The tube WIP inventories are smaller and lead times shorter. This is the result of the development measures carried out when the cell was set up. The measures have improved raw material quality, simplified material flows, improved tube handling and made the division of labour in the cell more flexible.
- The cell has a rotating spokesman. The spokesman is responsible for planning the detailed work scheduling in the cell. He also takes part in the weekly meetings which deal with the workshop work load, together with the production manager, foremen and other cell spokesmen.
- The cell holds monthly meetings which are also attended by the production manager and foremen. Under the production manager, these meetings go through the present and upcoming production situation, the overall cost budget, development measures in progress, upcoming holidays, etc. The meetings also make decisions about new development measures. The cell itself is responsible for some development measures, the cell support staff for others.

Cell operation is still far from the networking cell model set up as the goal for the project. This is not a surprise, as it was obvious when the project was started that a change in the mode of operation inevitably takes time. The company management nonetheless firmly believe that the company should start converting production to the cellular model and

altering the workshops' mode of operation in the manner tried out in the pilot cell. The development project involving the researchers has been adopted by the company as one of the three strategically most important goals for this year. The attitudes of the staff to the project have also been positive. For instance, the feedback questionnaire arranged in the pilot cell showed that everyone considered that the way in which the cell had been constructed was either 'extremely' or 'rather viable'.

REFERENCES

Adler, P.S. and R.E. Cole (1993). Designed for learning: a tale of two auto plants. *Sloan Management Review* **Spring/1993**, 85-94.

Alasoini, T., R. Hyötyläinen,. A. Kasvio, J. Kiviniitty, S. Klemola, K. Ruuhilehto, P. Seppälä, K. Toikka and E. Tuominen (1994). *Manufacturing change. Interdisciplinary research on new modes of operation in Finnish industry. University of Tampere.* Work Research Centre. Working Papers **48/1994**, Tampere.

Alford, H. (1994). Cellular manufacturing: the development of the idea and its application. New Technology, *Work and Employment* 9, 3-18.

Cole, R.E., P. Bacdayan and B.J. White (1993). Quality, participation, and competitiveness. *California Management Review* 3/**1993**, 68-81.

Harmon, R.L. (1992). *Reinventing the factory II. Managing the world class factory.* Free Press, New York.

Hyötyläinen, R. and M. Simons (1995). The network cell as a step to network factory. A topic article for the *IMSS-Book*. Manuscript.

Schonberger, R.J. (1986). *World class manufacturing. The lessons of simplicity applied.* Free Press, New York.

Tidd, J. (1991). *Flexible manufacturing technologies and international competitiveness.* Pinter, London.

Womack, J.P., D.T. Jones and D. Roos (1990). *The machine that changed the world.* Rawson., New York.

WORK PSYCHOLOGICAL TASK ANALYSIS FOR THE DESIGN OF SHOP FLOOR MANAGEMENT

Martina Zölch

*Work and Organizational Psychology Unit, Swiss Federal Institute of Technology (ETH),
Center for Integrated Production Systems, CH-8092 Zurich*

Abstract: The design of shop floor management which focuses on the joint optimization of human resources, technology and organization is based on decentralization of decision structures and functional integration, qualification of the employees and locally controllable computer-aided manufacturing execution systems (MES). Integrating these design issues is the only possibility to achieve the immense flexibility shop floor management requires. Business Process Re-engineering (BPR) methods promise this raise of flexibility but often neglect human resources which we see as the central key to organizational flexibility. Psychological task analyses offer to BPR an enhanced alternative. It adresses the fact that real persons have to cope with the tasks resulting from BPR and provide criteria to evaluate various design alternatives. One method, the "Contrastive Task Analysis", is introduced. How shop floor tasks are to be evaluated is explained. Lastly, a procedure is proposed which combines both process-orientation and work task focus.

Keywords: Business process engineering; co-operation; communication; human-centred design; production control; shop-floor oriented systems; socio-technical system design; task; work organization;

1. INTRODUCTION

Turbulent environments expressed by heterogeneous job structures, customer focus and shorter cycle times require increasing flexibility on the shop floor. Organizational approaches to meet with these demands consist for example in process orientation and introduction of work groups while technical support is mostly realized by so-called computer-aided manufacturing execution systems (MES) or finite-scheduling systems, in German "Leitstand".

However, function and technical-data orientation on the shop floor management dominates. This results in centralized decisions concentrated on a small number of employees, long decision paths and hence, a lack of possibilities for the majority of the workers on the shop floor to utilize their experience.

Additionally, the way computer-aided manufacturing execution systems are actually used in companies shows that central planning and control concepts prevail. Although praised as an aid for decentralized planning on the shop floor, the reality in the companies shows that an MES is, in fact often simply used as appendage of already implemented production planning and control systems and not as an independent planning system specifically for the shop floor (c.f. Köhler, 1990). Lastly the number of installed functions in an MES is often immense thus too complicated for the average worker to keep an overview.

The design of shop floor management which focuses on the joint-optimization of human resources, technology and organization is based on
* the re-integration of planning sequences for workers on the shop-floor,
* the qualification of the employees and the competent use of their abilities through an adequate design of work tasks supported by the required training measures, as well as
* new technologies, which are locally controllable and modifiable so the workers can keep the process of shop floor management under control.

Hence the realization of decentralized production and decision-structures are not be hindered.

This, however, does not simply occur by itself but requires, in addition to the above design principles, a precise understanding of the characteristics of shop floor management tasks, as well as the use of specific methods which cope with the demands of human, technology and organization (c.f. Ulich, 1994).

Reorganizing an enterprise, including the shop floor management, does call for a great variety of methods such as are included the MTO-analysis (orig. German: Mensch - Technik - Organisation) (c.f. Strohm & Ulich, in press). Design, in practice, draws on as many methods as is appropriate. However, one particular methodology, that of the psychological task analysis, is discussed with regard to the perspective of process-orientation.

2. CHARACTERISTICS OF SHOP FLOOR MANAGEMENT TASKS

Shop floor management includes planning, processing and monitoring job orders. Typical shop floor management tasks are
- utilizing resources
- taking on the responsibility of production time limits
- fine scheduling of various work orders
- monitoring the job orders
- monitoring the events on the shop floor
- database administration
- monitoring the material handling as well as
- drawing up statistics and evaluations.
(c.f. Gottschalch & Vöge, 1993)

Coping with disturbances such as machine down time, missing raw materials, cases of employee illness or express job orders are characteristic demands on people who manage the shop floor. An abundance of problems concerning the process flow, time limits and capacity, e.g. errors in PPS-data, overload of capacities, permanent time pressure due to delay, have to be anticipated. As a result, one of the main objectives of shop floor management consists of ensuring a smooth and orderly process flow. This entails keeping deadlines and high quality standards and production costs low in spite of the unfavourable conditions. Workers are therefore often confronted with contradictory demands which can have a stressful effect on them (c.f. Moldaschl, 1990).

In the case of heterogeneous job order structures and required customer focus, shop floor management occurs in very incalculable and uninfluenceable surroundings. Thus, situative action and particularly feedback-directed action is of vital importance (c.f. Schüpbach, 1994). It is necessary to actively interpret the planning data in each actual situation. Collectively remembered reference points, the avoidance of cognitive effort and a limited necessity to plan in advance seem to be crucial in coping with

shop floor management tasks (c.f. Mertins, Schallock, Carbon & Heisig, 1993; Planleit, 1994).

Based on the specific demands of shop floor management tasks outlined, it is obvious that an immense organizational and technical flexibility is a design objective. Because Business Process Re-engineering (BPR) methods target raising flexibility, the process-oriented approach is taken as a starting point also for the design of shop floor management tasks.

3. THE PROCESS-ORIENTED FOCUS

In modern management and production concepts, the process-oriented focus has replaced the conventional vertical view on an organization. Business processes are described as an informational or physical transformation and are performed by either human beings or technological resources. Time-based and target-oriented structuring of these processes leads to the formation of the overall process organisation. (c.f. Picot & Maier, 1993). In BPR projects, one of the main steps consists of establishing complete and unbroken chains of business processes with undivided responsibility. The identification and design of business process chains can be done according to products, targets groups or technologies. BPR usually also reduces hierarchy levels.

However, the current theoretical background reality of re-engineering projects can lead to big problems and risks. Particularly the development of the organization at the operative level is addressed insufficiently. The deliberate technical "top-down" approach of BPR attempts to sketch optimal business processes in a given way on a drawing board instead of carefully developing an organziation with participation of the employees. The fact that with re-engineering, as is the case with any other technical or organizational innovation, future work tasks with specific demands are created that persons have to cope with. This is also apparent with the newer BPR methods and software-tools which are on the market for specification and design of business processes (e.g. Scheer, 1994).

Certainly existing computer-aided tools and methods offer easy support for formal process specification even at high complexity (c.f. Scherer & Zölch, in press), but a criteria-based design of human, organization and technology is not supported adequately by this type of method. In contrast to the importance attached to identifying the genuine events of business processes and asking "when to do something" the creation of an organizational unit and a procedure of "how to cope with these events" is not formulated. This is noticeable because of the absence of criteria for defining and evaluating variations of business process chains. So the risk of splitting business process chains into those of complex and problematic nature and others which consists only of routine matters looms. Instead of traditional hierarchical structures a new form of establishing hierarchies enters unnoticed back in through the back door (c.f. Osterloh & Frost, 1994).

Because of the lack of content-related criteria and the missing relationship to work tasks, BPR-projects are liable to neglect irreplaceable human resources which are vital in coping complex and critical processes. Psychological task analysis offers an enhanced alternative.

4. FROM THE WORK TASK BOTTOM UP

Work psychological task analysis instruments like they have been developed in the German speaking work psychology for the production and clerical sector (e.g. Volpert & Oesterreich, 1991, Dunckel et al., 1993; Weik et al., 1994) make an important contribution toward the joint-optimization of human resources, technology and organization in advanced manufacturing systems. The action related work task - the heart of a sociotechnical system and point of intersection between individual and organization - forms the unit of analysis and focus for design efforts in this type of method (c.f. Ulich, 1994). From the work psychological perspective, the working task is not to be regarded as a function of a business process to be fulfilled by a technical system, but as a *task undertaken by a worker* who has to cope with its demands. From the goal of the working task and the working steps and the conditions to reach it, conclusions as to positive and negative psychological requirements for the worker during the working process are made and design proposals can be derived with the aid of this type of instrument.

One such task analysis instrument - the *Contrastive Task Analysis (KABA)* - originally developed for analysis, evaluation and design of computer-aided administrative and clerical work (c.f. Dunckel et al., 1993) - has also been successfully applied to production planning and control tasks. The evaluation of working tasks in this approach is based on human strengths which enable us, for example, to react flexibly in changing environments, with regard to goal orientation, object-relatedness and social nature of human action. Taking human potentials into account appropriately requires that the design of working tasks complies with the demands of work psychological criteria. Tasks that promote human potentials have to offer (1) a large decision latitude, including sufficient planning and decision-making requirements, (2) a temporal scope that allows the worker to structure his or her action with respect to time him- or herself, (3) a clear working context and the possibility to influence the working conditions, (4) a variety of jobs and methods of working, (5) communication and co-operation requirements as well as possibilities of direct communication. Additionally tasks should be (6) related to the material and social reality, (7) offer physical activity and (8) should be free of organizational or technical hindrances (c.f. Dunckel et al., 1993). During a design process, the KABA-method can be used to identify human oriented design potentials at a fairly concrete level based on a detailed recording of existing work tasks or prospective as achieved for example through a proposed business process lay out. The criteria profile of the work tasks can then be contrasted with possible impacts of an existing or planned technology to see if human potentials would be impaired.

5. HOW TO EVALUATE SHOP FLOOR TASKS?

In the following, the three main criteria of the "Contrastive Task Analysis" (decision latitude, communication and co-operation requirements and task-related work load) are explained with respect to shop floor management tasks. The fulfilment of the criteria in work and technical design supports the conservation of worker's experience as well as human and organizational flexibility, and is, therefore, a prerequisite for maintaining employee qualification and creating motivating and humane work places.

5.1 Decision Latitude

Reorganizing the shop floor management requires an idea of the extent to which various current work roles (from the production planner to the machine setter) are involved in doing shop floor management tasks as well as which amount of planning and decision-making requirements their tasks exactly incorporate and to which extent those are desired for the future for the various work roles.

Normally, the amount of planning and decision-making requirements for a single work role depends on the type of organization and how the planning control activities among the workers concerned are distributed. In addition, the extent of the competition among production resources and the fault liability of the production process have to be taken into consideration.

By means of the criterion "decision latitude", it is possible to discuss the amount of planning and decision-making requirements that the performed work tasks encompass. The range of the decision latitude is a result of the degree to which workers are able to act self-sufficiently at their work place and to make their own plans and decisions with regard to goals and the means for attaining them. Seven levels can be distinguished.

Table 1: Levels of the Criterion "Decision Latitude"

Level 1: Performing predetermined actions
Level 2: Determining a course of action
Level 3: Determining a course of action with consideration of the consequences
Level 4: Making a single decision
Level 5: Making several interconnected decisions
Level 6: Making decisions in several areas
Level 7: Deciding on new procedures
(c.f. Dunckel et al., 1993)

Assigning the analyzed tasks to one of these levels gives image of the job sharing concerning the shop floor management. In a further step of critical scrutiny, design potentials for reorganization can be identified. Organizational need for design is indicated if planning and decision-making requirements

- have low levels for tasks performed by operators on the shop floor,
- are concentrated on a small number of workers,
- are distributed over a large number of hierarchical levels, which usually results in long decision paths.

If also the introduction of an MES is planned reflection on the extent to which shop floor management tasks should be technically supported is required. It is imperative that the decision latitude that is organizationally conceded to the worker is not be restricted by a technical system. Also different levels of technical support can be distinguished:

Table 2: Levels of Technical Decision Support

Level 0: Decision is not supported
Level 1: Situative background information is displayed
Level 2: Various decision possibilities are shown (simulation)
Level 3: Decision recommendations are given
Level 4: The technical system makes decisions and follows through

Depending on the complexity of a particular task, various forms of technical support are conceivable. For the prioritizing and balancing of job orders on the shop floor, technical support is normally not necessary, while technical support on level 2 (various decision possibilities/simulation) or 3 (decision recommendations) could be useful for planning machine time, especially critical machines, rush orders, and the co-ordination of material flow. However, the final decision should remain with the user. From a work psychological viewpoint, technical support on level 4 should be foregone also because of limited decision latitude is associated with a low level of flexibility.

5.2 Co-operation and Communication Requirements

The manifold reasons for process disturbances, as well as the competition of departments for common resources that characterize shop floor management tasks demand a high level of communicative, co-operative interactions particularly with adjacent organziational units (c.f. Brinkop & Nullmeier, 1991).

However, not all employees have the same possibilities to participate at co-operation processes, which can often be traced back to an unequal distribution of planning and decision-making requirements. Thus, workers have little influence on the shop floor management and have little or no possibility to utilize their own experience. But exactly that is necessary for an efficient decentralization. Therefore, a further criteria to be evaluated is the extent to which the workers can co-operate and communicate in managing the shop floor.

In the analysis of task related communication, communication is not considered a mere exchange of information, but assumes that an exchange between partners takes place so that agreement can be reached and the task goal attained. Several levels of communication quality, dependant on the complexity of co-ordination and common decisions, are distinguished. These levels are similar to those concerning the criterion "Decision Latitude" (Table 1). A directly personal form of communication should be possible and unfavourable circumstances which may impair or disturb direct communication, like noise or hindrances in sight should be avoided.

Table 3: Levels of "Communication and Co-operation Requirements"

Level 1: Communication about deviations of predetermined actions
Level 2: Communication about how to determine a course of action
Level 3: Communication regarding the consequences of a particular course of action
Level 4: Communication in regard to making a single decision
Level 5: Communication in regard to making several interconnected decisions
Level 6: Communication about decisions to make in several areas
Level 7: Communication in regard to deciding on new procedures
(c.f. Dunckel et al., 1993)

Even if co-operation is supported by MES, the possibilities for direct communication between the users should not be limited by the system. Situative and real-time action using the resources of informal communication are essential. This calls for a technical system which is adaptable to the organizational form of co-operation.

5.3 Task-Related Work Load caused by Organizational and Technical Hindrances

Hindrances to work are usually caused by technical or organizational factors which determine the working conditions and complicate the achievement of the work result. In addition the worker does not have the possibility to overcome a hindrance although there may be organizational or technical solution. Examples of hindrances are *interruptions* caused by other persons, scarcity of resources, or technical faults. Lack of information, unreliable information, or information which is difficult to discern are also hindrances, so-called *informational impediments*.

Lastly hindrances encompass also *motoric impediments* caused by damaged, unsuited and complicated machinery (c.f. Dunckel et al., 1993).

Typical hindrances which occur in managing the shop floor are caused by
- inconsistent data
- different levels of parallel planning and lack of co-ordination
- insufficient data access
- time lag in receiving feedback from the shop floor as well as from the planning department
- creating a "data overload" in collecting and reporting back data which are not relevant for shop floor management
- unclear processing of PPS-data
- unclear responsibilities concerning data up-dates and decision-making
  and so on.

Hindrances have a potentially stressful effect on the worker because the task performance requires additional effort (for example an additional action may have to be recommenced following a "system breakdown") or she or he is forced to take a risk to achieve the desired result. Moreover, additional effort causes an increase in cycle times.

The design target is to remove hindrances by improving machinery and work environment as well as taking organizational measures by giving supplementary authority to solve the problems that lead to hindrances or support a strategy for coping with a hindrance far in advance. Moreover, an appropriate room for co-ordinating planning activities, adequate time set aside for this, as well as a reasonable monetary incentive for doing shop floor management tasks are prerequisites for success.

6. HOW TO COMBINE PROCESS-ORIENTATION AND WORK TASK FOCUS?

Through the combination of formal process-oriented methods with work psychological task analysis instruments, it is possible to counteract the above mentioned disadvantages of the process-oriented focus. The procedure proposed in the following combines both process-orientation and work task focus in the design process.

(1) Process flow analysis gives an initial overview on how shop floor management is organized. A process flow analysis preferably traces frequent, typical or critical job orders retrospectively, which are tracked through the single working stations. Utilizing fabrication documents, people who were involved in selected job orders are questioned with regard to operations performed and information used, as well as for necessary documents (for example warehouse receipts, PPS-print-outs, operation plans, ODA-data).

(2) The recorded data, interview statements and documents collected then serve as a basis in the second step of modelling the information process

with a formal process-oriented tool (like for example ARIS, see Scheer, 1994). With this kind of methods sequences of a business process are to be described as activities and linked to discrete events. Each activity is related to data and information necessary for execution. Additionally, each activity is assigned to its information resource, e.g., a computer system, as well as the affected organizational unit, e.g. departments or employees. The resulting information process model can later serve as basis for the design of a planned computer-aided information system (c.f. Scherer & Zölch, 1995; in press).

(3) In the third step, shop floor management tasks and the specific working conditions at key work places - from the production planner down to the machine setter and operator - are then to be analyzed and evaluated through the method of the "Contrastive Task Analysis". Two alternative situations, that of an existing system or that of a brand new system must be considered.

a. In the case of evaluating an existing system, so-called observation interviews are conducted. This means that the working person is questioned in detail about their work while a researcher observes his or her work activity. Work steps and conditions of working performance, particularly with respect to shop floor management, are recorded in detail as well as specific aspects which are necessary for the evaluation (e.g. decisions which must be made alone or in common, hindrances). In addition to a work psychological evaluation, this procedure implies the possibility for a validation of the documented information processes (step 2) in comparison with the results of the work task analysis.

b. In the case of a brand new system that is to be designed, formal sections of the intended business process should be combined with hypothetical working tasks performed by persons or groups. These scenarios can then be evaluated with the criteria of the "Contrastive Task analysis".

(4) If evaluation results show that human resources are not supported adequately with regard to the outlined criteria, proposals of task distribution are to be developed. For this reason, work tasks are divided up among process sequences. For example, process sequences which contain decision-making and communication requirements are to be identified and redistributed such that human potentials would be supported and a joint optimization of human, technology and organization will be reached. These design proposals can be iteratively evaluated by criteria of the "Contrastive Task Analysis" as well as by other relevant criteria. In this way, various design alternatives can be compared and discussed with employees in a participative process.

(6) In case of the implementation of an MES, the decision to which degree shop floor management tasks are to be supported technically must be made. For this purpose, based on the detailed work task records different scenarios of man-machine function-

allocation can be sketched out and evaluated with regard to positive (e.g. minimizing hindrances) as well as to negative impacts (e.g. reduction of planning and decision-making requirements) of the MES. The work psychological task analysis instrument "KOMPASS: - Complementary analysis and design of production tasks in sociotechnical systems" (see Weik et al., 1994) supplies further criteria for the evaluation of man-machine-function allocation (e.g. coupling, authority, flexibility, transparency and technical linkage) which can not be elaborated in the brevity of this contribution.

7. CONCLUSIONS

In designing the shop floor management, a joint optimalization of human resources, technology and organization is required. The design target of raising flexibility cannot only reached by optimizing the information flow processes in re-engineering business processes. The vital resource for flexibility are humans with their abilities to cope with turbulent, unforeseeable and co-operative demands in managing the shop floor. With the current methods offered by BPR, human resources are not considered adequately. Therefore, psychological task analysis instruments like the "Contrastive Task Analysis" form an important completion for the design of shop floor management tasks. They supply criteria, which consider human potentials and support the organizational development on an operative level and addresses the focus that concrete persons have to cope with the tasks to be designed by Business Process Re-engineering. In this way designed business processes can be evaluated with respect to their implications for qualified work tasks.

The detailed description of the work steps and the work conditions, one result of work psychological task analysis, provide a well-founded knowledge of the single work activities concerning shop floor management. It also supports a common understanding between designers and involved employees, which, inturn, supports the development of practical design alternatives.

Lastly, the development of practically oriented training measures can be devided from the results of task analyses. In comparing the actual qualification profile with the profile of demands for the future work tasks needs for qualification can be identified.

REFERENCES

Brinkop, T. & Nullmeier, E. (1991). Konkurrenz und Kooperation - Zwei Schlüsselkonzepte der Fertigungssteuerung. *ZwF, 86 (12)*, 597-601.

Dunckel, H., Volpert, W., Zölch, M., Kreutner, U., Pleiss, C. & Hennes, K. (1993). *Kontrastive Aufgabenanalyse im Büro. Der KABA-Leitfaden. Grundlagen, Manual und Arbeitsblätter*. Band 5. Zürich: Verlag der Fachvereine/Stuttgart: Teubner.

Gottschalch, H. & Vöge, M. (1993). Werkstattsteuerungen auf dem Prüfstand. *Technische Rundschau, 15*, 70-79.

Köhler, C. (1990). Der elektronische Leitstand - Befehlsempfänger der PPS oder Partner der Werkstatt? *VDI-Z, 132 (3)*, 43-49.

Mertins, K., Schallock, B., Carbon, M. & Heisig, P. (1993). Erfahrungswissen bei der kurzfristigen Auftragssteuerung. *ZwF, 88 (2)*, 78-80.

Moldaschl, M. (1990). Krankheit JIT - Therapie Leitstand. *VDI-Z, 132 (3)*, 40-43.

Oesterreich, R. & Volpert, W. (1986). Task analysis for work design on the basis of action regulation theory. *Economic and Industrial Democracy 7 (4)*, 503-527.

Osterloh, M. & Frost, J. (1994). Business Reengineering - Modeerscheinung oder Business Revolution? *ZfO 63 (6)*, 356-363.

Picot, A. & Maier, M. (1993). Interpendenzen zwischen betriebswirtschaftlichen Organisationsmodellen und Informationsmodellen. *Information Management 63 (3)*, 59-63.

Planleit (1994). *Leitstände für die Werkstattsteuerung*. Schriftenreihe der Bundesanstalt für Arbeitsschutz. Bremerhaven: Wirtschaftsverlag NW.

Scheer, A.-W. (1994). *Business Process Engineering: Reference Models for Industrial Enterprises*. Berlin: Springer.

Scherer, E. & Zölch, M. (1995). Nutzung humanorientierter Potentiale bei der Gestaltung von Geschäftsprozessen. *Management &Computer, (3) 1*, 35-42.

Scherer, E. & Zölch, M. (in press). *Design Activities in Shop Floor Management: A Holistic Approach to Organisation at Operational Business Levels in BPR Projects*. In: Re-engineering the Enterprise (J. Browne (Ed.)). London: Chapman & Hall.

Schüpbach, H. (1994). *Prozessregulation in rechnerunterstützten Fertigungssystemen*. Schriftenreihe Mensch-Technik-Organisation. (Hrsg.: E. Ulich), Band 4. Zürich: Verlag der Fachvereine/ Stuttgart: Teubner.

Strohm, O. & Ulich, E. (Hrsg.), (in press). *Ganzheitliche Betriebsanalyse unter Berücksichtigung von Mensch, Technik und Organisation. Vorgehen und Methoden einer Mehr-Ebenen-Analyse*. Schriftenreihe Mensch-Technik-Organisation. (Hrsg.: E. Ulich), Band 10. Zürich: Verlag der Fachvereine/ Stuttgart: Teubner.

Ulich, E. (1994). *Arbeitspsychologie* (3. Auflage). Zürich: vdf/Stuttgart: Poeschel.

Weik, S., Grote, G. & Zölch, M. (1994). KOMPASS - Complementary analysis and design of production tasks in sociotechnical systems. In: *Advances in Agile Manufacturing* (P.T. Kidd & W. Karwowski (Eds.)), 250-253. Amsterdam: IOS Press.

ACTION REGULATION, CO-OPERATIVE STRUCTURES AND TEAM WORK IN MANUFACTURING SYSTEMS

W. G. Weber

*Work and Organizational Psychology Unit, Swiss Federal Institute of Technology (ETH),
Center for Integrated Production Systems, CH-8092 Zurich*

Abstract: This contribution deals with one of our unit's current research projects. One central goal of the project is to utilise task-related concepts and methods of work psychology for evaluation of group work structures in the manufacturing area. Discussion will include how connections between concepts of the psychological theory of action regulation and the socio-technical systems approach can contribute to this aim. For example, the degree of collective autonomy, i.e. shared requirements for planning resulting from jointly-organised or executed work tasks, is analysed. Our investigation concept will be illustrated: the possibility of correlation between technological and organisational attributes of group work systems, structures of collective action regulation, and characteristics of the task orientation of the group members will be explored.

Keywords: Group work; self-regulation; socio-technical system design; communication; co-operation; human-centered design; quality of work life; production control; work organization; lean manufacturing

1. INTRODUCTION

There is a new boom of "team work" or "group work" in industry, which refers to group work on the one hand, and to discussion in the work sciences about advantages, problems and characteristics of humane group work on the other. In the tradition of the socio-technical systems approach, the terms "group work" and "work group" (cf. Susman, 1976; Alioth, 1980) as a special form of highly co-operative team work in manufacturing systems are preferable. The MIT study about Japanese "lean production" (Womack, *et al.,* 1990) strongly influenced this new wave of work organisation, but the statements concerning quality of work in "toyotistic" manufacturing teams are highly controversial (Berggren, 1992). The authors affirm, for example, that work in toyotistic production, even in a European context, would offer production assembly line workers "humanly fulfilling" activities and "creative tension" (Womack, *et al.,* 1990, pp. 103ff.). Among the features of the toyotistic "group work", however, is that the typical cycle time lasts "... from about one minute in a mass- or lean-production assembly plant", that "... a properly organized lean-production system   does remove all slack...". (p. 106), and that an assembly line without any buffer is the issue at all (p. 92). Among work psychological criteria, (see Ulich, 1989; 1994) such working conditions could hardly be called "humane" or "humanly fulfilling".

How could such a misunderstanding about the quality of the toyotistic work activities occur? The statements made by Womack, et al. (1990) about the humane quality of toyotistic teamwork are based on interviews conducted with managers and a few factory rounds. Neither psychological task analyses nor broader socio-technical system analyses were conducted, apparently. And just this, however, at the very least in a random sample among the factories examined, would serve to either empirically substantiate or reject the optimistic statements about the quality of working conditions. Whoever makes psychological evaluations and proposals for work design has a social responsibility: currently hundreds of thousands of work activities around the world are being restructured according to the lean production model.

2. THE SOCIO-TECHNICAL MODEL OF AUTONOMOUS WORK GROUPS

There are, indeed, "alternatives to lean production" (Berggren, 1992): Since the Sixties, and in both in Europe and the United States, many results have been gathered in research projects (due to the socio-technical systems approach) with regard to promoting personal development and self-enhancing experiences through work design in (semi-) autonomous work groups (Reviews: e.g. Hackman, 1986; Pasmore, *et al.*, 1982; Pearce and Ravlin, 1987; Ulich, 1994). The results are not homogenious. Not every case yielded the same relationship between the organisational characterisics of semi-autonomous groups and psychological - or business administration economics - outcome variables. Many of the sudies considered are case studies from which only a moderate number of generalisations can be drawn. Considering these limitations, medium and long-term *potential effects* of working conditions in semi-autonomous groups can be summarised globally as follows:

- Attitudes towards work change for the better
- Work motivation is increased
- Social and cognitive capabilities and skills are broadened (personality development)
- Absenteeism and employees' turnover are reduced
- The quality of work results is improved
- Productivity of work is increased, the cost of production is reduced
- Safety is increased

These results are often neglected in the current discussion. The socio-technical model of autonomous group work developed by psychologists, sociologists, economists and engineers views the company as an *open system* that defends itself in its turbulent context by self-regulation. (Emery, 1959; Pasmore, *et al.*, 1982; Susman, 1976). By means of the work technology used, the work organisation (technological subsystem) chosen and people (social subsystem), inputs (e.g. raw materials) are transformed into outputs (e.g. products for customers). The "occupational roles" join both subsystems together. Herewith, the principle of "joint optimization" requires that in the design of work, human skills and needs, the work organisation, and the technology should be chosen in accordance with one another. The enterprise system is divided into different *work systems*. A primary work system, e.g. a semi-autonomous group, is a unit of organisation that consists of independent work tasks, tools and workers. Technology as such is not a given, but an option; in other words, it can be designed according to human and organisational needs, as many examples demonstrate (Ulich, 1989; 1994).

For the existence of a company it is important for its management to develop mechanisms that can be self-regulating through flexible, variable, profitable, and - when necessary - innovative response to disturbances in the environment as well as in the manufacturing process. A central goal of socio-technical system design is to identify the so-called "key variances" on the production floor and to develop ideas which would help reduce and regulate these efficiently (cf. Emery and Thorsrud, 1976). Key variances are technology- or organisation-determined variations and disturbances on the production floor that crop up within a work system or between that work system and the others adjacent to it. The formation of (semi) autonomous work groups is a propitious way to regulate key variances directly at the place of their origin. The characteristics of autonomous work groups (synonyms: "self-regulating work groups", "self-managing teams") as representing the socio-technical perspective are described below (cf. Alioth, 1980; Berggren, 1992; Emery and Thorsrud, 1976; Susman, 1976; Ulich, 1994; Ulich *et al.*, 1973).

"(Semi-) autonomous group work" is a principle of work organisation and design that can be characterised as follows: several workers, in a spatially and organisationally *limited production unit,* share a *common task* that is divided into independent subtasks, and assume *shared responsibility* over the long term. The members of this production unit, this "work group", collectively determine ("collective self-regulation"), from a large to considerable degree, the *co-ordination of the work sequence and the allocation of jobs and tasks within their production unit* and the *input/output relation* ("boundary maintenance"). In doing so, every member can generally execute a variety of subtasks *("polyvalence")* and does so, depending on need (flexible job rotation). Inherent to this concept is the *principle of "job enrichment":* the work group is assigned structurally different work tasks, e.g. manufacturing planning and control, operating performance and quality assurance tasks. These are divided such that every member can execute challenging tasks repeatedly ("individual self-regulation").

In order for semi-autonomous group work to function, a few organisational prerequisites must be realised (cf. Ulich, 1994). (1) *Relatively independent organisational units:* the semi-autonomous work system must be technically and organisationally independent of the work systems adjacent to it. This means that the semi-autonomous group can exercise influence over when and in which order their input orders are taken up, and when these should be assigned to a nearby work system as output. Buffers, contrary to the lean production concept, are valid as inevitable and purposeful methods used to regulate the manufacturing sequence efficiently and to avoid mental overload; (2) *Relatedness of tasks within the organisational unit:* the work tasks which made up the work system must be linked in their content. The members of the group perform their subtasks with a relative autonomy; they must, however, verbally co-ordinate the sequence of the individual assignments with one another; (3) *Unity of product and organisation:* A product, or a spectrum of products inside the work system, must be as completely prepared, manufactured and tested as is possible. The members of the group make a contribution towards the step-by-step manufacture of the product, not only through their mutual planning and decision-making, but also through their various work activities. Every member's contribution and place in the manufacturing sequence is clearly recognisable. From this, the common task and the common product evolve both in material and psychological terms. Because of the collective decisions made by self-regulation, the semi-autonomous work group takes a *shared respon-sibility* for all of the activities and working cycles, regardless of who assumes a working phase for a special order. Internally, the semi-autonomous group can designate a "rotating" spokesperson. He or she also takes part in the production activities performed.

The two main criteria of work and organisational psychology for the evaluation of semi-autonomous group work are collective self-regulation and collective autonomy (cf. Alioth, 1980; Susman, 1976; Ulich, 1994). The term "self-regulation" is complementary to "autonomy" in as much as self-regulation presupposes autonomy. *Collective self-regulation* means the course of mutual action the work groups decide on in regard to the tasks and areas of autonomy entrusted to them or in regard to the principles of decision-making themselves. In question here are decisions directly linked to the manufacturing process, decisions, for example, about production planning and control within the work system itself, about the distribution of tasks, allocation of resources and regulation of inputs and outputs. These decisions can be taken more readily by the group than they might be from parties outside, as it is the group within whose territory the problems arise, depending on planning requirements, variations or disturbances. A high degree of autonomy and self-regulation at the group level does not necessarily mean, on the one hand, that every individual member of the group will profit automatically from collective decision-making. It is more likely to occur that one member, or a few group members, assume most of the decision-making competence over the long term. Such a form of group work is hardly classifiable according to work psychology criteria as "semi-autonomous". On the other hand, additional decision- making options can arise within the individual work activities which either do not need to be or cannot be used collectively. In the assessment of semi-autonomous group work, then, *"individual self-regulation"* also merits consideration.

The principle of *"polyvalent skilling"* means that every work group member masters several work tasks. *Job rotation among structurally diverse activities* (Ulich, *et al.*, 1973), or those which are different in regard to their promoting personality development, represent an important design concept of semi-autonomous group work. Over and beyond that, semi-autonomous group work, in contrast to individual work, offers many possibilities for "differential" and "dynamic" work design (see Ulich, 1989; 1994): By collective self-regulation, the group members can divide the subtasks among themselves according to their skills, their knowledge, their need for qualifying and their current situation.

Based on the characteristics of semi-autonomous group work described above, different instruments were, respectively are, developed at our institute which permit analysis and evaluation of the *"objective" level* of existing group work systems and their technical-organisational frameworks (Ulich, 1994; Weber, in preparation). They make it possible to differentiate personality-promoting forms of group work from the forms detrimental to personality. The results also make empirically grounded hypotheses possible about the *"subjective"* level of group work in actual practice. The stronger the characteristics of semi-autonomous group work, the more probable the evolution of a *shared task orientation* of group members over the middle- to long-term. By "shared task orientation" we mean (especially in reference to Ulich, 1994 and Emery, 1959; and to activity theory, e.g. Engeström, 1991) a work group whose members share insight and an attitude pattern, namely:

- to have and accept a common task and shared responsibility
- to give and receive communication, co-operation, support and understanding (in the sense of a mutually task-related and shared social perspective)
- to make a useful contribution for a product manufactured in the group work context
- to use and develop shared knowledge, skills and tools
- to positively influence working conditions in and around the work group

The "interiorisation" of the common primary task as opposed to simply isolated subtasks can lead each group member, as he or she engages in work activity, to consider what will impact the achievement of the entire group. This can be the case even if there is no actual link to subtasks he or she performed personally.

3. METHODS TO ANALYZE ACTION REGULATION AND CO-OPERATION IN AUTONOMOUS WORK GROUPS

From the beginning, the socio-technical system approach establishes in the work system - and not in the isolated task - the central unit of analysis and design of humanly fulfilling work activities. Stronger than in other work psychological approaches, aspects of task interdependence come to the forefront of analysis, and a co-operative effort of the work system members with the surrounding work systems ensues. Deficiencies of usable socio-technical methods for the analysis and design of work systems have been cited among the current representatives of the socio-technical approach (e.g. Pasmore, *et al.*, 1982). Many of the existing socio-technical checklists for the evaluation of work activities and tasks are useful, but particularly for the examination of subtasks per-formed individually in the group, they are imprecise.

Psychological task analysis instruments based on the *theory of action regulation* represent an important methodological supplement (e.g. Hacker, 1986; Oesterreich & Volpert, 1986; Volpert, 1992). Therefore, both approaches are suited to integration; they have a great deal in common theoretically (for more detail: Weber, in preparation). Human work activity is marked, for example, by being conscious, goal-directed, object-related and socially integrated.

One of the methods of task analysis is the VERA instrument (Oesterreich & Volpert, 1986). It can be used to analyse and assess requirements for planning and decision-making, the so-called "regulation requirements" that are inherent to the subtasks performed individually in a semi-autonomous work group. Accordingly, the VERA instrument can identify and assess important features of *individual autonomy* and *self-regulation*. It is a kind of empirically tested "observation interview": By means of a manual, working group members are carefully observed and interviewed. Work tasks can be assigned to five levels of requirements for planning and decision-making: the one most challenging and potentially the most personality-promoting level is Level 5. The lowest level 1 is assigned to tasks with no real demand for thinking. At every level, an additional "restrictive" step is defined (abbreviated; "R") which comes into effect when the processes for planning and decision-making are only required in part. The regulation levels are differentiated as follows (detailed description in Oesterreich & Volpert, 1986):

- Level 5: Establishing new working processes
- Level 4: Co-ordinating several working processes
- Level 3: Sub-goal planning
- Level 2: (Simple) Action Planning
- Level 1: Sensory-motor regulation

The VERA instrument is also able to assess opportunities for communication which are inherent to subtasks the work group members are individually assigned. *Task-related opportunities for co-operation and opportunities for communication* arise when the individual subtasks demand that a group member verbally co-ordinate certain work activities. The maximum VERA-K step, for parts of the task which promote communication, is determined for every individual worker. Seen from the psychological viewpoint, it is *not* a question of task-related communication, if information only travels one way. We speak of task-related communication only if both communicating partners have the same *de facto* rights in regard to co-ordinating certain operations. Communication is an exchange.

In addition, semi-autonomous group work is characterised by task - interdependent requirements of mutual evaluation, planning and decision-making processes. Mutual planning and decision-making for example, that is to say: who for what reason will assume what subtask at what time, make up part of the common "central task" of the group (see Kötter *et*

al., 1989; Kötter & Gohde, 1991) in other words, of the collective self-regulation (see above). The common central task connects the subtasks that make up the complete task of the group. The group members meet to distribute the subtasks that must be performed individually afterwards. In the analysis of literature and in our own case studies (Weber, in preparation), there were signs of varied *central task areas* among semi-autonomous work groups: (1) Planning the production and manufacturing sequences, (2) Allocation of personnel, (3) Job scheduling, (4) Development of improvement suggestions for technical and organisational problems, (5) Personnel development planning and need for qualifying, (6) Group decisions about self-management. Work groups are differentiated - among other criteria - by the number of these areas of collective autonomy which are delegated to them. To date, Ulich (1994) has proposed the most highly differentiated category scheme for determining degree of autonomy. It defines, in detail, the profile of the various decision-making opportunities that make up the common task areas.

In addition, the profiles of work groups can be distinguished from one another in the degree to which common planning and decision-making processes in the central task areas are challenging or meaningful. Dovetailed, shared regulation requirements can not be sufficiently analysed with the usual VERA instrument. For that reason, we are currently adapting the VERA steps model and combining it with Susman's (1976) and Alioth's (1980) classifications of "regulatory decisions". Susman and Alioth describe in detail the three classes of regulatory decisions which can be taken by mutual agreement by group members: "co-ordination" decisions regarding production planning and control, "allocation" decisions regarding resources at one's disposal, and "boundary maintenance" decisions (co-ordination of the inputs and outputs with adjacent work systems). Based on all the studies cited that relate to collective regulatory decisions, or to the concept of the common central task, it would seem advisable to so modify the VERA model so that *collective regulation requirements* could be analysed and evaluated with it. As long as several group members entitled with equal rights actually participate in the execution of regulatory decisions, it doesn't make sense to analyse each of the individual contributions separately. At a common weekly planning session of the manufacturing sequences (central task area No. 1), for example, the individual planning suggestions, ideas for solutions, etc. are subject to general discussion and modified. More and more, they lead towards a collective goal-action-program structure. From this, a (correctable) plan for manufacturing sequence, as well as plans for the utilisation of machinery evolve. Certain individual contributions are thus integrated, and others are rejected. In addition, computer-assisted knowledge storage banks and schemes of action, which were mutually developed by the group and are constantly accumulating, can be tapped. The *collective nature* of this regulatory process reveals itself true to form here, in as much as the group members can modify each other's suggestions with reciprocity and, again and again, realistic plans of action, decisions or suggestions for problem solving result The different members of the group motivate and complement one other ("resonance": cf. Volpert, 1992). The VERA step of this regulation processes, *locked* into themself, must be chosen for the group in its entirety. The result of this regulatory process, a "product", represents more than the "sum" of the isolated "parts". However, not even a modified model of the collective regulatory requirements is usable on every area of central tasks. The development of additional action psychology models is therefore necessary. It is for that reason that certain areas of central task in our study are evaluated exclusively according to what Ulich (1994) proposed: one of "activity-oriented categories for a profile of group work autonomy".

4. EMPIRICAL ILLUSTRATION

In conclusion, the use of the aforementioned model of collective action regulation in the examination of semi-autonomous groups should be demonstrated. This is only possible in part here, based on two exploratory case studies. One of the goals of our continuous research is to develop a *typology* of the structural characteristics of work groups in several areas of manufacturing. Among the characteristics for types of group work, there should be:

- the activity oriented categories of group work autonomy
- the step of collective regulation requirements of the central task or central task areas (VERA-CR step)
- steps of regulation requirements of included subtasks (VERA step)
- steps of communication requirements of included subtasks (VERA-C step)
- the possibilities for differential and dynamic work design within work groups.

Practical design proposals for step-by-step enlargement of the real group autonomy could be built on the foundation of this typology. Doing so should, on the one hand, spur the design of more personality-enhancing group work and, on the other hand, improve the efficiency of the work groups themselves. In Fig. 1 and Fig. 2, findings in relation to two analysed work groups are presented. Two of our (current) types of group work are represented by these work groups.

Fig. 1 shows a work group in a flexible manufacturing system ("FMS") for the sheet-working industry. The FMS Group includes a foreman as well as four operators. The four group operators alternate among the four work zones of the FMS weekly (systematic job rotation). While the foreman's tasks are not included in the job rotation, he does jump in, given a personnel shortage on the FMS. This organisational form effects a remarkable *polarisation*

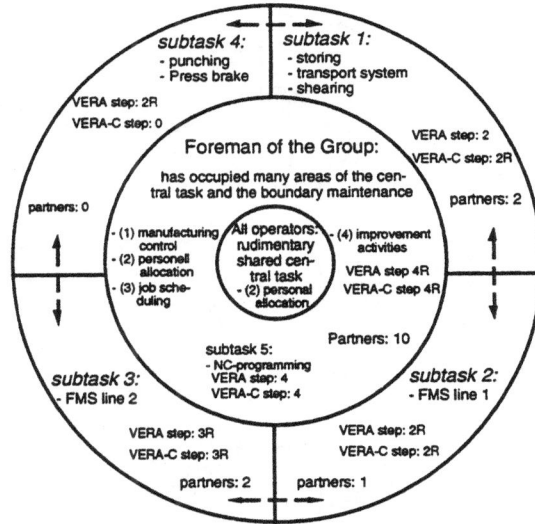

Fig. 1 Group work with rudimentary shared central task in a FMS

of *individual autonomy*. While the foreman's tasks (NC programming, production control, maintenance) require personally-challenging planning and decision-making skills (VERA step 4), the operators' tasks, with one exception, simply require mental processing of the foreman's directives in advance (Step 2R) or completion of them (Step 2). The notion of polarisation also marks the collective autonomy of the FMS group to a great extent. There is only a rudimentary shared central task. Primarily the foreman enjoys personally-challenging, task-related opportunities for communication (VERA-K step 4R). This form of "group work" has little to do with the socio-technical concept.

Fig. 2 represents a semi-autonomous work group in a flexible CNC production island ("PI") for mould manufacturing. It consists of eight skilled metal workers without a foreman. Rather than being strictly schematic, job rotation depends on the situation and is flexible in the sense of differential and dynamic work design (Ulich, 1989; 1994). Compared to the FMS group, PI operators have considerably more *individual autonomy*. All of the operators are performing challenging subtasks. They programme, for example, NC files or plan the operations on conventional machine tools, set up the machines and occasionally even maintain them. Therefore, planning and decision-making requirements in keeping with "skilled labour" characterise the activities performed (VERA level 3). *Communication* is also more strongly pronounced: Every work task requires collective planning, understandably at different levels (steps 2 to 4R). The number of communication partners for every operator in the group is much greater than in case 1 (averaging 6.9 versus 1.3). There is no foreman to block information channels; instead, a network communication structure extends even beyond the limitations of the group itself (boundary maintenance). Each group member is participa-

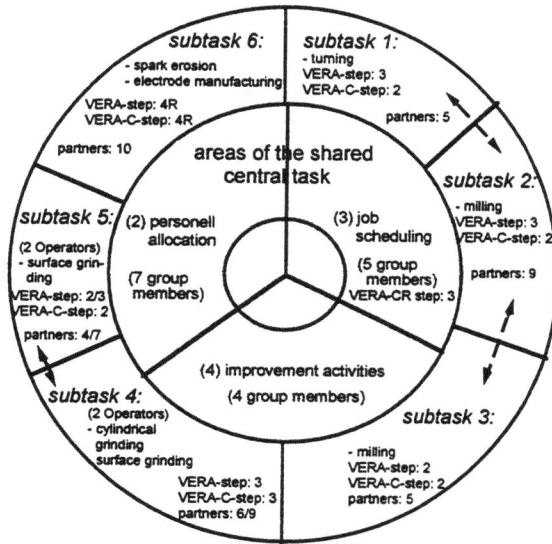

Fig. 2 Group work with distributed central task areas

ting in one central task area at least. The PI group represents autonomous group work to a considerably greater degree than the FMS group. Particulary in respect to several areas of the central task - personell allocation, job scheduling, improvement activities - the group respectively its subgroups take autonomous group decisions.

Meanwhile there is more and more empirical evidence that shared task orientation over from strong autonomy will considerably effect group cohesion, motivation and indeed performance (e.g. Emery & Thorsrud, 1976; Hackman, 1986; Pearce & Ravlin, 1987; Ulich, 1994). The aforementioned all agree that many specific consequences on or interactions among structural features and organisational conditions of group work, social psychological characteristics, moderating influences (e.g. stress, time pressure) and outcome criteria still deserve more exhaustive research. Some of these correlations will be examined more closely in our current research project "KOBRA" ("Co-operation in Computer-Assisted Work").

REFERENCES

Alioth, A. (1980). *Entwicklung und Einführung alternativer Arbeitsformen.* Huber, Bern.

Berggren, C. (1992). *Alternatives to Lean Production: Work Organization in the Swedish Auto Industry.* ILR Press, Ithaca, N.Y.

Emery, F. E. (1959). *Characteristics of Socio-Technical Systems* (Document No. 527). Tavistock Institute of Human Relations, London.

Emery, F. & E. Thorsrud (1976). *Democracy at work.* Martinus Nijhoff, Leiden.

Engeström, Y. (1991). Activity Theory and Individual and Social Transformation. *Activity Theory*, 4 (7/8), 6-17.

Hackman, J. R. (1986). The Psychology of Self-Management in Organizations. In: *Psychology and Work: Productivity, Change and Employment* (Pallak, M.S. & R. O. Perloff, Eds.), 85-136. American Psychological Association, Washington, D.C.

Hacker, W. (1986). Complete vs. Invomplete Working Tasks - A Concept and its Verification. In: *The Psychology of Work and Organization* (Debus, G. & H.W. Schroiff (Eds.), 23-36. Elsevier, Amsterdam.

Kötter, W. & H.E. Gohde (1991). Expertise: Fertigungsinseln - nur wirtschaftlich oder zugleich auch human? In: *Mit CIM in die Fabrik der Zukunft* (Fiedler, A. & U. Regenhard (Eds.), 179-248. Westdeutscher Verlag, Opladen.

Kötter, W., H.E. Gohde, & W. G. Weber (1989). Technological and Organizational Options for Skill Based Task Design in a Group Technology Project. In: *Skill Based Automated Production.* Preprints of the IFAC- / IFIP- / IMACS-Symposium, Austria, November, 15-17 (Kopacek, P., M. Moritz & R. Genser, Eds.), TS 12/1-TS 12/6. Austrian Center for Productivity and Efficiency.

Oesterreich, R. & W. Volpert (1986). Task Analysis for Work Design on the Basis of Action Regulation Theory. *Economic and Industrial Democracy*, 7, 503-527.

Pasmore, W., C. Francis, J. Haldeman & A. Shani (1982). Sociotechnical Systems: A North American Reflection on Empirical Studies of the Seventies. *Human Relations*, 35, 1179-1207.

Pearce, J. A. & E.C. Ravlin (1987). The Design and Activation of Self-Regulating Work Groups. *Human Relations*, 40, 751-782.

Susman, G.I. (1976). *Autonomy at Work.* Praeger, New York.

Ulich, E. (1989). Humanization of Work - Concepts and Cases. In *Advances in Industrial and Organizational Psychology* (Fallon, J., H.P. Pfister and J. Brebner, Eds.), 133-143. North-Holland, Amsterdam.

Ulich, E. (1994): *Arbeitspsychologie.* Verlag der Fachvereine, Zürich / Schäffer-Poeschel: Stuttgart.

Ulich, E., P. Groskurth & A. Bruggemann (1973). *Neue Formen der Arbeitsgestaltung.* Europäische Verlagsanstalt, Frankfurt/M.

Volpert, W. (1992). Work Design for Human Development. In: *Software Development and Reality Construction* (Floyd, C., H. Züllighoven, R. Budde & R. Keil-Slawik (Eds.), 336-349. Springer, Berlin.

Weber, W.G. (in preparation). *Handlungspsychologische Konzepte zur Analyse von Gruppenarbeit in der Produktion.*

Womack, J.P., D.T. Jones & D. Roos (1990). *The Machine that Changed the World.* Rawson, New York.

SPECIALIZED COMPUTER ARCHITECTURE FOR DYNAMIC CONTROL OF OPERATORS

Evgeny K.Pandov, Anelia Tz.Popandreeva

Technical University - Sofia 1156, Bulgaria

Abstract. The specialised computer architecture with module organization is designed for dynamic control of operators. This type of control is necessary for evaluation of the reliability of the system ˙man-machine˙. The specialized architecture consists of central computer, module for psychological testing, module for physiological examination and module for dynamic control .

Key words: computer architecture, control, medical applications.

1. INTRODUCTION

The professional psychological selection consists of examination of the condition of the candidate s for a specific job in terms of the most important physiological qualities needed for this job.

Accordingly to standard data a major part of all accidents connected with the process of working are function of the moment condition of the human. When analysing the data for the accidents it has been concluded that the people responsible for them are in fact perfectly suitable for the job they are doing. There is need for dynamic control of the reliability of the personnel due to the high degree of responsibility for the certain job which rises the requirements towards the psycho-sensor, psycho-motor and emotional qualities of the individual. It is possible to monitor all of these qualities and there is a serious change which should put in danger life of the operator a certain measures are taken immediately by the testing person until the causes of these changes are revealed.

The former testings of operators show the fact that best informative tests for evaluation of the moment state are:
1. The test for evaluating the mobility of the psychic proceses, switching over and distribution of the attention in the shortage of time. 1. The test for evaluating the mobility of the psychic proceses
2. The test for visual memory.
3. The test for condition motive reflexs.

1. The test for evaluating the mobility of the psychic proceses, switching over and distribution of the attention in the shortage of time. On the stimulus field with variable background color are appearing one or two digits, over which the tested person must make certain arithmetical operations depending on the background color and sign between digits.

2. The test for visual memory. In front of the tested person are visualized 7 digits for a fixed time. The tested person must memorize the digits and after that input them from the decimal keyboard.

3. The test for condition motive reflexs. On the stimulus field are visualized from twelwe light stimulus, organized in matrix (4 x 3). Te eleven combinations from the first series are visualized. They consists of two positive combinations on which the tested person has to react and two negative combinations on whiich the tested person must not respond. Each combination is visualized for three seconds, the pause is between one and two seconds. If the reaction is true the the exponation of the stimulus is terminated. If the series are positive and during the three seconds is not reaction the test counts the omitted reaction. If the tested person react on negative combination the reaction is registered as a wrong reaction.

2. ARCHITECTURE OF THE MICRO-COMPUTER SYSTEM

The microcomputer system for dynamic control of the operators consists of a computer IBM PC type and module for the psychological testing, module for physiological examinations and module for dynamic control . On the figure 1. is shown the block diagram of a similar specialized architecture.

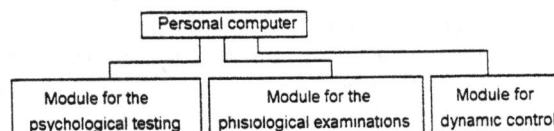

Fig.1. Mikrocomputer system for dynamic control
The specialzed architecture is organized as an open architecture.

The module for psychological testing consist of some tests realized as programs for the personal computer. For the punctually registration of the time of reaction the panel of the tested person is realised as a block connected to the computer by controller (figure 2).

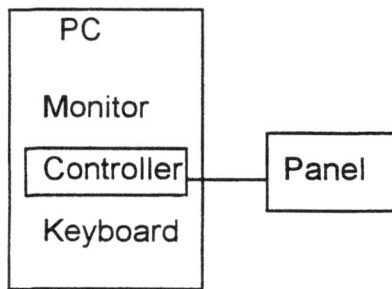

Fig.2 Module for psychological testing.

The modulee for physiological examination (fig.4) consist of a detector for registration of the pulse rate and respiration rate (1, 2), amplifiers (3, 4), modulators (5, 6) and controller (7) for the connection with personal computer.

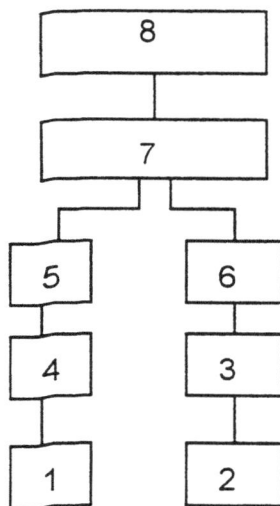

Fig.3 Module for phisiological examinations.

The module for dynamiic control (fig.4) consist of indicator field for visualization of the stimulus, speciialized keyboard for registration the persons reactions and specialized microcomputer . realized as one plate. This architecture is shown on the figure 4 and consists of : microcomputer (1). stimulus field A (2). stimulus field B(2), controller, decoder for the 7 indicators for the test 2 (4). decoder for monitoring the indicators 2 and 3 (5). decimal keyboard (6). block for realization the background

colors for the test 1(7). modulators for physiological signals (pulse rate and respiration rate)(8).The stimulus field A is a matrix of 4x3 lightdiodes and 3 segment indicators for test 1. The stimulus field B consists of 7 segment indicators for the test 2.

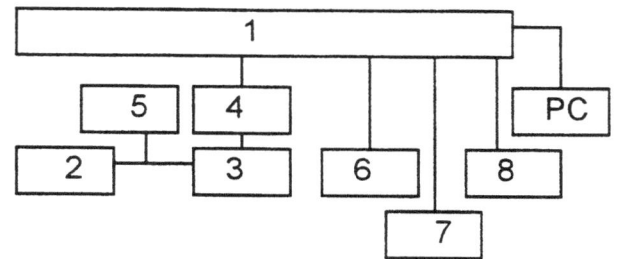

Fig.4 Module for dynamic control.

3. CONCLUSIONS

The specialized architecture for the microcomputer system is aimed at peforming dynamic control of the operators. It provides abilities for reorrgannizing the system. for capturing and automative calculation the results from the testings. for giving the sintezed information for evaluation of the state of the operator.

REFERENCES

1. Apostolov E., S.Minkov. Reliability of the system 'man-technics', Medicina i Fizkulutura. Sofia. 1980.
2. Minkov S., E.Apostolov. The professional fitnes of the man operator. Medicina i Fizkultura. Sofia. 1982.

SELF-SIMILARITY AS A NEW PRINCIPLE FOR INNOVATION PROCESSES

Giuseppe Strina

*Department of Informatics in Mechanical Engineering (HDZ/IMA),
University of Technology (RWTH), D-52068 Aachen,
and Institute for Industrial Cybernetics (IfU e.V.),
D- 45468 Mülheim/Ruhr, Germany*

Abstract: The so-called 'factory of the future' will be charaterized by an equilibrated and intelligent mix of technical, organizational and personal efforts. It will be more and more important for enterprises (and other organizations) to manage the parallelism of production process, innovation process and learning process. The paper shows for these processes how to use the fractal principle of natural sciences and how to how to use it for the management of innovation processes.

Key Words: Industrial production systems, human factors, socio-technical system, project management, employee participation

1. INTRODUCTION

Despite many advantages for enterprises *and* employees (increased productivity, higher motivation; (Martin, *et al.*, 1991), the diffusion of new concepts for technological, organizational *and* personnel innovation are progressing only slowly. In many cases the planning of automated systems concentrates on a few technical and economic parameters which automatically leave out of vision a wide spectrum of design options (Strina, *et al.*, 1993). In automation the paradigm of the human role is still the same: "The control system - including its assisting systems - acts, the human observes and supervises, he/she acts only in case of emergency." (Brandt, 1994)

This paradigm has to be replaced. It might be instructive to consider factories (and other organizations) as a socio-technical system, where at least three dimensions are important: technology (machines, buildings etc.), organization (explicit and implicit rules of working, information system etc.) and humans (employees, but also customers etc.). These subsystems are linked in complex and dynamic relations of actions and reactions. Fig. 1 shows a simple model of this view. It is intentional that the subsystems are drawn partly outside the main system; they are not fully within the main system (Strina and Hartmann, 1992).

From this point of view it becomes clear that it is not sufficient to optimize only the technical system (and perhaps a little bit of the organizational system). It is only what may be called 'a sub-optimum improvement of a polyparameter problem'. Instead of considering the whole problem, this strategy begins at the easy angle where the world seems to be deterministic, linear and measurable. The deeper dimensions of real-life problems are complexity, dynamic

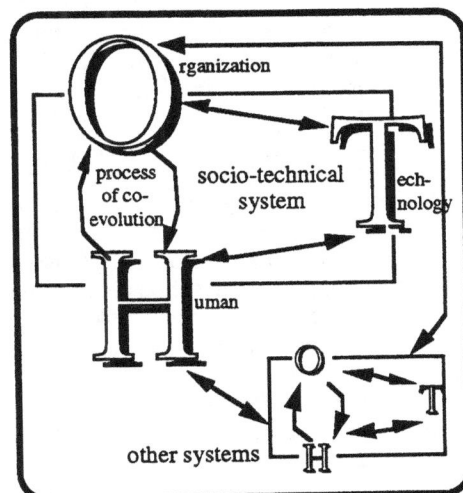

Fig.1: Simplified model of socio-technichal system

changes and non-linearity. They remain out of view. This is the reason for many difficulties of enterprises to respond to the increasing market instabilities.

If in a scientific discipline the problem solving process comes to a standstill it is a proved method to look around how other disciplines with similar problems are developing answers. Concerning themes like complexity, instability and non-linearity, especially natural sciences (e.g. physics and biology) and human sciences (e.g. sociology) have made progress. Examples are chaos theory and autopoiesis theory (Strina 1996). The present paper shows, as an example for the application of these modern theories, how to make use of genuine concepts of the fractal theory. To do that it might be instructive to refer to the mathematical definitions and to extract from them the basic characteristics which can be used for the application on the management of innovations (chapt. 2). In a second step, these characteristics are translated into guiding principles for organizational innovations, not only for their structure but also for their processual aspects (chapt. 3). Finally a case study shows the practical importance of this approach (chapt. 4).

2. SOME ASPECTS OF FRACTAL'S THEORY

In the recent history of sciences there was a parallel evolution during the last two decades in those disciplines that are occupied with systems of high complexity and dynamics. There was an increasing awareness that sciences (and with them our 'normal' concept of the world) had considered only a special part of nature. It had left out of vision those system behaviours that are characterized by instability, turbulence and non-linearity (Briggs and Peat, 1989). The new Chaos Theory teaches us that phenomena like intermittence, strange attractors and others are much more frequent than we thought, even in daily life.

It seems to be possible to cautiously transfer these concepts - among others - into considering a social system, e.g. a factory or any other organization (Isenhardt, 1994; Henning, 1993). Under certain conditions of outside influence and/ or inner uncertainty, the system may become unstable. It may switch to completely unexpected states (Brandt, 1994). In those situations chaos theory may offer new explanation possibilities.

This theory contributes not only to explanation, but also to design. This might be its contribution for management of production systems. An example is the fractal geometry (Mandelbrot, 1982), one of the mathematical languages to describe phenomena of chaos. Scoped as one of the basic principles of nature, self-similarity can serve as a guiding principle in factory organization. There exist already some applications (e.g. Warnecke, 1992); but they are mainly metaphoric, without taking in consideration the basic mathematical content of the fractal theory.

The first definition was given by Mandelbrot; it uses the Hausdorff-Besicovitch-Dimension (HBD) as a measure of the unregularity of a given quantity. The main assertion of this definition is that the HBD of a fractal is always greater than its topological dimension (Mandelbrot, 1992). Usually the HBD is a fraction, but there are also cases of fractals with dimensions that are integers. (e.g. the so-called devils stage or the Brown movement; (Peitgen, et al., 1992). Therefore the term 'fraction dimension' is misleading. It better should be spoken of a 'fractal dimension'.

The advantage of the Mandelbrot definition of fractals is his validity for the *majority* of all fractals. The main disadvantage, however is that it is far too difficult to calculate the HBD and with it the fractal dimension, beside the fact that it is not valid for *all* fractals. Therefore scientists have looked for alternative definitions.

One of these alternative definitions is the way Peitgen et al. tried to define fractals. Their approach is based on the fact that for many fractals characteristic transformation operations can be found to copy them on themthelves (ibid.). These operations are: scaling up or down (S), rotation (R) and translation (T).

$$T(R(S(F_{n-1}))) = F_n \qquad (1)$$

It can be shown that for many fractals a characteristic combination of these operations exists so that the result F_n of these operations is again the original curve F_{n-1}. This characteristic combination is called the 'Hutchinson operator' (W):

$$W(F_{n-1}) = T(R(S(F_{n-1}))) = F_n = F_{n-1} \qquad (2)$$

This concept is illustrated in Fig. 2. The original curve F_{n-1} is the Koch curve. The transformations, i.e. scaling down by 1/3, translation (w_1 and w_4) and rotation/translation (w_2 and w_3), lead to a curve F_n, that is identically with the original curve F_{n-1}.

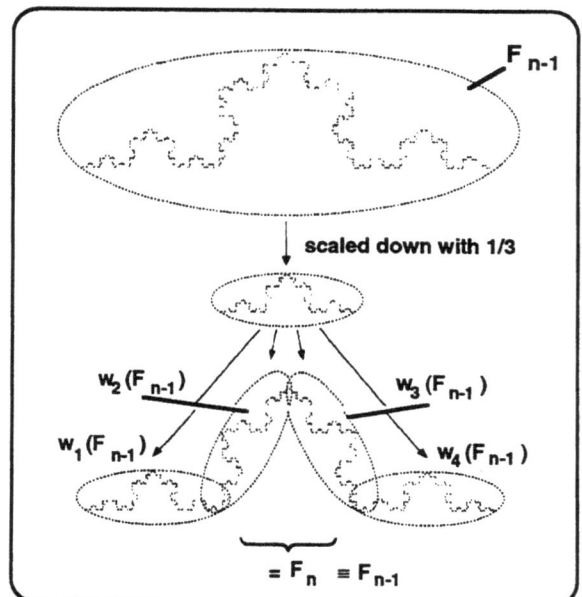

Fig. 2: Transformation of the Koch Curve
(see Peitgen, et al., 1992)

These expositions show that the definition approach in (2) mathematically explains the characteristics of self-similarity. It shows at the same time that this definition is not valid for all fractals either. The main advantage of this approach lies in the fact that it gives an idea of how we can conceptualize the construction of fractals: as an endless recursive iteration. It shows furthermore that fractality is often a phenomenon of line borders of quantities (e.g. in Fig. 3, the 'Mandelbrot Quantity' in Fig. 3; see Henning, 1993).

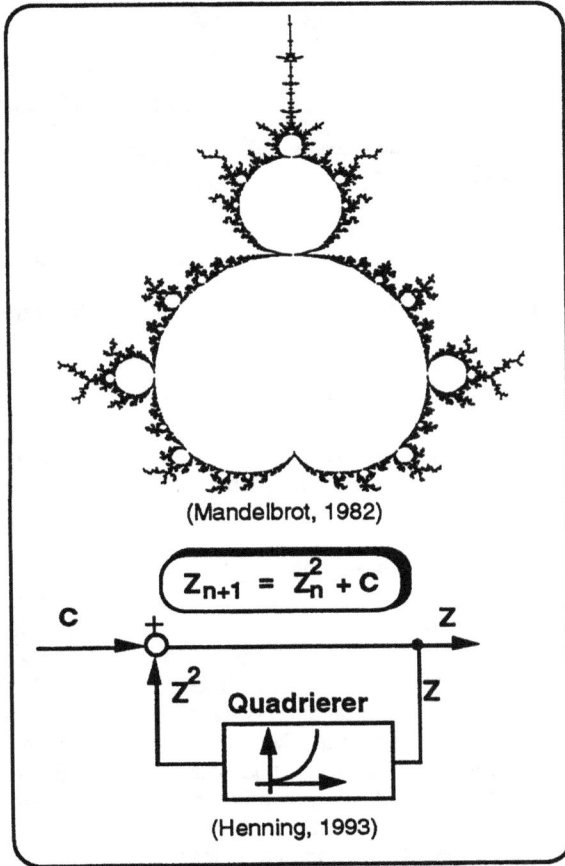

$$Z_{n+1} = Z_n^2 + C$$

(Mandelbrot, 1982)

(Henning, 1993)

Fig. 3: The Mandelbrot Quantity and its recursion (see Mandelbrot, 1982, and Henning, 1993)

It seems to be impossible to find a definition that covers all possible fractals. Therefore, Falconer draws the conclusion that it might be better to characterize fractals through a list of properties; hence, fractals more or less own some of these properties (Falconer, 1993). Combining this method with the aspects discussed above, it leads to the following **definition of fractals** (Strina, 1996):

1) They have a microstructure on many scales.
2) They are splitted structures and the adequate measure for this is the fractal dimension.
3) In many cases they show the property of self-similarity. Thereby they represent a characteristic relation between a part and the whole.
4) They can often be viewed as 'limit objects':
   • as a limes of an endless and recursive iteration,
   • as an endless line border of a limited quantity,
   • as a border phenomenon between the well known topological dimensions.

The application of these theories will not be quantitative - perhaps this kind of transfer can never be made. The qualitative adaptation brings new ideas to theory and design of organizations. Therefore it is 'allowed'. Some consequences have to be considered in design processes as set out in the chapters below.

3. SELF-SIMILARITY IN INNOVATION PROCESSES OF PRODUCTION SYSTEMS

It is one of the secrets of the Japanese success that in a factory the following concepts have to be 'implemented': the concept of continuous improvement (Imai, 1986) and of managing innovation processes. Enterprises try to find new answers for changing system environments. This is one of the characteristic features of a living and surviving organization. One condition for this is the ability of finding structures that are strong enough to guide the innovation process, and that are weak enough to be abandoned when they are no longer needed.

Concerning the principle of fractality we find several levels where it can be applied. Firstly we have the aim of the whole innovation process: to build up decentralized subentities where every subunit represents in a special way the whole system (Warnecke, 1992). But actually it is not fractality in the narrow sense: the principle of fractality corresponds to looking from macrocosmos to microcosmos along an infinite number of scales. Therefore we speak here only of the 'self-similarity' in the factory (Henning, *et al.*, 1994). Secondly we have the structures during the innovation process. Here project management is to be considered. According to the Chinese wise old men: the way is the aim. Thus self-similar structures can also be implemented in this process. Even in the different teams of a project management we find a reproduction of the whole enterprise, not only in terms of departments (as in a 'matrix organization'), but also in terms of participation of the different hierarchy levels. Fig. 4 shows a project structure that tries to implement this concept.

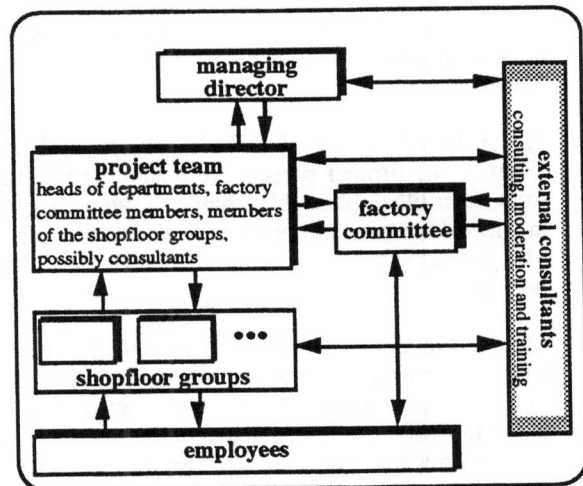

Fig. 4: Project Organization

The third application level is a special design approach. Not only the project organization may be the object of a self-similar structuring; also the project process can be structured in a self-similar way. Fig. 5a shows an arrangement of the five typical steps of any system design concept: aim planning, concept development, formation, evaluation and final assessment. The connections between them demonstrate that such a design process is an iterative process. Only the last step, the final assessment, is made at the end of a project and therefore is occuring once on the general level of the project process. Each of the other four project steps of the innovation process can be described in more detail by four further steps.

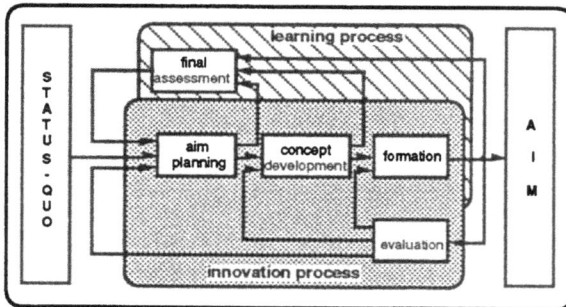

Fig. 5a: The basic structure of a project process

In each step of the innovation process in Fig. 5a the same division into four steps can be found. For the 'aim planning' step - as shown in the 'enlargement' of figure 5b - it means: after an aim planning step 'first problem and aim definition' (e.g. 'What are the main problems to be solved and what are the aims?') a 'concept development' step may follow (e.g. information of the employees and implementation of a 'round table'). After a formation step (agreement on aims, e.g. 'What are the visions for the enterprise for the next ten or twenty years?'), an evaluation step follows ('What can we do ourselves and where do we need help of outside consultants?'). The other project phases may be detailed in the same way.

Self-similarity can be found in each of these steps as shown. This process of enlargement can be described 'ad infinitum', because in any human act we can find these elements. That is why we can call it rightly a fractal.

The main advantage of this design approach is as follows: the project phases are no longer seperated from each other (first this, than this, after this etc.); in each phase we find elements of the other phases. The process appears more like a network than a linear process.

The experiences of persons who have been confronted with this design approach correspond to this complexity: project management approaches that refer to the complexity of the change process are more difficult to accustomed, but they help to increase the sensitiveness and the awareness for this copmlexity of the process.

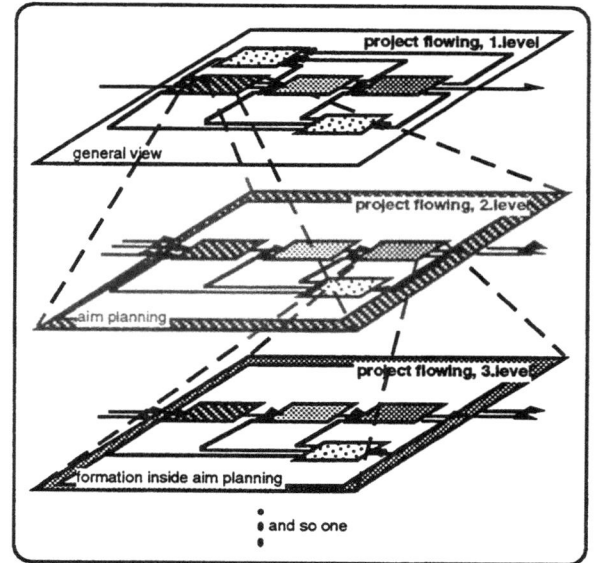

Fig. 5b: The project process in its fractal structure

4. A CASE STUDY

The results discussed above will be supported by experiences of a casestudy, where the design approach was used.

Process: In a large enterprise (about 900 employees; products: cement-mixers), the aim of the innovation process was to introduce production islands. On the manager level the project aims and the assessment of the resources were agreed on. The management decided to contact consultants who were specialized in 'process consulting'. With these consultants they repeatet the agreement on aims. They decided to begin with a 'pilot group' before reorganizing the whole shopfloor. The reason was to gain first experiences with how to choose people to cooperate in teams on the shopfloor.

On the project team, the departments concerned (production, planning, quality control etc.) management and the factory committee were represented. This project team together with the consultants and the shopfloor representatives made the selection of the pilot group. In talks with the shopfloor they found out that - for a real experience to be transferred on to the whole plant - it was necessary not to assemble an 'olympic team' where only the best workers are chosen, but to find a 'cross-section group' where several competence and qualification levels are represented. In this way they learn to manage the problems that appear during the implementation process of group working.

Evaluation: In the process of this project the concept of self-similarity and fractals can be seen:
• In the project development, the iteration steps can be found that are not linear in the narrow sense. This corresponds to the fractal project process.
• In the different project groups (project team, pilot group, shopfloor group), the members tried to implement self-similar elements. It corresponds to the

fact that the vision of the future organization is strongly influenced by its social context: only by a self-similar approach it can be guaranteed that 'corporate identity' is experienced by the employees, not only forced upon them by the management.

After a pilot project time of about nine months, the enterprise began to transfer the results and experiences to other departments. Some essential steps of the process had to be repeated with each new group of employees getting involved. Now working groups are progressing successfully in most of the departments of the enterprises.

5. CONCLUSIONS

We need new perspectives and ideas for the manufacturing and production systems of the future. Here new concepts developed in other disciplines can help us to find ideas.

REFERENCES

Brandt, D. (1994). Automation in Manufacturing, Control versus Chaos? Proc. Fourth International Conference on Human Aspects of Advanced Manufacturing and Hybrid Automation, Manchester (GB), 06.-08.07.1994

Briggs, J., Peat, F.D. (1989). Turbulent Mirrow. Harper & Row, New York, 1989.

Falconer, K.J. (1993). Fraktale Geometrie. Spektrum, Heidelberg, Oxford, 1993.

Henning, K. (1993). Spuren im Chaos. Olzog Verlag, 1993.

Henning, K., Strina, G. and Wollenweber, D. (1994). Die selbstähnliche Fabrik. In: Scheel, J., Hacker, W., Henning, K., Enderlein, K. (Ed.). Fabrikorganisation neu beGreifen. TÜV-Verlag, Köln, 1994

Imai, M. (1986). Kaizen.

Isenhardt, I. (1994). Komplexitätsorientierte Gestaltungsprinzipien für Organisationen - dargestellt an Fallstudien zu Reorganisationsprozessen in einem Großkrankenhaus. Augustinus, Aachen, 1994

Mandelbrot, B. (1982). The Fractal Geometry of Nature. San Fransisco, 1982

Martin, T. , Kivinen, J., Rijnsdorp, J.E., Rodd, M.G. and Rouse, W.B. (1991). Appropriate Automation - Integrating Technical, Human, Organizational, Economic and Cultural Factors. *Automatica* **27**, 6 (1991), 901-917.

Peitgen, H.-O., Jürgens, H., Saupe, D. (1992). Bausteine des Chaos: Fractale. Springer-Verlag, Berlin, New York, 1992.

Strina, G. (1996). Application of Self-Similarity and Self-Renewal Principles on the Management of Innovations in Small to Medium Sized Enterprises. VDI-Verlag, Düsseldorf, 1996 (to be published) [German].

Strina, G., Hartmann, E.A. (1992). Komplexitätsdimensionierung bei der Gestaltung soziotechnischer Systeme. In: Henning, K. and Harendt, B. (Ed.): Methoden und Praxis der Komplexitätsbewältigung. Dunker & Humblot, Berlin, 1992, 169-181.

Strina, G., Süthoff, M., Grinda, S. and Brandt, D. (1993). Automation without Organizational Development Won't Do. IFAC Congress, Sydney (Australia), 18.-23.07.1993, **7**, 339-343.

Warnecke, H.-J. (1992). Die Fraktale Fabrik. Springer-Verlag, Berlin, New York, 1992.

WORK PSYCHOLOGICAL ISSUES OF RESTRUCTURING A SUPPLIER FOR INTERORGANIZATIONAL COOPERATION

Olga Pardo

Swiss Federal Institute of Technology, ETH Zürich
Work and Organizational Psychology Unit
CH-8092 Zürich

Abstract: Within the scope of a research project, an electronics supplier was analyzed and design proposals suitable for interorganizational cooperation were derived. Work psychological analyses revealed that the task structures did not promote the utilization and enhancement of skills nor did they provide flexibility, a prerequisite for meeting increasing customer demands. The design proposals made, focus on appropriate training for the creation of holistic tasks and the enhancement of the workers' polyvalence.

Keywords: Computer-aided manufacturing; manufacturing systems; personnel qualifications; socio-technical system design; training; work organization.

1. INTRODUCTION

Many small and medium size companies face a competitive environment with heightened, rapidly changing market demands. The total cycle time along the value chain is a crucial competitive factor. The demand for more complex, customized, high-qualitiy products at low cost calls for flexibility in internal processing, i.e. efficient utilization of different types of resources, as well as in external interfacing, i.e. the interaction with customers and suppliers (D'Ambrogio, 1992). Thus, the optimization of processes is a critical design issue for many companies.

The view on processes has been recently expanded to include interactions at the company's boundaries: processes are understood as part of a larger value chain comprising suppliers and customers. In fact, many companies have been building up interorganizational relationships. The emerging cooperative relationships can be characterized as intermediate coordination arrangements between market and hierarchy (Williamson, 1991; Barney and Ouchi, 1986). They are based on mutual trust and entail equal status of the companies involved (Smith Ring and Van de Ven, 1994).

2. INTERCIM: THE DESIGN OF A COOPERATIVE RELATIONSHIP

The project InterCIM focuses on the logistics issues of the cooperation between a medium size manufacturer in the machine tool industry (AGIE) and two small and medium size suppliers in the electronics (Juri Elettronica) and metal-working (Franzi SA) industries (Frigo-Mosca and Alberti, 1995). InterCIM was launched at the Institute of Industrial Engineering and Management (BWI) at the Swiss Federal Institute of Technology and was jointly carried out with the Swiss-Italian regional CIM-center (CIMSI) and the work and organizational psychology unit (IfAP) at the Swiss Federal Institute of Technology. This paper addresses work psychological aspects connected with the restructuring of the electronics supplier (Juri Elettronica) for interorganizational cooperation.

Juri Elettronica, manufactures cables and different types of assembled electronic circuit boards. 33 of its total 39 employees are semi-skilled, mostly foreign-speaking female workers. At the time of the study, AGIE, a leader in the market of electro-erosion machines, had already introduced logistics concepts such as JIT (just in time) and Kanban and had restructured its manufacturing departments as production cells. In order to reduce total cycle times and time-to-market, AGIE manifested a keen interest in intensifying the relationship with the electronic circuit boards supplier Juri Elettronica. In this context, AGIE reinforced the quality requirements and decided to charge the supplier with purchasing tasks which had been carried out by AGIE so far. According to AGIE's "free-pass" policy, the supplier was supposed to deliver the required components directly to the customer's production line. Furthermore, the "free-pass" policy required that at least 97.5% of the components delivered by its suppliers be error free. In addition, AGIE decided no longer to purchase and provide the supplier with electronic circuit boards and components, but to assign this task to its supplier. The supplier was expected to be able to purchase and keep on stock materials and components autonomously. Obviously, increased flexibility and quality was required in order to meet these specific demands. With Juri Elettronica, work structures were required which were suitable for ensuring quality and efficiency in terms of costs and total cycle times in the long run to meet these arising customer demands.

3. THE WORK PSYCHOLOGICAL PERSPECTIVE

The work psychological perspective is based on the socio-technical systems approach in which systems consist of two jointly designed and optimized subsystems, the technical and the social subsystem (Ulich, 1994). Work-oriented design concepts focus on the joint design of the three aspects human - technology - organization, i.e. the use and development of skills, the implementation and utilization of technical systems and organizational design (Ulich, 1994). This holistic conception of CIM (computer integrated manufacturing) is termed MTO-model (orig. German: Mensch - Technik - Organisation). The hypothesis is that work-oriented design concepts contribute to humane working conditions as well as to economic efficiency. Techni-cally-oriented concepts, on the other hand, primarily address the implementation of technical systems and their components and regard organizational and human ressources aspects as secondary or negligible. Based on the MTO-model, a MTO-analysis procedure has been developed and empirically tested at the Work and Organizational Unit of the Swiss Federal Institute of Technology (Strohm and Ulich, in press). In the case study presented here, MTO-analysis steps have been applied.

Interorganizational relationships can affect techno-logy, e.g. specific machinery to be purchased by a supplier or the implementation of EDI (electronic data interchange), they can also affect organizational design, e.g. the redistribution of functions and tasks, and qualifications, e.g. know-how transfer for co-design of products. Cooperative relationships between manufacturer and suppliers imply that both reconsider their processes. Functional integration, a key issue for the design of manufacturing structures, affects all the companies involved in the interorganizational relationship. The redesign of processes also implies that work tasks have to be redefined. Furthermore, the distribution of tasks between the companies involved in the value chain may need to be reconsidered and redesigned. Undoubtedly, the design of interorganizational relationships simultaneously affects questions of implementation and utilization of technical systems, of use and development of skills, and of work organizational design (MTO-model). For Juri Elettronica, the emerging interorganizational relationship was coupled with new job demands due to the assignment of purchasing and stock-keeping tasks as well as increased quality requirements.

4. WORK PSYCHOLOGICAL ANALYSES

In work psychology, the work tasks are the center of interest. The goal of work psychological research is to analyze, evaluate and design work structures compatible with criteria on humane work conditions. In the following, three methods for work psychological analysis, evaluation and design are presented and applied to the present case study.

4.1 Condition-related work analyses

The unit of condition-related work analysis is a subtask which in combination with other subtasks related to each other constitutes the task of a socio-technical unit, e.g. a department. According to the MTO-investigation procedure (Strohm and Ulich, in press), work analyses include both condition-related and subjective analyses.

Condition-related work analyses focus on the structural features of a task and its execution conditions; the task is analyzed and evaluated irrespective of the person actually performing the task. The goal of condition-related work analyses is to evaluate tasks regarding features that potentially promote and those that impair the development of the worker's personality.

Features that promote the development of the employee's personality include evaluation criteria for humane work based on the action regulation theory (e.g. Bergmann and Richter, 1994) as well as those related to features of work design (Ulich, 1994). According to Ulich (1994), the enhancement of these features of work design is a prerequisite for task orientation and the generation of intrinsic work motivation.

Features that impair the development of the employees' personality are defined as regulation hindrances in terms of work conditions leading to work load.

Instruments for the condition-related analysis of tasks include the VERA work analysis tool (Oesterreich and Volpert, 1991) and the KABA task analysis tool (Dunckel et al., 1993). Regulation hindrances can be identified by means of the RHIA instrument (Leitner *et al.*, 1987).

The VERA, KABA and RHIA work analysis instruments, all based on the action regulation theory, were utilized and partially adapted for the analysis of production tasks (Zölch, 1993). The VERA work analysis tool (Oesterreich and Volpert, 1991) was used in this study for the evaluation of tasks with respect to planning and decision-making requirements. For communication requirements, the corresponding section of the KABA task analysis tool was utilized and adapted for production tasks (Zölch, 1993). Furthermore, the identification of regulation hindrances on the basis of both the RHIA instrument and the KABA task analysis tool was included in the adapted work analysis tool.

Planning and decision-making requirements are evaluated by means of assigning the task to a specific VERA-step. VERA-steps denote the amount of planning and decision-making requirements that a task comprises (see Fig. 1.). They refer to the extent to which a worker can autonomously establish goals and determine the means for the achievement of the work outcomes. Planning and decision-making requirements reach from the execution of pre-defined steps, VERA-step 1, to comprehensive planning as described in VERA-step 5. For each VERA-step a corresponding restricted step (e.g. 1R) is defined which characterizes partial fullfillment of the corresponding VERA-step. The amount of planning and decision-making requirements indicates the extent to which a task contains features that promote the development of the worker's personality in terms of potentially utilizing and enhancing the worker's skills (Weber, in press). The lower the VERA-step, the more a task is considered to be partialized in terms of a tayloristic work design (Oesterreich and Volpert, 1991). From a work psychological view, tasks promoting the worker's personality are ensured from regulation requirements on VERA-step 3 upwards (Weber, in press).

Evaluation of Tasks
Planning and decision-making requirements

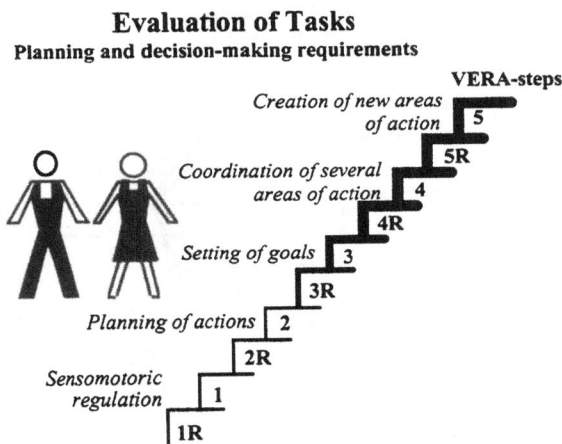

Fig. 1. Evaluation of tasks with respect to planning and decision-making requirements. VERA-steps denote the amount of planning and decision-making requirements.

Evaluation of Tasks
Communication requirements concerning ...

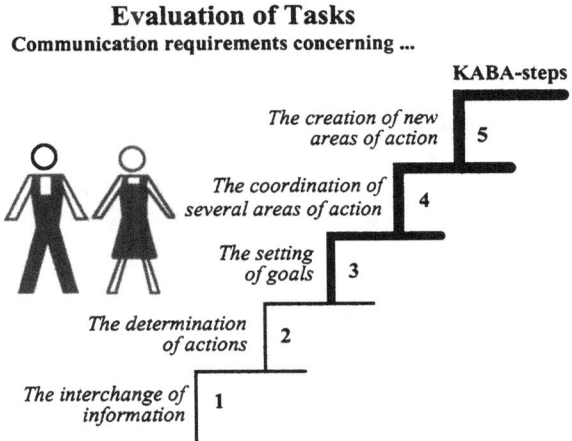

Fig. 2. Evaluation of tasks with respect to communication requirements. KABA-steps denote the type of work-related communication.

Communication requirements of a task refer to the extent to which a task requires work-related communication concerning planning steps or determinations (Dunckel *et al.*, 1993; Zölch, 1993). Work-related communication implies that at least two persons engage in communication for coordinating the execution of a task or single task steps. A task is evaluated with respect to communication requirements by assigning the task to the appropriate KABA-step. Five KABA-steps reflect the range of communication requirements (see Fig. 2.).

Regulation hindrances not only impair the workers' well-being, they also can result in long-term psycho-physiological diseases. Regulation hindrances comprise regulation impediments and regulations overload. Regulation impediments directly affect the execution of a task. They become apparent in recurrent events or conditions that render a task difficult or interrupt its execution, entail additional effort to perform a task or lead to risky actions by ignoring safety measures (Leitner *et al.*, 1987). Typically, regulation hindrances imply additional effort for the execution of a task. The higher the additional effort and the more time impositions have to be considered, the more a task potentially involves psychological work load (Weber, in press).

Regulation overload does not directly interfere with the execution of a task, but accumulates and has an effect on the worker's attention or concentration over time and thus excessively strains the action regulation. Regulation overload includes working conditions with tasks that lack planning and decision-making processes, that are repetitive and require continuous processing of visual information. Further forms of regulation overload comprise time pressure and task-unrelated negative ambient conditions such as noise, temperature, and chemical vapours.

For condition-related work analyses in this study, production tasks in both the electronic circuit boards assembly and the cable manufacturing departments were chosen (Fig. 3).

Process Flow

Electronic Board Assembly

- incoming inspection
- *preparation and distribution of materials*
- *pre-assembly*
- insertion:
  - *manual insertion*
  - *semi-automatic insertion (insertion machine)*
  - *automatic insertion (SMD-technology)*
- *soldering*
- *tests of soldered contacts/rework*
- final assembly
- *quality control*
- packing and shipping

Cable Manufacture

- incoming inspection
- cutting of cables
- *cable manufacturing*
- *quality control*
- packing and shipping

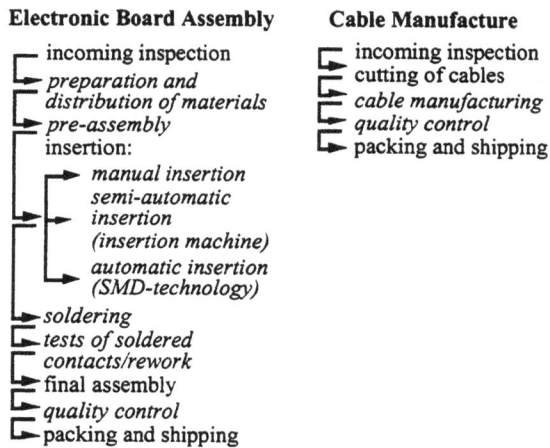

Fig. 3. Process Flow in the electronic circuit boards assembly and cable manufacturing departments. (Tasks chosen for the condition-related work analyses and evaluations in italic letters.)

The tasks attain VERA-steps between 1R and 2R for planning and decision-making requirements (Table 1). These results indicate that the tasks do not comprehend opportunities for the workers to utilize and further develop their skills. The execution of tasks assigned to VERA-steps 1R and 1 does not require planning; the work steps leading to the work outcome are routinized. For tasks at the VERA-step 2R, the sequence of operations leading to the work outcome is predetermined, the actions have to be brought to mind, though.

Table 1. Evaluation of tasks.

Tasks	Planning and decision-making requirements VERA-step	Communication requirements KABA-step
Preparation and distribution of materials	2R	1
Pre-assembly	1	1
Manual insertion	1R	1
Semi-automatic insertion	1	1
Automatic insertion	2R	1
Soldering	1	1
Test of soldered contacts/rework	1R	1
Quality control electronic circuit boards assembly	1	1
Cable manufacturing	2R	1
Quality control of cables	1	1

Communication requirements are notably low throughout all tasks analyzed (Table 1). Work-related communication is restricted to receiving and transmitting information, i.e. exchange of information within a largely routinized sequence of work steps.

Regulation hindrances are found with all tasks analyzed. For the majority of the tasks analyzed, the execution of a task is interrupted by fellow-workers enquiring for information on orders, by rush orders or by machine breakdowns. Most strikingly, half of the tasks show ergonomic deficiencies and negative ambient conditions.

4.2 Subjective work analyses

Employees' perceptions and redefinitions of their work situation can differ substantially from 'objective' features of work as analyzed and evaluated by way of condition-related work analyses. Accordingly, a subjective work analysis was conducted in form of a written survey based on the questionnaire for the subjective work analysis on salutogenesis by Rimann and Udris (1993). The questionnaire comprises scales on salutogenetically relevant work features which are associated with the design of personality promoting work (Rimann and Udris, 1993; Fig. 4.). The employees were asked to state the degree to which they agreed with a variety of items related to these scales. With a return rate of 61%, 20 completed questionnaires were collected for the analysis. Obviously, most work features rather are not or partially agreed on (Fig. 4.).

Subjective Work Analysis

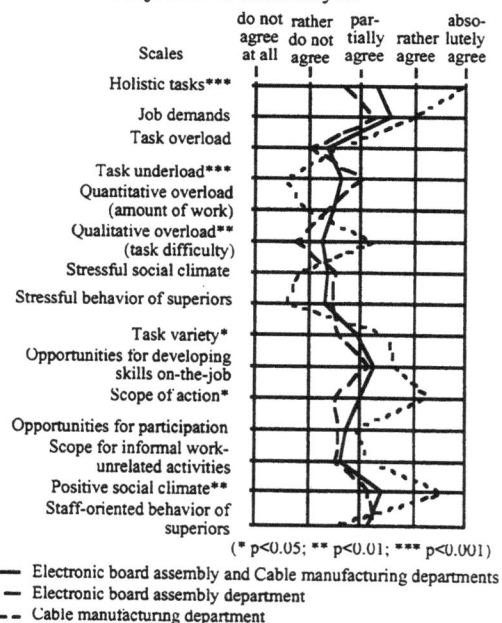

Fig. 4. Results of subjective work analysis questionnaire (n=20).

However, both production departments differ significantly with respect to the employees' perceptions of holistic tasks (t=-5.83, df=8.75, p<.001), task underload (t=4.99, df=10.43, p<.001), qualitative overload (t=-4.08, df=12, p<.01), task variety (t=-2.51, df=10.18, p<.05), scope of action (t=-2.77, df=12, p<.05), and positive social climate (t=-4.62, df=9.77, p<.01). Especially the results on task underload and qualitative overload reveal a subjectively perceived underutilization of skills in the electronic circuit board assembly department. In contrast to this, in the cable manufacturing department the tasks are perceived as holistic, rather difficult, and varied. There are also more affirmative answers with respect to scope of action and positive social climate. These results are striking because in the cable manufacturing department the range of attained VERA-steps (1 and 2R) only slightly differs from that in the electronic circuit board assembly department (between 1R and 2R).

4.3 Analysis of polyvalence

In order to gain an insight into the distribution of tasks and their evaluation according to the condition-related work analyses, the polyvalence of the employees, i.e. the level of multi-skillfulnesss, was analyzed according to the MTO-investigation procedure (Strohm and Ulich, in press). Polyvalence was studied by means of a function-task-tool matrix.

Socio-technical systems comprising many work tasks that are carried out in turns and contain relatively high amounts of regulation requirements according to VERA- and KABA-definitions, make a substantial contribution to the local self-regulation of arising variations and obstructions (Weber, in press). Thus, employees who are qualified for performing various demanding subtasks within a socio-technical system - multi-skilled workers - are a prerequisite for flexibility. Such work systems meet the requirements for the joint design of the three aspects man - technology - organization (MTO-model).

The analysis of polyvalence revealed a relatively large number of workers qualified for various tasks and able to substitute each other in the electronic circuit board assembly and cable manufacturing departments (Table 2., cf. also Fig. 3). Many workers are qualified for and frequently or occasionally perform several tasks. However, only few persons are qualified for performing tasks such as quality control, preparation of materials, NC-programming of the SMD machine, soldering tasks, test of soldered contacts, inquiry of material delivery times, administrative tasks, production planning and fine scheduling, writing of production documents, warehouse management and transport of products, and communicating with suppliers or customers. Strikingly, tasks with planning and disposition elements excluding the administrative and the quality control tasks, are assigned to the few male persons, including the head of production, the head of the cable manufacturing department, and the NC-

programmer. The polyvalence found could not be concluded to ensure flexibilty due to the restricted potential for substitution.

Table 2. Analysis of polyvalence.

Tasks	Workers qualified for and performing tasks	
	frequently	occasion-ally
Incoming inspection	1	2
Preparing materials	1	5
Preparing components	7	8
Distribution of materials	3	-
Pre-assembly	8	11
Manual insertion	12	9
Semi-automatic insertion	3	1
Automatic insertion	2	3
NC-programming	1	1
Soldering	1	1
Test of soldered contacts	1	-
Soldering rework	4	3
Final assembly	1	3
Board assembly quality control	1	-
Packing, shipping	1	-
Inquiry of material delivery times	1	-
Cutting of cables	2	-
Cable manufacturing	3	8
Cable quality control	1	1
Administrative tasks	1	-
Production planning and fine scheduling	1	-
Training	3	-
Writing of production documents	1	-
Communication with suppliers or customers	1	-
Maintenance	1	-
Supervision, production problem solving	1	-
Warehouse management and transport of products	1	-

5. DESIGN PROPOSALS AND CONCLUSIONS

The findings show that the tasks performed in both production departments require a very low amount of planning and decision-making. They cannot be termed suitable for promoting the development of skills. Furthermore, work-related communication requirements are mostly restricted to the exchange of information within pre-defined sequences of work steps. These findings indicate that the workers only need to be familiar with their segment of the process flow - there is no need for coordinating their actions with preceeding and subsequent production units, let alone with external persons. Despite some differences found between the departments analyzed, the workers' perceptions and redefinitions in general reflect the work structures evaluated by way of condition-related work analyses. Analyses of qualifications and tasks performed by the employees reveal a considerable number of polyvalent female workers on the shop-floor level, i.e. several workers are qualified for a number of tasks. However, most female workers perform execution tasks, quality control tasks are assigned to few female workers only. Planning and disposition tasks are performed by few male superiors and specialized workers.

The planned inclusion of purchasing tasks may add to this bottleneck situation. It is essential to distribute planning and disposition tasks as well as quality control tasks among the workers. Furthermore, a broad distribution of processing know-how is needed in order to comply with increasing quality demands. Present qualification deficiencies prevent the assignment of holistic tasks comprising planning elements, execution and self-inspection to the workers. Qualification deficiencies reduce the potential for self-regulation in the production departments which is needed to increase in flexibility in order to, in turn, meet customer demands.

Training of the female workers in order to assign them holistic tasks suitable for the utilization and enhancement of their skills is particularly important. For the foreign-speaking workers, basic language skills are required enabling them to understand production documents and communicate with internal and external persons. Multi-skilled workers qualified for autonomously making decisions related to the processing of manufacturing orders contribute to internal flexibility. The existing multi-skillfulness with respect to execution tasks is an appropriate starting point for this type of changes. The design of holistic tasks implies that disposition elements such as fine scheduling and quality control tasks, formerly assigned to single persons, be re-distributed and thus re-integrated into the scope of more work positions.

Furthermore, a work redesign that disregards the wage scheme is likely to fall short of its expectations. Qualified work and multi-skillfulness should be supported by an appropriate wage scheme corresponding to the workers' multi-skill levels. Pay-for-knowledge or "polyvalence wage systems" (Ulich, 1994) are suitable for this purpose.

REFERENCES

Barney, J.B. and W.G. Ouchi (1986). *Organizational Economics*. Jossey-Bass, San Francisco.

Bergmann, B. and P. Richter (Hrsg.) (1994). *Die Handlungsregulationstheorie*. Von der Praxis einer Theorie. Hogrefe, Göttingen.

D'Ambrogio, F. (1992). CIM-Aktionsprogramm im Dienste der Zulieferer. *Technische Rundschau*, 45, 98-100.

Dunckel, H., W. Volpert, M. Zölch, U. Kreutner, C. Pleiss and K. Hennes (1993). *Kontrastive Aufgabenanalyse im Büro*. Der KABA-Leitfaden. Grundlagen und Manual. Schriftenreihe Mensch-Technik-Organisation (Hrsg. E. Ulich), Band 5a. vdf, Zürich; Teubner, Stuttgart.

Frigo-Mosca, F. and G. Alberti (1995). JIT-Deliveries for a Small Electronics Supplier: A Case Study in Switzerland. In: *46th International Industrial Engineering Conference Proceedings*, 21-24 May 1995, Nashville, Tennessee, USA (Institute of Industrial Engineers (Ed.)), pp. 293-298. Norcross, Georgia.

Leitner, K., W. Volpert, B. Greiner, W.-G. Weber and K. Hennes, unter Mitarbeit von R. Oesterreich, M. Resch and T. Krogoll (1987). *Analyse psychischer Belastung in der Arbeit. Das RHIA-Verfahren*. Verlag TÜV Rheinland, Köln.

Oesterreich, R. and W. Volpert (1991). *VERA Version 2*. Arbeitsanalyseverfahren zur Ermittlung von Planungs- und Denkanforderungen im Rahmen der RHIA-Anwendung. Institut für Humanwissenschaft in Arbeit und Ausbildung, Berlin.

Rimann, M. and I. Udris (1993). *Belastungen und Gesundheitsressourcen im Berufs- und Privatbereich*. Eine quantitative Studie. Institut für Arbeitspsychologie, ETH, Zürich.

Smith Ring, P. and A.H. Van de Ven (1994). Developmental Processes of Cooperative Interorganizational Relationships. *Academy of Management Review*, 19, 90-118.

Strohm, O. and E. Ulich (Hrsg.) (in press). *Ganzheitliche Betriebsanalyse unter Berücksichtigung von Mensch, Technik und Organisation*. Vorgehen und Methoden einer Mehr-Ebenen-Analyse. Schriftenreihe Mensch-Technik-Organisation (Hrsg. E. Ulich), Band 10. vdf, Zürich; Teubner, Stuttgart.

Ulich, E. (1994). *Arbeitspsychologie*. 3. Auflage. vdf, Zürich; Schäffer-Poeschel, Stuttgart.

Weber, W.G. (in press). Analyse und Bewertung von Produktionstätigkeiten. In: *Ganzheitliche Betriebsanalyse unter Berücksichtigung von Mensch, Technik und Organisation*. Vorgehen und Methoden einer Mehr-Ebenen-Analyse. Schriftenreihe Mensch-Technik-Organisation (Hrsg. E. Ulich), Band 10 (Hrsg. O. Strohm and E. Ulich). vdf, Zürich; Teubner, Stuttgart.

Williamson, O. (1991). Comparative Economic Organization. *Administrative Science Quarterly*, 36, 269-296.

Zölch, M. (1993). *Manual VERA/RHIA/KABA für die Produktion*. Unveröff. internes Manuskript. Institut für Arbeitspsychologie, ETH Zürich.

THE RELEVANCE OF
COMPUTER SUPPORTED COOPERATIVE WORK
FOR ADVANCED MANUFACTURING

Thomas Schäl

RSO SpA, Via Leopardi 1, I-20123 Milano
HDZ/IMA, RWTH Aachen, Dennewartstraße 27, D-52068 Aachen

Abstract: Computer Integrated Manufacturing (CIM) is faced with issues which are crucial to Computer Supported Cooperative Work (CSCW). However, despite the large amount of work on CIM and its obvious links to the CSCW field, this domain is almost totally absent in the work of the CSCW community. Therefore, this paper wants to open the discussion about the relevance of CSCW in manufacturing and to combine new concepts for cooperative work with requirements for information system design in production. Production related models, e.g., an order-driven mode of coordination, can be augmented with mechanisms of interaction for coordination, collaboration and codecision and the design of customer/supplier chains. It is suggested, as one example, to use workflow management technology in advanced manufacturing.

Keywords: computer supported cooperative work (CSCW), groupware, business process reengineering (BPR), human centred CIM (HC-CIM), production planning and control (PPC), concurrent engineering, process automation, workflow management technology,

1. THE NEED FOR COMPUTER SUPPORTED COOPERATIVE WORK

1.1. Mechanisms for interaction and coordination in advanced manufacturing

In the current business environment, manufacturing enterprises have to cope with shorter product life-cycles, roaring product diversification, minimal inventories and buffer stocks, extremely short lead times, shrinking batch sizes, concurrent processing of multiple or different products and orders, etc. (Schäl, 1991; Schmidt, 1991). A work organisation operating in this environment cannot rely on advanced planning for task allocation and task articulation. The compilation of tasks into jobs must allow for a high degree of flexibility, and in order to be able to adapt to unforeseen contingencies, task articulation must basically rely on local control. In any case, *coordination* of activities based on the production of parts and inventory is not possible any more.

A feature of the emerging work organisation in advanced manufacturing has been discussed by Cummings & Blumberg (1987) concerning the *self-regulating workgroup*, and by Kern & Schumann (1984) concerning the *multi-functional workers*.

However, cooperative work relations in advanced manufacturing are not limited to a *group* or a *team* responsible for, e.g., a particular shop. Cooperative work relations embrace the entire enterprise, from marketing to delivery, from design to final assembly. As manufacturing involves the unpredictable interaction of many persons and functions, the reciprocal interdependency of the different actors and units is very high. Therefore, the various categories of workers – product designers, production engineers, programmers, supervisors, operators, etc. – are highly dependent on each other in order to keep the factory running. Accordingly, the whole social network of actors in the manufacturing enterprise has to be able to adapt dynamically to changing conditions in the environment; the entire system must react simultaneously and cooperatively.

The traditional *mechanism for interaction* in manufacturing has been the *master production schedule*, a coordination mechanism based on *forecasts* and *standard lead times* for the various parts involved in production. The underlying concept relies on the believe that forecasts are accurate. Coordination among functional units and departments is mediated by plans or other organisational procedures. As a consequence, *direct cooperation* occurs only in case of *exception handling*.

Exceptions, however, are the norm in manufacturing. These are, e.g., shortage of material, delays, faulty parts, variations in component properties, design ambiguities and inconsistencies, design changes, changes in orders, cancellation of orders, rush orders, defective tools, software incompatibility and bugs, system failure or crash, machinery breakdown, changes in personnel, illness, etc.

As a response to the turbulent manufacturing reality, *feed-forward oriented coordination mechanisms* have been introduced. Manufacturing companies are changing towards an *order-driven mode of coordination*. *Kanban* or *just-in-time* systems are of that kind of mechanism. These production related models can be augmented with *mechanisms of interaction* for *coordination*, *collaboration* and *codecision* in manufacturing and the design of *customer/supplier chains* for the material exchanges and the linking of *production islands* or *semiautonomous working groups*. Beside the need to understand cooperation in manufacturing, the need for computer support becomes evident.

1.2. Computer Integrated Manufacturing

The strategic means to meet these challenges include advanced manufacturing technologies and information systems, summarised under the concept of *Computer Integrated Manufacturing* (CIM). The ambition of the efforts in CIM is to link and fuse the diverse information processing activities of the various manufacturing functions, such as design and process engineering, production planning and control, sales, purchasing, distribution, accounting etc., into a unitary information system. A CIM system embracing these information processing activities on a company-wide scale should be seen as a unified database system facilitating and supporting the horizontal and hierarchical, direct and indirect, coordinated and collaborative dependencies of a heterogeneous ensemble of distributed persons involved through all functions of manufacturing. This multi-divisional information system should allow to reduce development and production time which leads at the end to a reduced lead-time (Eversheim & Brachtendorf, 1987). CIM is thus faced with the issues which are crucial to CSCW (Schmidt, 1991). However, despite the large amount of work on CIM and its obvious links to the CSCW field, this domain is almost totally absent in the work of the CSCW community (Bannon & Schmidt, 1992). This paper wants to open the discussion about the relevance of CSCW in manufacturing and to combine new concepts for cooperative work with requirements for information system design in production. It is suggested, as one example, to use *workflow management technology* (Schäl, 1995) in advanced manufacturing based on *production islands* (Schäl, 1991).

1.3. Production Planning and Control

Most of the commercially available *Production Planning and Control* (PPC) systems have been developed in the 70s. They have been designed according to a Tayloristic, centralised organisational model. This results in systems which have a deterministic view about the working processes and information flows in production companies. The working processes, as supported by these systems, are rigidly controlled and fed back to the central control units of the company. The WZB (1989) argues that the failure of Tayloristic CIM systems and especially of PPC components is due to the increased personal communication and coordination in companies which are not sufficiently reflected by PPC software developers (Hildebrandt & Seltz, 1989). The study shows that the introduction of information systems with their internal control mechanisms and company specific functionalities can only be understood in relation to the social organisational of work in the specific context of the domain.

The example of the German mechanical industry shows that manufacturing is not based on a Tayloristic large scale production principle, but on differentiated production plans, personal control, intrinsic motivation, mechanical workers with experienced skill and expertise, informal cooperation and trust among different division involved in production. The missing reflection of the socio-technical setting results in a partial use of PPC software in the sample organisations analysed by Hildebrandt & Seltz (1989). The macro planning functionality of PPC systems survives, while the micro planning is not considered at all. As the decentralised shop floor planning is still done by improvisation, the printed job lists from the system are generally thrown away. Therefore companies which recognised this lack of support in general purpose PPC systems have to develop the micro planning for their own. This in-house software modules reflect EDP-logic and decentralised self-regulation of work in an appropriate combination (WZB, 1989).

Management in small batch production is the ongoing effort to manage chaos (WZB, 1989). This results from the reciprocal relation between a necessary chaotic coordination of actions on the one hand, and the attempt to plan and control according to rules and norms on the other hand. This tension between plans and improvisation is however accepted reality. In advanced manufacturing, PPC thus requires horizontal and direct cooperation across functional units and professional boundaries which have not been taken into consideration in present PPC mechanisms and systems, as well as in other CIM components. This paper develops a model to work in these realities based on workflow management for shopfloor PPC, which integrates the planned progress in production

68

with the continuous management of chaos and break-downs by interpersonal communication.

1.4. Concurrent Engineering

But not only the shop-floor is effected by the changing environment of production companies and possible implementation of computer systems in the networked production company. Increasing product complexity, competition and tighter time-to market deadlines has a great impact also on design engineers. Most manufacturing companies organise their design efforts as a kind of pipeline with the results of one step in the design process being 'thrown over the well' to the next step (Malone et al., 1993). This traditional serial design generates costs for making a change which increase by an order of magnitude for every step in the cycle.

Recently, most manufacturing companies have come to the believe that there is a much better way to organise the design process: the new concept is called *concurrent engineering* where all functions are performed concurrently and iteratively by design teams with specialists from all the relevant functional areas (Carter & Baker, 1991). Design teams will turn to a multidisciplinary approach where all disciplines involved traditionally in different phases of the product development will work together concurrently to define and develop the product. The team will work together to make the right design decisions from the very beginning, when mistakes are less expensive and easy to fix. These new engineering teams will need some changes in their supporting tools and systems, adjustments in team management and the way engineers interact with each other. The incubation period of modern product design is reduced due to the introduction of computer technology in concurrent engineering. Because of this, product life-cycles are reduced dramatically. For example, in automobile production major re-designs of engines, bodies, axles and brakes used to occur every six or seven years. Today, however, product life-cycles are cut down to half the time, that is, to about three or four years (Gunn, 1987).

The key to successfully design teams is to emphasise *team*. Therefore one of the challenges for concurrent engineering is getting engineers to work well in teams. Also universities and other higher education institutions should reflect this issue in their curricula to educate prepared persons for new organisational settings (Schäl, 1989).

The emergence of concurrent engineering teams will introduce also new disciplines to the traditional ones into the design cycle. Some other disciplines will merge or change its mission. The discipline which will become one of the most important, is to understand the customer's requirements and to guarantee that everyone is participating in the engineering process. Implementing concurrent design without capturing the customer's requirements is useless, because the team might produce a better, cheaper product which does not meet the customer's needs.

The design process will start with capturing requirements from the customer in a method which is easy to communicate and easy to refer back to as more detailed design is done. Today design teams often do not have this effective and efficient way to capture and record the *customers requirements* (Maliniak, 1991).

Another important discipline which will appear in concurrent engineering is *business process re-engineering* and the resulting use of *workflow management technology*.

2. WHAT IS COMPUTER SUPPORTED COOPERATIVE WORK (CSCW)?

The decade of the '80s saw a massive rise in the importance of personal computing which was followed by a trend towards networking all the end-computing devices. This was the context for the birth of the area of *Computer Supported Cooperative Work* (CSCW). The term CSCW was used to describe the topic of the interdisciplinary workshop in 1984, intending to discuss on how to support people in their work arrangements with computers (Greif, 1988). Since then CSCW has emerged as a new interdisciplinary forum for research into the issues central to the design, implementation and use of technical systems which support people working cooperatively (Bannon et al., 1991).

Four years after the first workshop in relation to the second CSCW conference, Greif (1988) defines CSCW in a book of readings on the topic as computer support for group work. The idea of *group* as the main unit to focus on is common to several approaches in CSCW and directly related to the term *groupware*, i.e., software for groups (Johansen, 1988). Some groupware definitions take a workspace-oriented view where groupware is a support for people engaged in a common task by an interface to a shared environment (Ellis *et al.*, 1991). Groupware is often regarded as the technology-driven development of applications which represent a natural extension of software for workstations interconnected via a local area network, a development of *Office Automation* ideas and a significant extension to *Office Information Systems* (De Michelis, 1990). Another term related to the field of CSCW is *Workgroup Computing* which refers to the support provided by networked micro-computers. A definition of CSCW is given by Bannon & Schmidt (1991):

Computer Supported Cooperative Work (CSCW) is an identifiable research field focused on the understanding of nature and characteristics of cooperative work with the objective of designing adequate computer based technologies to support such cooperative work.

This definition focuses on the understanding of *cooperative* in order to support the work by taking into consideration the results gained in the design process before applying information technology. The following chapter gives an example of the potential of CSCW in decentralized CIM environments.

3. HUMAN-CENTRED COMPUTER INTEGRATED MANUFACTURING

3.1. Esprit Project 1217 (1199)

The human-centred approach in the development of technology was researched in the ESPRIT-project 1217 (1199) to demonstrate that CIM, under the philosophies of human centredness, is significantly different to a traditional Tayloristic division of labour. The human-centred approach is focused on people working within and managing manufacturing cells. The development of multi-skilled personnel is a key concept: Much of the production planning and scheduling activities that take place in offices are transferred to the shop floor (Schäl, 1991).

A working prototype of such a human-centred production cell was developed by a company based in London. This cell is to produce high-frequency connectors. The company specialises in the manufacturing of precision connectors for communication signals at frequencies up to 46 GHz. The connectors can be described as consisting of an outer body and a central pin which is surrounded by a PTFE insulator. The dimensions of the parts are of some millimetres. The whole variety of the company's production is approximately 44000 items. Batch quantities are mostly in the range of 200 to 1000 parts.

The process for the production of the final connector consists of several main steps. The metal and plastic parts are turned on a lathe (30 minutes for a batch of 250 parts). Thereafter main production steps are second operations, heat treatment, gold plating, inspection and assembly.

3.2. Design of a cellular factory

The author suggested an overall factory model for a cell-based production of the connectors under the underlying philosophies of Group Technology, Period Batch Control and Optimised Production Technology (Schäl, 1991). The model of the factory consists of two cell types with different skill-levels and a constant number of persons required. The cells with the higher skill-level (machining cell) produces finished parts which are later assembled in the second cell (assembly cell) to the final products. Three machining cells and one assembly cell work together within a product family.

Machining cell: The order of a connector consists of the pin, insulator and body. All operations and processes to finish these parts are done within the machining cell. The machining cell consists of four lathes and one multi-purpose machine for second operations. The required number of operators is six.

Assembly cell: In the assembly cell machined components are assembled. The cell dispatches the finished connector. The required number of operators is four persons for six assembly and despatch working places.

The plating of the connectors is done in a centralised process. The manufacturing cells could not plan their throughput if they do not know the maximum time required in the process cell which should be reasonably short. Assuming that subcontracting usually occurs at the same time, e.g., at the middle of a period, an organisation by *Smoothed Period Batch Control* (Schäl, 1991) helps to overcome the requirement of infinte capacity. The capacity peak is smoothed by adjusting the initial dates of periods for manufacturing cells on the timescale.

3.3. Information and control

Information flow and control have to change in this factory. Especially the middle management will loose responsibilities because some of their tasks will be done on the shop-floor level. The planning scenario looks as follows: The material manager receives all orders from customers. He identifies the product families and sorts them by groups. The resulting job-list is transferred via the computer-network to the cells. People view this list which is in a random form. Subsequently an operations list is scheduled with assistance from the system and/or manually. Basic rules for an optimised sequence of orders are available in the cell-software. It is the choice of the operator which method he wants to use. After a first draft of macro-scheduling he calculates the estimated makespan with a rough cut program. The result is the base for feedback to the material manager. The cell will reject the planned load if it would be excessively overloaded. The material manager has then to cut down the job-list to a reasonable load for the cell. If the cell would be underloaded the workers can ask for more jobs to do in that period.

Once the load is in the range of 100 per cent the cell members microschedule the operations list. They check and adjust all the used data; special attention is paid to the set-up times. These are important for small-to-medium sized batch production because they reduce the utilisation of the machinery. The complexity of the products makes it difficult or even impossible to program the scheduling. The system can assist the operator by providing several methods which can be chosen to get a starting sequence. This might be altered by the operator.

Once the operator thinks that he has found the best sequence he simulates it on the computer. These results are more precise than the rough-cut makespan. The operator informs the materials manager whether he agrees with the allocated job-list, or whether he wants to reduce it again or desires to have some more jobs to do in that period.

The material manager at factory level cannot control the decisions in the cell. The material manager's task is the balancing of orders over the cells available. If a cell rejects a planned job he would have to find a different cell to do the job.

3.4. Workflows for shop-floor PPC

The negotiation among the material manager and the cell operator can be modelled as a cutomer/supplier relation in a workflow (see Fig. 1).

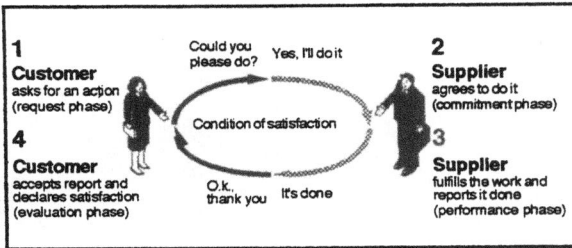

Fig. 1: Workflow describing customer/ supplier relations (elaborated from Keen, 1991; Medina-Mora et al., 1992)

This relation among the material manager and the cell operator is difficult to support in conventional CIM environments. However, this more informal discussion anout workloads and commitments is in the scope of support tools for cooperative work. One such support can be seen in workflow management technology (Schäl, 1995) or other groupware applications. Thus, CSCW offers a new understanding of cooperative work settings and a complementary set of applications in addition to CIM environments. The discussed needs to support cooperative work in advanced manufacturing and the short example for improving shopfloor PPC with workflow management technology might be a first step towards the linking of the application domain of manufacturing and the CSCW research community.

REFERENCES

Bannon, L.; Robinson, M.; Schmidt, K. (eds). *Proceedings of the Second European Conference on Computer-Supported Cooperative Work*. Kluwer Academic Publishers, Dordrecht/ Boston/London

Bannon, L.J.; Schmidt, K. (1991). CSCW: Four Characters in Search of a Context. In Bowers, J.M.; Benford, S.D. (eds) (1991). *Studies in Computer Supported Cooperative Work. Theory, Practice and Design*. North-Holland, Amsterdam: 3-16

Carter, D.E.; Baker, B.S. (1991). *Concurrent Engineering: The Product Development Environment for the 1990's*. Addison-Wesley, Reading, Massachusetts

Cummings, T.; Blumberg, M. (1987). Advanced Manufacturing Technology and Work Design. In Wall, T.D.; Clegg, C.W.; Kemp, N.J. (eds). *The Human Side of Advanced Manufacturing Technology*. Wiley, Chichester: 37-60

De Michelis, G. (1990). *Computer Support for Cooperative Work*. Butler Cox Foundation, London

De Michelis, G. (1994). Computer Support for Cooperative Work: Computers between Users and Social Complexity. In Bagnara, S.; Zucchermaglio, C.; Stucky, S. (eds). *Organizational Learning and Technological Change*. Springer, Berlin

Ellis, C.A.; Gibbs, S.J.; Rein, G.L. (1991). Groupware: Some Issues and Experiences. *Communications of the ACM*, **34(1)**: 39-58

Eversheim, W.; Brachtendorf, T. (1987). PPS - ein zentraler Baustein für CIM. *Industrie-Anzeiger*, 109-19: 50-59

Greif, I. (ed.) (1988). *Computer-Supported Cooperative Work: A Book of Readings*. Morgan Kaufmann Publishers, San Mateo, California

Gunn, T.G. (1987). *Manufacturing for Competitive Advantage: Becoming a World Class Manufacturer*. Ballinger, Cambridge, Massachusetts

Hammer, M.; Champy, J. (1993). *Reengineering the Corporation: A Manifesto for Business Revolution*. Harper, New York

Hildebrandt, E.; Seltz, R. (1989). *Wandel betrieblicher Sozialverfassung durch systemische Kontrolle? Die Einführung computergestützter Produktionsplanungs- und steureungssysteme im bundesdeutschen Maschinenbau*. Sigma, Berlin

Johansen, R. (1988). *Groupware: Computer Support for business teams*. The Free Press, New York

Kern H.; Schumann, M. (1984). *Das Ende der Arbeitsteilung? Rationalisierung in der industriellen Produktion*. Beck, München

Malone, T.W.; Crowston, K.; Lee, J.; Pentland, B. (1993). *Tools for inventing organizations: Toward a handbook of organizational processes*. CCS Technical Report #141, MIT, Massachusetts

Schäl, T. (1989).*The need for non-engineering subjects in common core curriculum of engineering education and the Aachen model*. In Proceedings SEFITALIA89, Naples, 17-20 September: 297-302

Schäl, T. (1991). Menschenorientierte CIM-Konzepte für die Flexible Fertigung. *Werkstattbericht*, **100**, Ministerium für Arbeit, Gesundheit und Soziales, Düsseldorf

Schäl, T. (1995). *Workflow Management Technology for Process Organizations*. PhD-Thesis, HDZ/IMA, RWTH Aachen

Schmidt, K. (1991). Computer Support for Cooperative Work in Advanced Manufacturing. *International Journal of Human Factors in Manufacturing*, **1(4)**: 303-320

WZB (1989). EDV im Maschinenbau: Systemische Rationalisierung. *WZB-Mitteilungen*, **46**, Dezember: 7-10

ORGANIZATIONAL DEVELOPMENT IN HOSPITALS - EXEMPLIFIED BY THE INTRODUCTION OF PROFIT-CENTER-STRUCTURES

I. Isenhardt, J. Grobe, U. Steinhagen de Sánchez

Department of Informatics in Mechanical Engineering (HDZ/IMA), University of Technology, (RWTH), D-52068 Aachen, Germany

Abstract: Large hospitals are highly complex systems in terms of dynamic interrelations between people, the organization they work in and the technology they work with. Such complex social systems must constantly develop and redesign themselves to meet all requirements of their task. The Department (HDZ/IMA), particularly deals with organizational development and the reorganization of large hospitals as one of ist areas of activity. For several years the HDZ/IMA has been implementing in two large hospitals a modern decentralized organization based on participation and supported by decentralized concepts of technology.In this report, some experiences of re-organzing two large hospitals are described.

Keywords: Hospitals, Socio-technical systems, systems design, organizational development, decentralization

1. INTRODRUCTION

For the aim of organizational and technical changes, the question of power needs to be considered: large complex organizations exert power both within themselves and towards their environment in society. They tend to grow in a way which resembles cancer growth. They continuously increase size and influence even if in the end, they may destroy themselves through their growth. But they are unable to see their growth process as their path into destruction. Examples of this cancer growth may be the end of communism in East-Germany and Eastern Europe, but also the bankruptcies of large enterprises and the decline of acceptance of political power in many countries. These processes are threatening the fabric and the future of our societies.

One way to counteract this danger is to implement strategies of decentralization and subsidiarity through all layers of organizations. This paper discusses two out of several research and developent projects of the HDZ/IMA to change complex organizations toward decentralized structures. Several change strategies have been developed and tested in these projects. They integrate both organizational and technological changes. Some of the largest German enterprises and hospitals have been involved into these projects. These projects are examples of succesful cooperation of university with non-university institutions and industry. The projects discussed here deal with the organization of two large hospitals.

2. PROBLEM DESCRIPTION

Large hospitals are highly complex systems in terms of dynamic interrelations between people, the organization they work in and the technology they work with. Such complex social systems must constantly develop and redesign themselves to meet all requirements of their task. The university hospitals to be discussed comprise each about 1500 beds, more than 5000 employees and about 3000 students of medicine. There is a high standard of medical care associated with a high degree of specialization. All parts of these hospitals are accommodated in modern buildings.

Although these hospitals have got a high reputation in public opinion - because of their superior medical standards - there are some problems which according to the management of the hospitals need to be solved. These are described as follows:

- The flow of information and material does not always work to full satisfaction of staff and patients.

- Many employees are not satisfied with their working conditions and the organizational structures.

- The costs for treatment and services are too high and there exist hardly any awareness of the costs involved on the employees' side.

- The cooperation between various sections of the hospitals is not satisfactory.

- The aim of customer-oriented services and of flexibility in solving problems is not always achieved.

- In public opinion criticism of the non-medical services for patients seems justified.

In the hospitals conflicting views exist on the need for change. Therefore, the Department of Informatics in Mechanical Engineering (HDZ/IMA), University of Technology (RWTH) Aachen, has been asked to apply new approaches to redesigning the organizational and technical structures of the hospitals.

The department's approach to organizational development is based on the cybernetic approach which is presented in the next paragraph.

3. THE OSTO - APPROACH

As a method to analyze, redesign and monitor a system the OSTO approach has been used (Hanna, 1988; Rieckmann and Weissengruber, 1990; Henning and Marks, 1990; Marks, 1990).

OSTO stands for "open, socio-technical-economic system", i.e. the open system comprises social and technical as well as economic components. It builds on the socio-technical system theory which was considerably influenced by members of the London Tavistock Institute of Human Relations.

OSTO is particularly suitable for organizations and focusses on analyzing high-quality work processes which are strongly inter-related. They are largely the result of a combination of high level of technology and high quality standards. OSTO understands systems as "living systems" (open cybernetic systems). Such systems include human beings with their processes of work and life. Feedback processes stabilize and renew these open living system (fig. 1).

The system transforms input to output through a transformation process. The mission is the reason for the existence of the organization. It represents the unwritten contract between the system and its environment. However, all processes within the system only become core processes, if they are oriented towards the mission. In the long run, living systems cannot survive without reason for existing or core processes.

Moreover, it is necessary for survival of the system that the purpose of the system is future-oriented. Only then, the members of the organization maintain both motivation for and identification with the system. The system can maintain its acceptance in a wider social context.

Further elements of cybernetic systems are feedback loops which are either negative or positive (weakening or enforcing). They give to the system the qualities of stabilization and renewal. In this context the organization of feedback processes is an important managerial task with regard to survival of the system.

The system can be characterized by certain terms as follows. The goals of an organization are linked to the mission. They describe, what is to be achieved by the system. Strategies are the means to achieve both objectives and the mission of the organization.

The system structure conducts the transformation process. The system behaviour (culture) makes the transformation process visible. It is determined by objectives, strategies and structure of the system.

The concept presented here is suitable to analyze organizations and thus, to prepare organizational change in the system. Interventions can be planned and system responses can be analyzed.

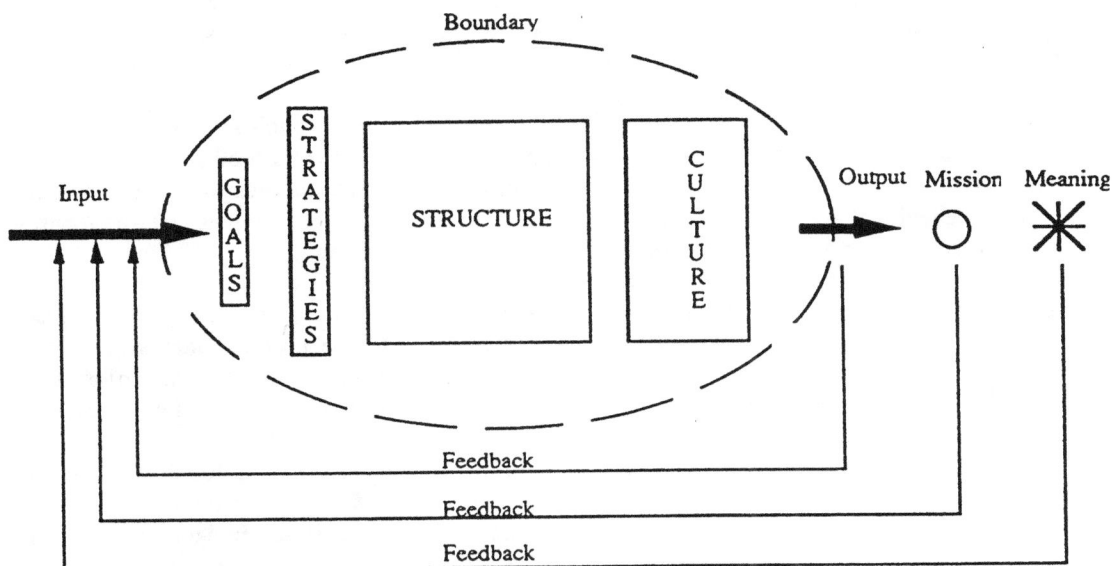

Figure 1:   OSTO approach, a recognized model of Systemic Management

4.   DIAGNOSIS THROUGH THE OSTO-APPROACH

The following statements on output, culture and structure of the system are the results of continuous observation and analysis of communication and information flow, opinion polls (including patients and staff), interviews of individuals and groups, and observation and interpretation of reactions towards interventions of the department (HDZ/IMA).

The results of the analysis are correlated with the contract of the system with its environment, namely the mission. According to statements of staff and management a hospital today has got the following mission:

The existence of the university hospital is based on the need of

- patients for high medical quality standards and financially reasonable medical and non-medical care,

- current and future medical staff for general and specialized professional qualification and further education,

- society for financially reasonable medical research with high quality standards,

The goals and strategies of a system should ideally represent the realization of its mission, its reason for existence. For the hospital this would mean an orientation of all subsystems of the organization towards this mission. In order to check this, it is necessary to consider the system outputs. The department HDZ/IMA, however found outputs like e.g. high costs, patients dissatisfied with the service, waste, overworked staff, but also highly qualified doctors and nurses.

These outputs are again attributed to a certain behaviour (culture) of the system. Culture includes any observable behaviour of the system, of its members and subsystems, e.g.:

- The behaviour in working and problem-solving situations (hardly any awareness of costs);

- Rules and norms (e.g. strong orientation towards job descriptions);

- Management culture (decisions have to pass many hierarchical levels, employees do not like making decisions for themselves);

- Working atmosphere (strongly competitive thinking, hardly any teamwork, some employees feel discriminated against because of low-reputation jobs);

- Feelings and attitudes of staff (aggression against superiors, low motivation, high acceptance of technical equipment, also a strong sense of duty towards patients).

- Reliability of technical equipment (e.g. high reliability of electronic data processing (EDP)).

System output and behaviour are regarded as consequences of the structure or design elements of the system, where these are in turn interpreted as system strategies and objectives put into practice. This structure of the system is shown in Figure 2. The design elements of a system are linked with each other by dynamic and complex interrelations.

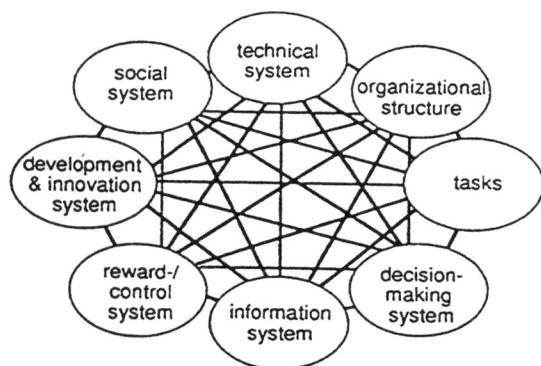

Figure 2:   Structure of an organization

Here only a few illustrative results of the analysis of the existing organizational structure in the hospitals can be mentioned:

- A functional principle of organization prevented task-oriented acting and thus flexible problem solving. This led to frequent conflicts between the various sections of the hospital (organizational structure).

- In many areas the decision structure is determined by a centralized information system, which thus impedes an unbureaucratic behaviour (information system, decision-making system).

- A hierarchical decision structure prevented members of staff to make decisions in their section, although this would correspond to their qualifications (decision structure, social system).

- Some of the organizational sections were too large. This overtaxed the executives. It also had the effect that decisions, tasks and orders were not transparent for the employees. Frustration,

lack of motivation and lack of identification of employees with their work were the consequences (organizational structure, decision-making system, social system).

- Mistakes were often punished, or disguised and presented as mistakes of other sections. This, too, led to conflicts between the various sections and groups of the hospitals (reward/control system, social system).

- There was not much feedback taking place within the system, e.g. on the quality of work or on costs (reward/control system, information system, development and innovation system).

- Further education, the basis for development of special and key qualifications, was rarely supported due to small budgetary provisions (development and innovation system, social system).

These points show that some elements of the organizational structure are not oriented towards the mission but towards maintaining the power of some executives or that of whole sections. They are the main obstacles against the de-centralization into small organizational sections; or improved cooperation between various sections and transparency in making decisions for the organization as a whole, as demanded by many employees. These changes would mean a possible shift in hierarchy and power structures, which obviously is not wanted by everybody in the hospitals.

Still there was a number of employees in different levels and sections of the two hospitals who were motivated and prepared for new ways of cooperation and problem solving. A further advantage of the system was and still is its capability to mobilize hidden capacities in order to manage highly critical situations. On this basis, the staff of the hospitals got very strongly involved in the changes envisaged, across all levels of employment.

In the following paragraphs the experiences of the research team of the HDZ/IMA are briefly described related to the two different hospitals mentioned.

One of these hospitals shows a paricularly strong centralized organizational structure which is supported by being housed in only one huge building and by a thoroughly centralized information and transportation system.

The other hospital superficially shows a decentralized structure. It is housed in about 60 different buildings. linked by more than 20 km of inner-urban roads. The underlying organizational structure, however, is also strongly oriented towards centres of power. Decisions are made top-down. The information system is also highly centralized.

5. APPROACHES FOR REORGANIZATION

For both hospitals, the diagnosis of the systems has shown that the official goals and strategies, oriented towards the hospitals' mission, are not reflected in the structure to a sufficient extent.

These following changes were initiated by the HDZ/IMA in cooperation with administration and employees:

- Qualification of executives in seminars on systemic management (along with other executives from industry);

- Introduction of project work: various innovation tasks were solved by teams working on certain projects. These teams consisted of members from different sections of the hospital, (e.g. introduction of a new cost assessment system of the outpatient department; reorganization of nursing care; designing new decentralized technological components of the information network etc);

- Establishment of service departments within the system which provide internal interdepartmental and customer-orientated services for various subsystems;

- In one of the two hospitals, establishment of a department for continuous controlling of the orientation towards goals and tasks as well as the development and innovation of the system (department of "Development and Controlling");

- Also in one of the two hospitals, decentralization of the administration (change from five to nine departments) and of nursing services (change from one to four relatively autonomous nursing departments);

- Decentralization of several services for the nursing service (integration of supply assistants and secretaries into the ward teams);

- Job rotation of assisting staff between various departments and sections;

- Introduction of the concept of technological pilot experiments, dealing mainly with decentralized information processing and transport;

- Providing transparency of the relationship between service and costs in various sections of the hospital (e.g. introduction of a system of itemized costs for the outpatient department).

6. APPROACHES FOR RE-DESIGNING TECHNOLOGY

In this paragraph, the development of some technical componentes of the information system in one of the two hospitals is described in some more detail. This example has been chosen because it shows the close interrelation of technological change and organizational development.

Based on the research of the HDZ/IMA team, the following conditions were identified for one of the hospitals as necessary for effective teamwork between medical doctors and nursing personnel:

- Hospital department teams should operate on the basis of decentral revenues and costs considerations.

- Revenues and costs should be allocated over the decentral units; however decentral performance should be measured and accounted for.

7. CONCLUSION

The outcomes of these two projects - and several other projects of similar kind as implemented by the HDZ/IMA - have shown that it is possible to change large complex organizations like such hospitals or industrial companies. In order be successful it helps to have available:

- an analysis and innovation strategy which is based on participating observation, empathy and creativity

- a team of young, independent and committed 'change agents'- students and young research staff who may act as catalysts for changes which people within the organization are eager to introduce,

- patience, hope, the determination not to give up despite difficulties and antagonisms - and the willingness to accept a low profile and visibility so that changes occur out of the organization itself rather than being introduced from outside,

- and the support of those in power.

REFERENCES

Hanna, D P. (1988); Designing Organisations for High Performance, Reading, Massachusettes.

Henning, K. and Marks S. (1989); Application of Cybernetic Principles in Organizational Development, 2nd International Symposium on Systems Research, Informatics and Cybernetics, Baden-Baden.

Henning, K. and Marks, S. (1992); Kommunikations- und Organisationsentwicklung, Vorlesungsmanus-kript, 2. überarbeitete Auflage, Aachener Reihe Mensch und Technik, Aachen.

Marks, S. (1991); Gemeinsame Gestaltung von Technik und Organisation in soziotechnischen kybernetischen Systemen, VDI-Verlag, Düsseldorf.

Rieckmann, H. and Weissengruber P. H. (1990); Managing the Unmanagable? - Oder: Lassen sich komplexe Systeme überhaupt noch steuern? - Offenes Systemmanagement mit dem OSTO-Systemansatz, in: Kraus, H.; Kailer, N.; Sandner, K. (Publ.): Management Development im Wandel, Wien.

Isenhardt, I. (1994): Komplexitätsorientierte Gestaltungsprinzipien für Organisationen - dargestellt an Fallstudien zu Reorganisationsprozessen in einem Großkrankenhaus, Aachener Reihe Mensch und Technik, Aachen.

INFORMATION MANAGEMENT FOR INTEGRATED SYSTEMS IN PROCESS INDUSTRIES

Peter Loos

Institut für Wirtschaftsinformatik - IWi (Institute for Information Systems)
Universität des Saarlandes, 66041 Saarbrücken, Germany
e-mail: loos@iwi.uni-sb.de, http://www.iwi.uni-sb.de

Abstract: The paper proposes the integration of information systems in process industries, like logistic systems, engineering systems and process control systems, by the use of integrated process and product models. After an overview of integration requirements in process industries, it presents an approach to integrate material and process information and to manage them by the use of graphical modeling techniques. The approach uses concepts like hierarchisation and business process orientation to support the integration. Graphical representations for the models are shown and their application in a graphical recipe editor are described.

Keywords: Business Process Engineering, Chemical Industry, Industrial Production Systems, Information Integration, Information Systems, Organizational Factors, Production Control

1. INTRODUCTION

In contrast to discrete manufacturing - the typical application area of Computer Integrated Manufacturing (CIM) - the process industries, to which the chemical industry as well as food industry, rubber industry and paper industry belong, have not been thoroughly researched yet. Therefore there is still a great potential for information systems integration. While in discrete manufacturing concepts exist to link CAD and CAM, similar concepts for process industries are still missing. The same is true for linking production planning to process execution.

Until now various information systems supporting these areas operate independently and only have very weak connections. Since they were developed independently, they have their own internal logic and structures. Interfaces are realized by exchanging papers and manual notes.

However, since a business process flows through various departments it has to be processed by several information systems. In order to get full advantage of business improvement the integration of different information systems can be an important instrument for business process reengineering.

Because of special production conditions in process industries, concepts for business process design and integrated information management have to be adapted to the specific requirements (Luber, 1992; Schürbüscher, et al., 1992; Loos, 1995), for example:

- Physical and chemical attributes of the materials to be processed and their end-products must be considered within the framework of material management, e.g. fluctuating quality in mineral, vegetable or animal raw materials for primary production, limited shelf-life or special storage conditions and restrictions.

- Production and manufacturing processes often result in the creation of co-products. This is especially true for primary production. The co-products of such analytical material conversion processes, which could be either primary, co- or secondary products as well as cyclic material flow, e.g. catalysts, lead to a complex, internal and excessive material composite with corresponding effects on production and material planning.

- Environmental protection determination and hazardous material regulations have a direct influence on production planning and the execution of logistics processes. Thus, simultaneous production or storage of two certain products could be ruled out.

- Frequently, functions for the calculation of input-output relationships or production times are non-linear (e.g. non-linear substitution of raw materials) or parameters which cannot be determined in advance (e.g. influence of weather on the production process).
- Production processes often have narrow time restrictions and complicated equipment which depends on the sequence of products.
- Even more so than with discrete manufacturing, production data must be logged and archived. This concerns not only the quality of the product, but rather the quality of raw materials and the logging of the manufacturing processes.

2. INTEGRATION REQUIREMENTS

There are several information systems that deal with information on products and processes which have to be integrated, e.g. systems for process development with process definitions, production planning systems with master materials and rootings, process control systems with recipe descriptions, laboratory information and management systems or systems for hazardous materials and environmental protection.

According to the Y-Model for CIM developed by Scheer (1994a), figure 1 shows the information systems indicated as CIP - Computer Integrated Processing (Polke, 1989; Eckelmann, et al., 1989) for production in the process industries (Loos, 1993).

2.1 Focus of Information Systems

Information systems can be characterized by various criteria describing the main focus of the systems and the conditions of ther use:

- Planning-oriented systems versus execution-oriented systems
  Within the information systems functions can be distinguished which are more planning-oriented on the one hand and which are more execution-oriented on the other hand. In figure 1 the planning-oriented systems are shown on the upper part, the execution-oriented systems are shown on the lower part. Planning-oriented systems support the functions for preparing the production in a long and medium time horizon. Results of these functions are released orders and process descriptions, which are processed by execution-oriented systems.
- Logistic-oriented systems versus engineering-oriented systems
  Logistic-oriented systems mainly deal with material, inventory and production order. They are one part of the enterprise-wide logistic management. Engineering-oriented systems deal with the technical aspects of products, processes and resources. In figure 1 logistic-oriented systems are depict on the left side, the engineering-oriented systems are depict on the right side.

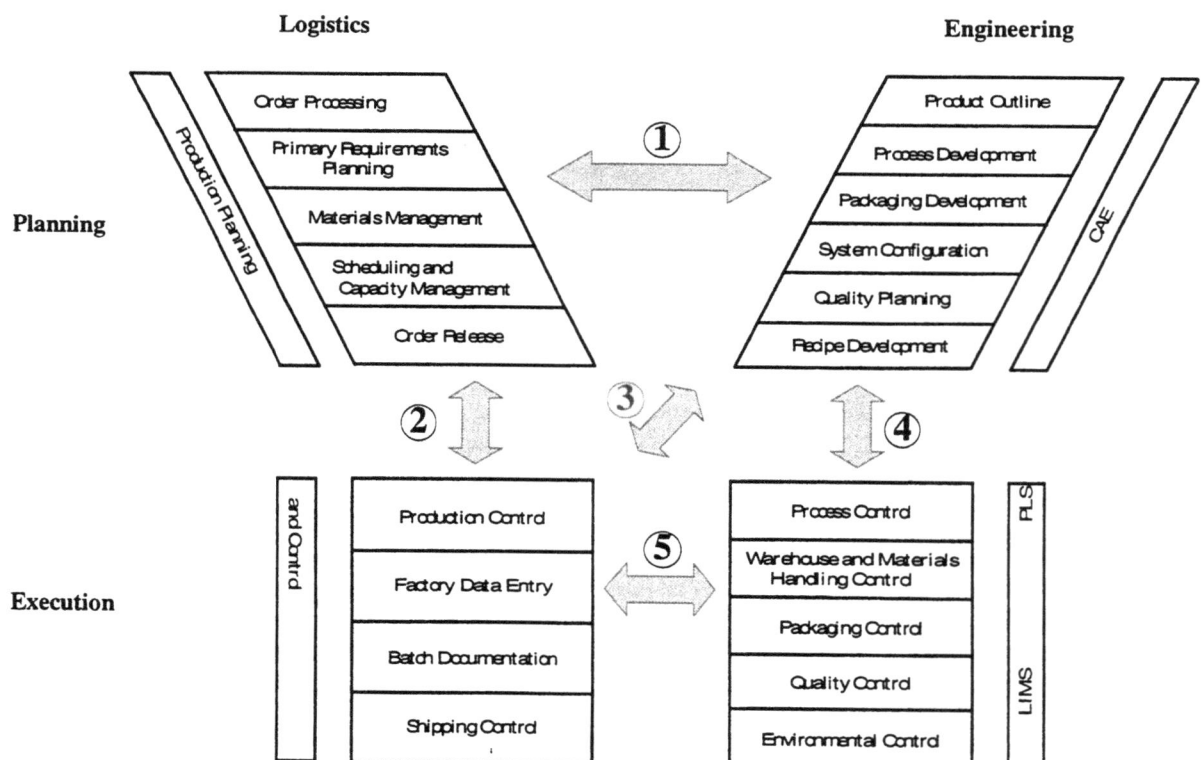

Figure 1: CIP Model with Information Flow

- Transaction-oriented systems versus real-time oriented systems
  Transaction-oriented systems are used in interactive or batch mode. The transactions are controlled by the users. Although performance is an important issue, data processing in transaction-oriented systems is not time-critical. In logistic-oriented systems user transactions last seconds or minutes and can easily be implemented with relational data bases. Real-time oriented systems are used to control equipment and processes. Data processing is critical and has to be guaranteed in a given time frame.

With these different requirements and conditions it is obvious that with today's information technologies it is not sensible to try to cover all the CIP functionality for all products and processes in one single information system. An overall database which stores all information does not seem to be suitable either.

2.2 Information Flow between Information Systems

However, information has to be exchanged between the various systems with the flow of business processes. The flow of information has mainly to do with product and process description. In process industries, production process and product information is usually stored in recipes. They contain information about required raw materials, ingredients and their quantities as process input, intermediate products, finished products, the by- and co-products and the waste, environmental impact and hazards, required production resources like equipment or human resources and detailed process descriptions.

Now follows a rough overview of the main information flow (a description of information flow from and to recipe development is given in Loos, et al., 1994)

- Material and production planning has to decompose the primary requirements which result from customer order and sales forecast into production and purchasing orders. It has to take material inventory into consideration. Among others, information about the structure of materials and the quantity ratio between subordinated and superordinated material is needed. This information is usually defined in bills of materials. However, this information is based on the product development, a function of engineering systems on planning level, and is enlarged with logistic related information.
  Furthermore it requires rough information about the production process, e.g. process times and resources to allocate, for the capacity requirements planning. This information also origins in the engineering systems on planning level, in the equipment construction and in the system configuration. This information flow is indicated with ① in figure 1.
  The result is a medium-term production plan and production orders.

- Logistic-oriented systems on execution level, especially the production management and control (sometimes called MES - Manufacturing Execution System), have to schedule production orders and allocate the required resources in a more detailed way. They need the logistic process description (information flow ②) and some more precise technical description (information flow ③)

- For process execution process control systems need information about the technical capability of the equipment (information flow ④) and about the production orders which have to be produced (information flow ⑤). Form the point of view of process control, two committees have defined and classified terms of process execution, i.e. the SP88 committee of ISA (SP88, 1994) and AK 2.3 of NAMUR (NE33, 1992). For the process description they distinguish between different generations of recipes. A process describes the general procedure for the production of a particular substance. General recipes specify these processes and contain additional information about materials, intermediate products, finished products and their respective quantity ratios as well as generic requirements to the capabilities of the production equipment. Thereupon, they are transformed into site recipes, master recipes or basic recipes by adding further information about the process step to be made and the equipment to be used. While the recipe types mentioned before are considered to be master data, control recipes are generated for production orders and always refer to actual batches. They contain times, quantities and qualities of the planned production order or actual data about terminated orders respectively. The generation of a recipe reflects the process of development.
  In addition to the generations, the components of recipes are differentiated according to their granularity for the process description. The process of a recipe thus divides into partial recipes for the description of a process section according to the NAMUR recommendation, e.g. sulphonation. These can consist of multiple operations which form the smallest independent executable units, e.g. dosage H_2SO_4. Operations, on the other hand, are composed of different functions, e.g. temperature at 140°C. Functions are realized via programs in automated systems which are facility-specifically implemented and referenced and parameterized from recipes. The SP88 draft follows an analogous concept. The components of process description are classified here in Unit Procedure, Operation and Phase.
  Although the recommendations take the material side into consideration, there are no suggestions about how to describe material flow and logistic aspects.

3. INTEGRATION OF PROCESS AND PRODUCT INFORMATION

3.1 Principles for Integration

For integration of process and product information some general principles should be followed:

- By designing integration, business process orientation of function and organization has to be taken into consideration (Scheer 1994a). Functions should not be designed in isolation, but as part of a business process. Information technology enables to re-integrate functionality, which had been decomposed according to Taylorism.

- The same information structure should be used in various information systems and for various information objects (in (Becker, 1991), these common information structures for the CIM concept are called data structure integration). Common information structures not only simplify the information exchange between different systems, but also enable the reuse of software functions.

  The combined description of products and processes is an application of common information structure, this means material and material relations on the one hand and operations and operation relations on the other hand can be described as net with knots and edges in the same structure.

  If it is not suitable to have common information structures, references between objects in different information systems can establish the connection. The NAMUR recommendation proposes a reference between the function in the recipe and the technical function realized at a specific equipment. By that, process description in recipes can be stored independently from a specific realization in an equipment module.

- Information can be hierarchically decomposed and detailed in order to cover the specific requirement of various functions. As proposed in the recommendations mentioned above, process descriptions in recipes are decomposed in three levels. The advantage of decomposition is that each application can use the granularity of information which is required, e.g. the medium-term capacity management usually requires information only on partial recipe level, short-term scheduling requires information on operation level for detailed resource allocation and process control requires information on the most detailed level.

With regard to the requirements of process and product information and to the principles of integration, an approach for modeling of production processes has been developed (Loos, et al., 1994).

3.2 Approach for Modeling of Production Processes

The information which should be modeled can be distinguished using different criteria. They lead to distinct views of recipes:

- Material flow view: The material flow shows the material input and the ratio of input materials, the output material like finished products, by- and co-products and waste, and the intra-process material which is created by a process activity and consumed by a successive process activity. The material flow can be depicted with the material-consuming and material-creating process activities.

- Process view: This view depicts process activities and their relationship on different levels of the recipe. Relationships can result from sequencing, i.e. a process may have one or more predecessors and successors. Furthermore relationships can also result from alternative process activities which can substitute one or more other process activities if several production paths exist. Besides the availability of input materials as a precondition for the execution of a process activity other events can be required such as particular weather conditions, defined states of other processes etc. Materials can be regarded as special pre- or post-events of an activity in the process view.

- Time-related view: While the process view represents the logical dependencies between activities, this view deals with those coming from temporal relationships. Temporal dependencies are relevant for control recipes in course of scheduling. In this context Gantt-charts which depict activities on a time axis are helpful. The time axis can bear different scales (from minutes to months).

- Resource view: In this view the increase and decrease of resources are of interest. This corresponds to an inventory view when dealing with consumable resources as material, and to a resource occupation view when dealing with capital goods as equipment. The resource view is of importance to order-related data (to master data to a lesser degree) so that it can be used for control recipes.

These views can be applied to different recipe levels. Hence, the material view is useful at the recipe level, partial level or operation level. The time-related view is also helpful at the phase level.

For the visualization of recipe structures and their dependencies, graphical representation is used to depict the views. For the resource view histograms can be used. Implementing time-related views with Gantt-charts has been successfully demonstrated for discrete production in Leitstand contexts.

The integrated modeling of products and processes can be done with the help of material process chains. They cover the material flow view and process view mentioned before.

3.3 Material Process Chains

Figure 2 shows the graphical representation of a material process chain of a recipe. Materials are represented as octagons and process operations as soft rectangles. Flows of material which connect materials and operations are depicted by using pointed arrows and connectors for unification and ramification. The example shows a recipe for the production of M5 in which two alternative production paths exist. The first path results in the production of Material M2 using material M1 in operation OP1. M2 is used as an input for operation OP2 which also produces an input for operation OP3. At the same time processing of the input material M3 by operation OP4 takes place. This operation produces the co-product or waste product M6 (10% of the quantity of the input material M3) as well as an intermediate product (90% of M3's quantity) which is processed in OP3 together with the output of OP2 resulting in the end product M5. Moreover, the controlling events of the process are shown. These events are the preconditions like 'process release' for the start of operation OP1 or 'C1'

for the start of OP4 (e.g. atmospheric humidity of less than 90%). The resulting event 'batch finished' indicates the completion of the process after operation OP3.

Furthermore, two alternatives are shown in the graphical representation. On the one hand, operation OP5 can substitute OP2 which is expected in 10% of all cases. On the other hand there is a 20% probability of using operation OP6 instead of OP4. In contrast to the first alternative there is a change in the flow of materials, since M7 replaces material M3 and no production of M6 takes place. The definition of alternatives in the recipe makes sense due to the fact that in production control there is the need for knowing alternative paths in order to have disposing capacities, i.e. to execute operations on a different unit and to react in real time to changes in actual production data.

Material process chains cover the modeling of products and processes mentioned above which were indicated as material flow view and process view. A process within a material process chain can be de-

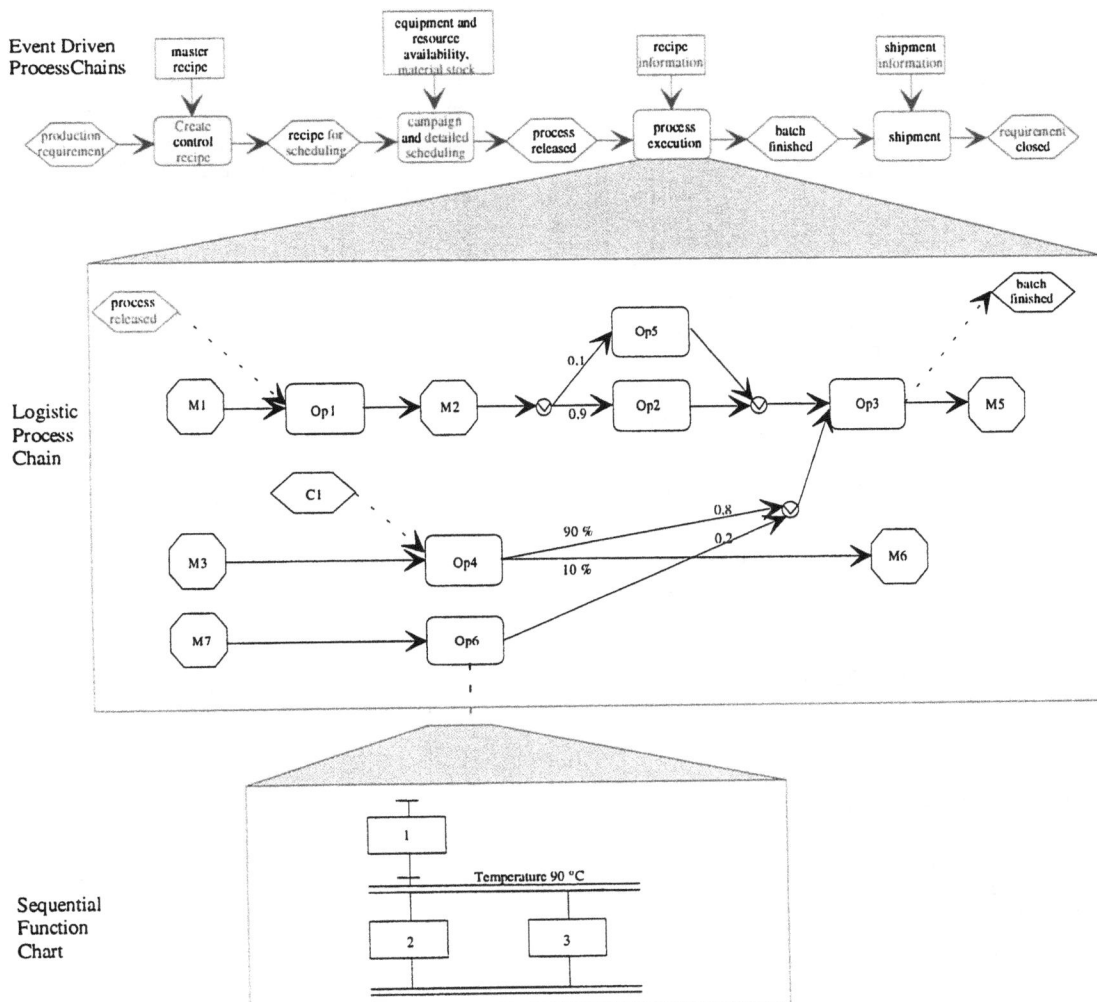

Figure 2: Material Flow and Process Model

83

composed by another material process chain for a more precise definition. With that, hierarchization can easily depict. Within the prototype GRACE, a graphical recipe editor (Loos, 1993), material process chains are implemented for the management of recipes. In order to depict the time-related and resource aspects mentioned above other views are also implemented. Figure 3 shows a screen layout for the time-related view and the resource view of a material process chain. For the management of recipe, the material flow view is mainly used for the upper level (bill of materials or partial recipe level), while the view shown in figure 3 is usually used for the more detailed description (operation or phase level).

Material process chains can be regarded as a decomposition of a production process within a business process. Figure 2 indicates that by a superior business process which is modeled with the ARIS-method Event-driven Process Chain (Scheer, 1994a; Scheer, 1994b). It is also shown that other descriptions can be referenced from the smallest process action in the diagramm, e.g. a sequential function chart that decribes the control logic realized in an equipment function.

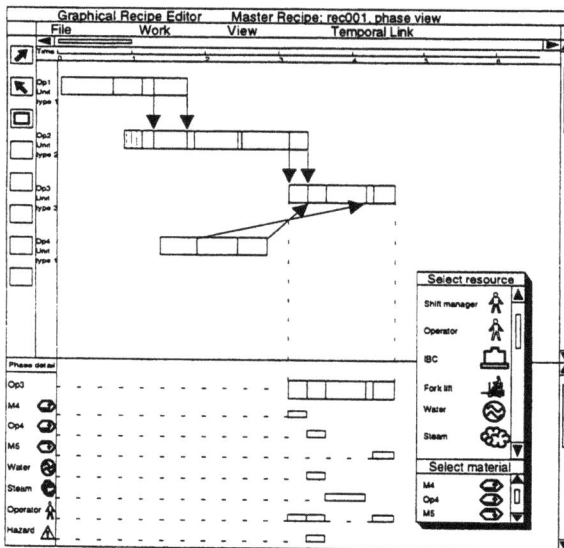

Figure 3: Time-related View in GRACE

REFERENCES

Becker, J. (1991). *CIM-Integrationsmodell*. Springer, Berlin et al.

Eckelmann, W.; Geibig, K.-F. (1989). Produktionsnahe Informationsverarbeitung - Basis für CIP. *CIM Management* **5(1989)5**, 4-9.

SP88 (1994). *ISA-dS88.01 Batch Control, Part 1: Models and Terminology*, Draft 12, (International Society for Measurement and Control).

Loos, P. (1993). *Konzeption einer graphischen Rezeptverwaltung und deren Integration in eine CIP-Umgebung*. In: Publication of the Institut für Wirtschaftsinformatik, paper 102 (Scheer, A.-W. (ed.)), Saarbrücken.

Loos, P. (1995). Production Management - Linking Business Applications to Process Control. In: *Proceedings of the World Batch Forum 1995* (Newtown Square, PA, May 22-24, 1995), 2.1-2.16.

Loos, P., Scheer, A.-W. (1994). Graphical Recipe Management and Scheduling for Process Industries. In: *CIMPRO'94 - Rutgers' Conference on Computer Integrated Manufacturing in the Process Industries*, (Boucher, T.O.; Jafari, M.A.; Elsayed, E.A. (eds.)), Piscataway, PA, 426-440.

Luber, A. (1992). How to Identify a True Process Industry Solution. *Production and Inventory Management* **12(1992)2**, 16-17.

NE33 (1992). *NAMUR-Empfehlung Anforderungen an Systeme zur Rezeptfahrweise*, (Standardization Committee for Measuring and Control Engineering in the Chemical Industry).

Polke, M. (1989). CIP in der Verfahrensindustrie. *CIM Management* **5(1989)5**, 34-35.

Scheer, A.-W. (1994a) *Business Process Engineering - Reference Models for Industrial Enterprises*, 2nd edition, Springer, Berlin et al.

Scheer, A.-W. (1994b). ARIS Toolset: A Software Product is Born. *Information Systems* **19(1994)8**, 609-626.

Schürbüscher, D.; Metzner, W.; Lempp, P. (1992). Besondere Anforderungen an die Produktionsplanung und -steuerung in der chemischen und pharmazeutischen Industrie. *Chem.-Ing.-Tech.* **64(1992)4**, 333-341.

ORGANISATIONAL CONCEPTS FOR PRODUCTION PLANNING AND RESOURCE ALLOCATION IN A MULTI-NATIONAL PHARMACEUTICAL ENTERPRISE

Silke Hübel and Jürg Treichler

Institute of Industrial Engineering and Management (BWI)
Swiss Federal Institute of Technology (ETH), Zurich

Abstract: The production of active substances in pharmaceutical enterprises sometimes involves different production sites. This situation complicates the development of an optimal production plan and results in long lead times and large buffer stocks. Therefore, new solutions for the production planning tasks have to be derived. The MRP II concept, basically developed for discrete manufacturers, seems to be a useful basis. But several adjustments, especially for the capacity planning processes, have to be made. In this article, a modified planning architecture and supporting organisational structure are introduced.

Keywords: Process Industry, MRPII, Master Production Schedule, Capacity Planning, Information Systems, Multi-National Enterprise

1. INTRODUCTION

The globalization of world markets has compelled Swiss multi-national pharmaceutical enterprises in recent decades to found and acquire new production sites in different countries around the world.

There are various reasons for this expansion. Existing trade barriers forced the companies to found production facilities in their most important markets. Moreover, there are many legal restrictions including very restrictive environmental requirements for the construction of new facilities, e.g. in the area of Basle the realisation of a new factory lasts about 10 years including the legal approval processes. Owing to this fact new factories were built in countries with fewer restrictions. Another reason is the expensive workforce in Switzerland. Various mergers and acquisitions have also resulted in additional production sites.

Furthermore, these companies face changing market conditions which force them to improve their operational effectiveness of manufacturing and distribution. On the one hand these new situations are caused by:

1. worldwide cost-reduction programs in health care

2. discontinuing patents

3. more competitors entering the markets (generic companies) with lower costs and prices

On the other hand increasing regulatory limitations (e.g. the validation requirements of the American Food & Drug Administration FDA) require an extensive quality assurance in the total pharmaceutical manufacturing process and increased efforts in registration. These regulatory limitations made for customer's safety result in higher costs for the pharmaceutical companies. This tendency is shown by the fact that manufacturing costs in

85

pharmaceutical production have risen from 10-15% to around 20-25% cost of sales since the 1970's (Byrne 92).

In view of this new situation the pharmaceutical companies are developing strategies to improve the efficiency of the workforce and the invested capital. However, there are different strategies depending on the different production stages. The production processes of pharmaceutical enterprises are normally separated into three production stages. The first stage includes the synthesis steps from raw material to the active substances. In the second stage the active substances are transformed into dosage forms. The last stage is the packaging of dosage forms into finished products.

In this article we will concentrate on the production of active substances. This production stage is characterised by complex syntheses and lengthy lead-times. The main objectives are to ensure product quality and safety of supply. Because a production plant for active substances requires high capital investment the selection of the production location is strongly influenced by tax incentives and infrastructural considerations such as energy supply and a well-educated workforce. Expensive manufacturing equipment and regulatory requirements lead to a strategy of largest possible concentration (Fischer 92). This means that most intermediates and active substances are only produced at one or two locations. But this concentration of specific equipment and, therefore, synthesis steps at one location can result in enormous problems for the production planning process. In the following we will describe these problems, taking Ciba-Geigy as an example, and present a possible solution.

1.1. The production of active substances at Ciba-Geigy

The pharmaceutical division of Ciba-Geigy produces approximately 150 active substances, of which some have cumulative lead-times of up to 2 years. This requires a make-to-stock production and leads to high safety stocks to buffer the uncertainties of demand. Further uncertainty, and therefore buffer stocks, are added by the distributed production system.

1.2. The distributed production system

In Ciba-Geigy the transformation from raw materials to active substances is called a manufacturing pipeline. A pipeline consists of various stages which

eventually leads to one or more active substances. The pipeline structure can be diverging as well as converging. This means some pipelines result in more than one active substance, others combine more than one intermediate at one stage. Fig. 1 shows an example of some rather complex manufacturing pipelines.

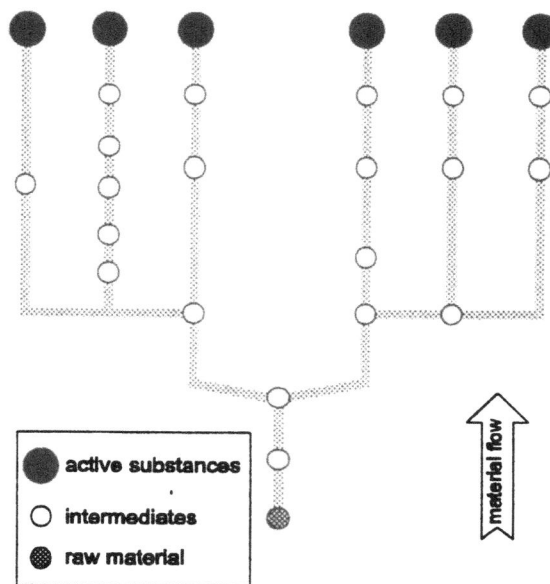

Fig. 1. Example of manufacturing pipelines (taken from Grüninger, 1992)

The different synthesis steps for these pipelines are produced in different volumes and require different equipment. For example, some steps are produced in large volume in dedicated mono-lines, others have less volume and could be produced in multi-purpose equipment. Therefore, various locations and even third parties have to be involved. Fig. 2 shows the pipelines with their different production sites, which are indicated by different shadings of grey.

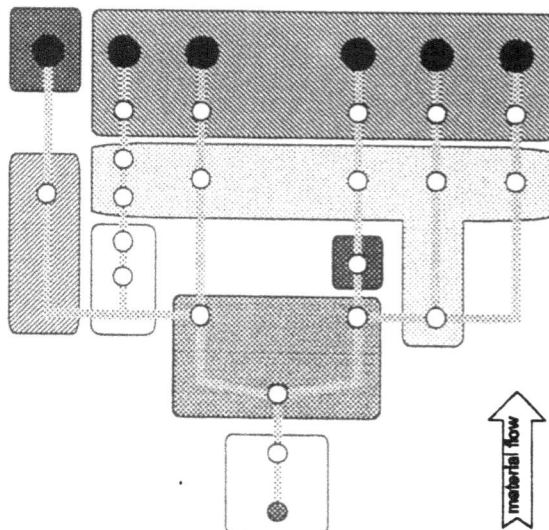

Fig. 2. Example of manufacturing pipelines with their production sites

The distributed production system is characterised by a customer-supplier relationship between production sites. For some large-volume products the production process involves three or four different sites in several countries. The autonomous production planning process at each site complicates an effective production planning. Each site makes its own plan which is optimised for that specific site - but the coordination between the sites suffers. This results in enormous buffer stocks and causes a further extension of lead times.

To address these problems, a joint project of Ciba-Geigy and the Institute of Industrial Engineering and Management (BWI) at the ETH was started, which aims at the formulation of an adequate planning architecture and the evaluation of a supporting computerised tool.

2.   THE NEW CONCEPT

The production plan of a plant has to consider the three opposing parameters - stock, capacity and lead time.

The improvement of one parameter may deteriorate the performance of the other parameters (see Fig. 3).

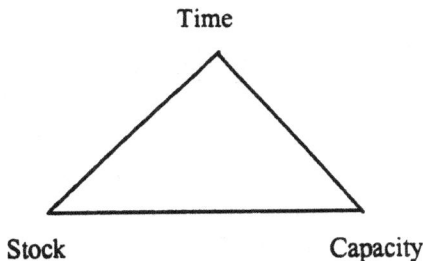

Fig. 3. Planing constraints

In the case of distributed production sites the local optimal plans could conflict with the global optimum.

The production planning process has to ensure that an optimal plan over all production plants can be developed. Several optimal plant production schedules (PPS) normally do not result in a global optimum. Three requirements are derived:

1. The production plan of each production site has to reconcile with the global goals.

2. The view of all pipelines or all plants respectively has to be ensured.

3. An arbitrator has to make strategic decisions in the case of disagreement.

An optimal plan covers the demand with low stock, high capacity usage, low production cost and 100% adherence to the shipping dates. The proposed new concept covers two aspects:

1. A production planning process which allows the generation of an "optimal" plan.

2. An organisational structure supporting the production planning process.

2.1.   The production planning process

The proposed planning process (see Fig. 4) is based on the manufacturing resource planning (MRP II) concept (Wight, 1974), which was developed initially for discrete manufacturing environments.

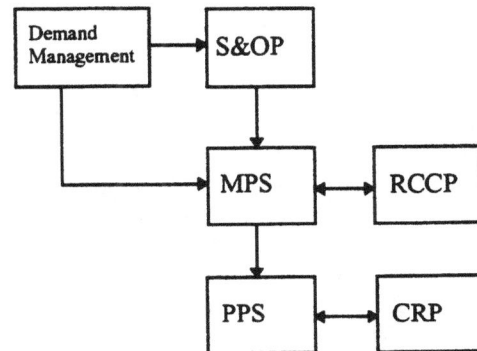

Fig. 4. Proposed planning process

Demand management encompasses forecasting, order entry, other customer-related activities and establishes the link to the marketplace. The sales and operation planning (S&OP) provides key communication links from top management down to manufacturing. It determines the basis for focusing the detailed production resources to achieve the firm's strategic objectives. The heads of marketing and production participate in the S&OP meetings (Vollmann, et. al., 1992). Up to this level the tasks for pharmaceutical companies are in accordance with the traditional MRPII concept.

The master production schedule (MPS) determines the basis for making customer delivery promises, attaining the firm's strategic objectives as reflected in the S&OP and resolving trade-offs between manufacturing and marketing. The rough-cut capacity planning (RCCP) helps to develop a MPS, considering the aggregated capacity constraints of all plants. The horizon of the MPS is three years based on the long lead times of the active substances. After defining the MPS the plants validate the MPS with a plant production schedule (PPS). The PPS has a horizon of 1 to 2 years. If it is not possible to generate a valid PPS for a plant a new MPS process starts with the new restrictions of the production sites. Fig. 5 shows a graphical representation of the planning process flow.

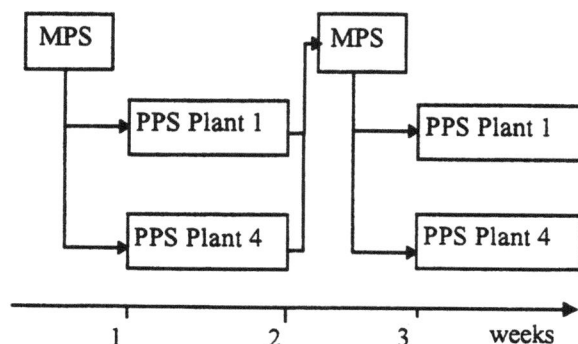

Fig. 5. Planning process flow

After a maximum of 1 month a valid MPS should be available. Otherwise, the arbitrator has to take a decision. For example, in the case of a capacity conflict, the arbitrator has to decide which shipping date or amount has strategic importance and which date can be moved or which delivery amount can be split. To minimise the changes of the MPS the very first MPS should always be an optimal plan. This depends on the quality of the RCCP.

The PPS is a schedule which has to consider finite capacity. As mentioned before the horizon is 1 to 2 years which is caused by the long lead times. That is why the capacity requirement planning (CRP) in this context is a finite scheduling task.

This planning process should normally be repeated 3 to 4 times per year. It is necessary not only to construct the MPS but also to process actual transactions and modify the plans. In case of unpredictable occurrences in production sites, like machine breakdowns or serious quality problems, which cannot be solved locally, the MPS has to be changed. This also holds for dramatic demand changes.

These requirements differ in various aspects from the traditional MRPII concept, especially for the RCCP and the CRP processes. In discrete manufacturing environments the capacity planning task is usually performed taking infinite capacity into account. This is justified because lead times and setup times are quite short and therefore the machine loading and sequence planning are short-term problems and performed only several weeks in

advance. Standard software for scheduling systems is available to support these tasks. For pharmaceutical production the planning with finite capacities has to be done almost one or two years in advance. Even the RCCP should consider finite capacity, albeit in an aggregated form. Here no adequate tools exist.

2.2. Organisation structure

The following organisation structure was defined to support the production planning process. First, two definitions:

1. A person who has the view of one or more pipelines is called a pipeline-manager (PM). Pipeline-managers typically have a background in logistics.

2. A plant scheduler (PS) is responsible for the PPS and has a background in chemical engineering.

All plant schedulers and pipeline-manager are members of the planning group (see Fig. 6).

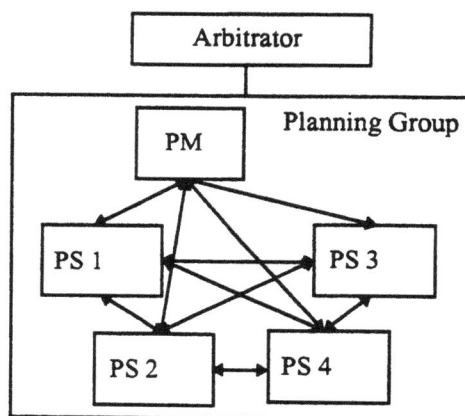

Fig. 6. The planning group

The planning group holds the responsibility for the stock, the adherence to the shipping dates and the capacity usage. To ensure that the plant schedulers are interested in a global optimum the incentives depend 50 percent on the global and 50 percent on the local goals. Global goals are low stock, low production cost, short lead times, etc. Local goals are roughly the same but could include quality and internal stocks. Unknown capacity breakdowns or quality problems are considered by involving the plant scheduler in the MPS process. The pipeline problems are taken into account by the pipeline manager. The pipeline-managers give information on problems incurred by local changes in the production plants. They can check with the master production schedule (MPS) the coverage of the actual demand and make suggestions for the next steps if demands or dates change.

Such a structure is quite different to the traditional separation between logistic functions (pipeline planning) and the production functions (plant scheduling). Here a different attitude is demanded to overcome the functional oriented behaviour of the people involved in the planning process.

A problem which has not been addressed until now is the involvement of third parties where the production orders for one company compete with those of others. Perhaps new concepts for customer supplier relationships could be considered (Alberti and Frigo-Mosca, 1995). Possible solutions are mutual insight into the production plans. The necessary pre-requisites are given because relationships between pharmaceutical companies and third party suppliers are normally long-term partnerships.

2.3. Requirements

The fact that the very first generated MPS should be a valid plan and that within 1 week the plan should be confirmed leads to some specific requirements.

1. The used aggregated capacity model in the RCCP function must be correct.

2. The multi-purpose production sites have to use a tool to quickly verify the relevant data of the MPS. The mono-line production sites may not have this problem - scheduling is usually simple for these sites.

3. The MPS has to consider the cost and stock level constraints to generate an optimal plan.

4. Changes in the capacity or the cost structure of a site, planned shutdown times, vacation or changes in the operation plans, can have an impact on the RCCP. So the used RCCP model has to be immediately updated.

5. A fast and flexible method validates the MPS in the production sites.

To satisfy these requirements a tool has to be used to generate an MPS. No manual method can optimise a plan with such complexity in reasonable time. Also a multi-purpose production site can only fulfill the requirements with a software tool to simulate various work allocations.

In the project with Ciba-Geigy scheduling tools were evaluated to test their usefulness for the RCCP process. But the long time horizon and the huge quantities of data lead to run time problems for the existing software.

3. CONCLUSIONS

Pharmaceutical companies have some elements in their production planning process which differentiates this kind of industry from others. The traditional MRPII concept could be used as a basis but some changes have to be introduced.

Particularly, the RCCP process has to imply a finite capacity planning for the generation of effective production plans. Additionally, an adequate organisational structure has to be implemented to overcome the functional oriented behaviour of the people because different production sites with local interests are involved. Much better productions plans could be generated this way and therefore reductions in lead times and buffer stocks could be obtained.

REFERENCES

Alberti, G. and F. Frigo-Mosca (1995). Advanced Logistic Partnership: Neugestaltung der Beziehungen zwischen Kunden und Lieferanten.. *io Management*. **64: 1/2**, 67-72.

Byrne, F. (1992). The challenge to manufacturing - Setting the challenge. In: *Challenging traditional remedies. Conference Proceedings. Coopers & Lybrand. 8 - 9. December 1992.* Basle, Switzerland.

Fischer, B. (1992). Developments in European manufacturing strategies. In: *Challenging traditional remedies. Conference Proceedings. Coopers & Lybrand. 8 - 9. December 1992.* Basle, Switzerland.

Grüninger, A. (1992). Managing complexity - A case study in pipeline planning. In: *Challenging traditional remedies. Conference Proceedings. Coopers & Lybrand. 8 - 9. December 1992.* Basle, Switzerland.

Vollmann, T.E., W.L. Berry and D.C. Whybark (1992). *Manufacturing Planning and Control Systems.* Business One Irwin, Illinois

Wight, O.W. (1974). *Production and Inventory Management in the Computer Age.* Cahners Books International, Boston.

APPROACHES TO COMPLEXITY AND UNCERTAINTY OF SCHEDULING IN PROCESS INDUSTRIES:
PROCESS REGULATION IN HIGHLY AUTOMATED SYSTEMS

Eric Scherer

Institute for Industrial Engineering and Management (BWI)
Swiss Federal Institute of Technology (ETH) Zurich

Abstract: In process industries functional relations between the different organisational levels are difficult to define. While it is possible to describe the relationship of resources considered at the various levels and thereby form an integrated information system, it is nearly impossible to achieve a full functional integration within a computerised system. This leaves a gap in functionality between planning oriented business levels and execution oriented process levels. In practice this gap has been closed by human experts at plant level. With new computerised systems on the market it is necessary to reassess the role of the human experts at plant level and see how they can interact between production planning and scheduling systems and process control systems.

Keywords: Process industry, production management, process control systems, scheduling, resource allocation, hybrid systems.

1. INTRODUCTION

To reduce complexity in production planning hierarchical, multistage planning concepts are used. In process industries, this is often presented as a five level pyramid (Eckelmann & Geibig 1989, Fig. 1.).

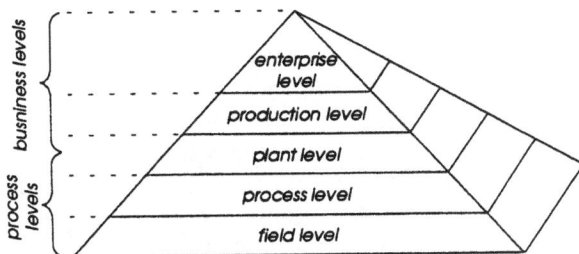

Fig. 1. Five-level organisational model for process industries (adopted from Eckelmann & Geibig 1989).

1.1. Business level and process level systems

Thereby scheduling systems usually form the front end of a production planning and scheduling systems (PPS). Derived from the business strategy and usually based on a MRP II related approach it is anticipated to achieve an optimal plan of resource utilisation, i.e. capacities and material measured in times and quantity (e.g., McKaskill 1994a). This pretty much covers the enterprise and production level as indicated in Fig. 1. In continuos and discontinuous production as found in process industries mainly highly automated manufacturing systems are utilised at the technical process levels (Schumann et al. 1994, 552ff). Thereby the production process is permanently controlled and regulated by process control systems (PCS) (Clevermann 1991). PCS control a number of resources necessary for production and regulate their relation. Utilisation of PCS in process industries is very high and can be found in one type or the other in virtually any enterprise (Schumann et al. 1994, 558).

91

1.2. Integration of organisational levels in planning

Informational Integration. In practice currently a gap exists in computerised information systems between process control systems (PCS) at the process level and production planning systems (PPS) at business level(s). PPS systems currently utilised in process industry mostly have been designed for discrete manufacturing and lack the ability to model resource relations as found in process industries, e.g., synthetical bill-of-materials or couple production (Hofmann 1992, Allweyer et al. 1994). New efforts in software-system development try to develop more suitable PPS systems. Therefore system architectures are defined intending to bridge the gap between business applications and process control to form an overall production management framework (e.g., Loos & Scheer 1994). The main focus for integration thereby is on information. This is useful since the very basic resource models utilised at each organisational level within an enterprise are principally the same. Achieving full informational integration ensures that all data is related throughout the different levels in aggregated or disaggregated form respectively and the material flow can be fully monitored and optimised accordingly.

Functional Integration. Since planning is the major activity at business levels, PPS systems are based on a planning paradigm of predictability and optimisation. They usually tend to lack efficiency in the real time environment of complex technological production processes. In process industries production processes form a complex and chaotic system with unpredictable and often not even stochastic behaviour. This real-time situation leads to a concept of job scheduling and execution as a mainly reactive task triggered by various events, e.g., changing process parameters, varying product quality, facility breakdown, lack of production resources or any other disturbances. This problem has been known in discrete manufacturing for a long time and resulted in several approaches to close the gap and form closed control loops based on realistic assumptions about system behaviour. Still the complexity of production in discrete manufacturing lies in the product and production structures allowing various different solutions for job routings and sequences. In process industry complexity is mainly generated by the single production process itself and the relation of resources and process parameters varying while the process is in progress (see Hofmann & Scherer 1994).

1.3. Problem statement

In process industries functional relationships between the different organisational levels are difficult to define and therefore make it difficult to form a fully integrated, computerised systems. This leaves the second gap of functionality between planning-oriented business levels and execution-oriented process levels still open. In practice this gap has been closed by human experts at plant level. Due to their experience and expertise they are able to foresee disturbances and react accordingly. Utilising knowledge-based methods several researchers and software engineers tried to solve the problem and generate a fully automated and integrated system without taking human expertise at plant level into full consideration (e.g., McKaskill 1994b). Most of these approaches do reduce the complexity of the scheduling problem for the system operator through simulation techniques but ignore the uncertainty of the current situation in a real-time environment. So it is usually not only difficult to predict the occurrence of a specific event as trigger to a new scheduling procedure but it is notably impossible to know and describe the current situation to a computer-based system within any reasonable period of time. This makes it necessary to reassess the role of the human experts at plant level and see how they can interact between PPS systems and PCS while taking into consideration the existence of more and more integrated information frameworks.

2. REQUIREMENTS FOR SYSTEM DESIGN: COMPLEXITY AND UNCERTAINTY

The two main parameters for system design at plant level are complexity and uncertainty. They form the basic requirements to be meet with system design. Complexity thereby refers to the problem of finite-scheduling and allocation of resources as enforced by the complexity of the production structures itself. Uncertainty describes the problem of continuos job execution in a highly automated production system with several disturbances possible.

2.1. Complexity: production structures in process industries

To provide a closer insight into the problem, some basic characteristics of production in process industries will be shortly described: process structures, resource relations and facility layout.

Process structures. Compared to discrete manufacturing several differences can be found in process industries (also see Schürbüscher et al. 1992). These namely are

- no routings and bill-of-materials but recipes (Fig. 2),
- by-products and recycling of resources,
- converging and diverging material relations,
- non-linear demand relation,
- unstable interim products and flow resources,
- fixed sequence of production processes, i.e., campaigns,
- inflexible facilities.

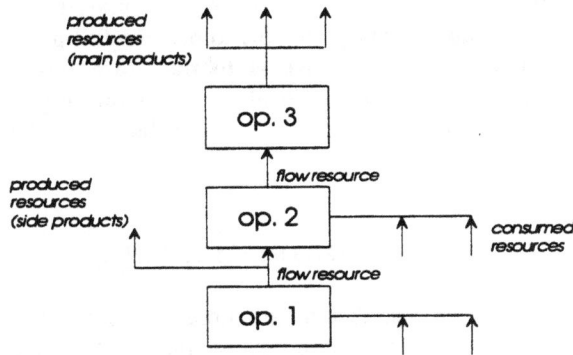

Fig. 2. Principal description of a production process by integration of resources and operations.

Resources. Resources present objects with a somehow stable character. Resources are related to operations or processes (see Fig. 2.) and are required, consumed, or generated by them (Hofmann & Scherer 1994)[1]. Many different kinds of resources are utilised in process industries and have various characteristics to be described. Resources controlled during production in process industries are multifarious compared to discrete manufacturing, e.g., capacities, space, material, operators, energy, etc. Basis features are the (1) the possibility to stock a resource, e.g., energy cannot be stocked, and (2) the possibility to consume a resource, e.g., space is required but cannot be consumed. Fig. 3. gives a principle example of resources related to process.

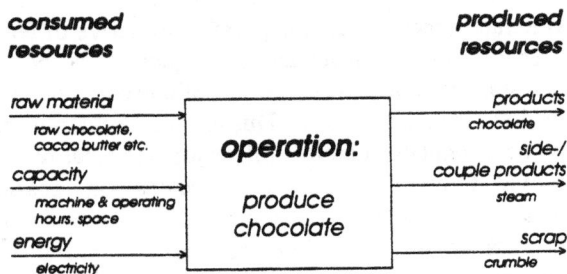

Fig. 3. Example of resources necessary to perform a process.

---

[1] an operation can be described as a single process but is also part of a process on an aggregated level.

Facilities. In process industries, the layout of the plant in most cases is determined by the product or a class of products, i.e., facilities are designed to produce a closely related group of products. Facilities and plants are delimited by the maximum line flow and the capacity of the bottle neck operation. The facilities are very specialised and change over times are rather high. Therefore the emphasis of mid-range planning lies on optimal capacity through ideal sequencing of production campaigns. At the short-range scheduling level it is the main focus to allocate all necessary secondary resources and determine what auxiliary operations are necessary to ensure a high quality product and a secure process. Thereby it is impossible just to operate the facility but to regulate the process continuously.

2.2. Uncertainty: Order processing in highly automated systems

Through finite-scheduling systems all necessary resources are allocated. These systems mostly are based on simulation techniques and determine all resources necessary, e.g., tank space (see Jänicke 1994). Once a schedule is established describing all resources and basic relations necessary for job processing, a job can be released for execution. Now it is the task of the PCS and the human operator to ensure optimal processing. Therefore the originally established schedule has to be altered according to certain events occurring. Such events are breakdown of facilities, delays in delivery, lack of quality, missing personel, etc., and cause unavailability of one or more resources. The occurrence of a specific event thereby is unpredictable and results in looses of time, resources and product quality. Possible losses are consumed within the schedule by inclusion of safety margins. These margins result from approximations based on previous runs of a specific process. In case of a very high statistical availability of resources at a specific date the completion of the overall job at scheduled date is usually very uncertain. Fig. 4. gives an mathematical example, while Fig. 5. indicates the results of a statistic survey of facility availability in industry.

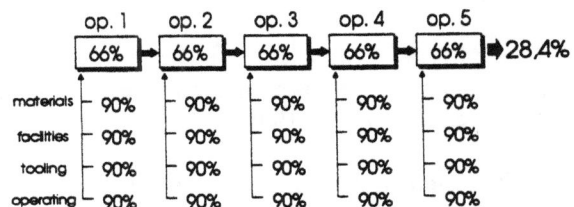

Fig. 4. Schedule uncertainty at unpredictable resource availability

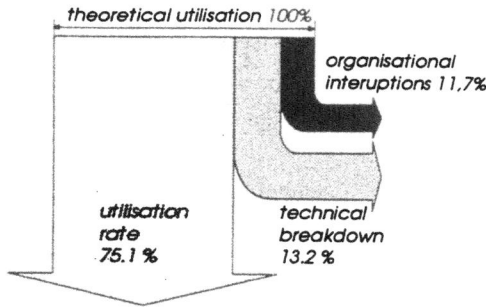

Fig. 5. Facility utilisation rate and availability in highly automated manufacturing (average of several surveys 1989/90, source: Wiegershaus 1991)

3. PROCESS REGULATION AND HUMAN SKILLS

In an highly automated system environment it is necessary to regulate the production process permanently. Since no full integration between business and process levels can be achieved within a fully computerised framework, it is the task of the human experts at the plant level to fill the gap. In process industries the high degree of automation lead to the creation of a new type of production worker: the system regulator. System regulators are not directly involved in the production process itself but control the behaviour of a system, i.e., a production facility. It is the task of a system regulator to adjust the actual system behaviour in accordance to the anticipated target (see Schumann et al. 1994, 86f). Fig. 6. gives an comparison between different degrees of work at plant level in different types of industries.

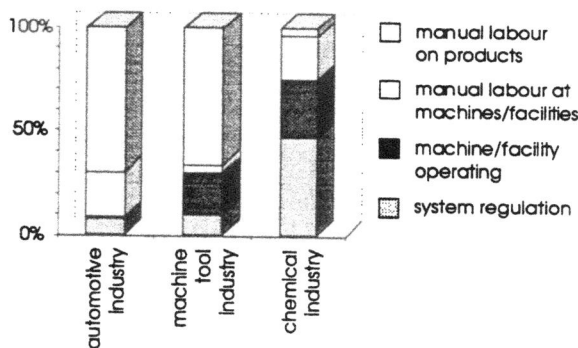

Fig. 6. Work characterisation at plant level (Source: Schumann et al. 1994, 575)

It can be assumed, that human acting is goal-oriented, object-related, social and flexible. Humans can adopt strategic goals as set by business level strategies for their own acting. Object-related cognition and thinking easily reduces complexity and makes it possible for humans to understand certain situations fast and implicitly. Flexibility allows a human to adapt his method of working to changing situations without explicitly being advised to do so. The high share of system regulators at plant level in process industries, i.e., 47 %, indicates the high degree of expertise and qualification common in process industries. By providing system regulators with further autonomy they are able to become full process regulators, i.e., their task is to transform business level plans into process level processes and regulate the transformation accordingly (see Schüpbach 1994).

4. HYBRID SYSTEM LAYOUT

Based on the described assumptions it is necessary to design a hybrid system layout that fully utilises computer technology to form an integrated information system as well as utilises human potentials to close the functional gaps. For such a complementary system it is necessary to determine the degree of automation and support is useful for the human process regulator.

Due to the described complexity of production in process industry it is the task of a finite-scheduling system to provide the process regulator with a possible schedule for all resources and inform him about possible difficulties, e.g., missing materials or possible side products. Additionally, the scheduling system has to coordinate the resources produced and required by several parallel processes throughout time, e.g., parallel processing of energy producing and energy consuming processes. Still it is the objective of the scheduling system to create possible and probable schedules easy to understand and to be carried out by the process regulator. Through simulation techniques it is possible to derive different scenarios and indicate the consequences of certain decisions without taking the risk of insecure processes or low product quality. During job execution the process regulator controls all major technical data as provided by the PCS and decides on possible measures to achieve the technical objectives of the process as well as the objectives set by the PPS system. These objectives can be contradicting, since it may be possible to carry on with the execution of a single process after adjustment of certain process parameters but impossible to meet the objective set by the PPS system. Fig. 7. gives an possible layout for the proposed hybrid system.

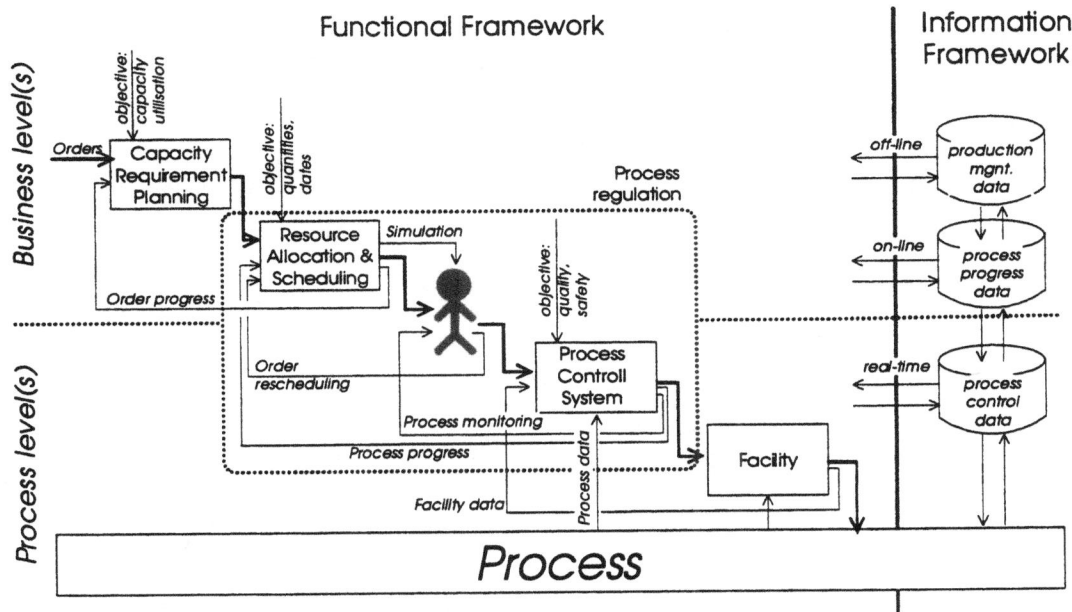

Fig. 7. Hybrid system layout for process regulation

5. CONCLUSIONS

The proposed system layout closes the functional gap between PPS and PCS and leads to a fully integrated system architecture. Whereas any computerised system needs humans mainly as operators the proposed approach sees the human experts in the role of process regulators as a core part of the system. A human process regulator can provide the system with further flexibility and adaptability.

To ensure an optimal system behaviour further work is necessary. One main topic is the efficient and ergonomical design of the local work environment for the process regulator fully utilising computer aided simulation and monitoring.

REFERENCES

Allweyer, Th., Loos, P., Scheer, A.-W. (1994): An empirical study on scheduling in the process industries. *Veröffentlichungen des Instituts für Wirtschaftsinformatik No. 109*. Universität des Saarlandes, Saarbrücken.

Boucher, T.O., Jafari, M.A., Elsayed, E.A. (eds.) (1994): *Proceedings of Rutgers' Conference on Computer Integrated Manufacturing in Process Industries*. East Brunswick, April 25-26, 1994.

Clevermann, K. (1991): Produktionsleitsysteme als CIM-Baustein für die Verfahrensindustrie. *Sonderteil in Hanser-Fachzeitschriften* 3, 45-48.

Eckelmann, W., Geibig, K.-F. (1989): Produktionsnahe Informationsverarbeitung - Basis für CIP. *CIM Management* 5:5, 4-10.

Loos, P., Scheer, A.-W. (1994): CAPISCE - A system architecture for production management in process industries. *World Batch Forum*, Phoenix AZ, March 1994.

Hofmann, H. (1992): PPS - nichts für die chemische Industrie? *IO Management* 61:1, 30-33.

Hofmann, H., Scherer, E. (1994): POA - A Process-oriented approach to modelling production structures. in: Boucher et al. 1994. 472-486.

Jänicke, W. (1992): Computergestützte Apparatebelegung für Mehrzweckanlagen. *Chem.-Ing.-Tech.* 64:4, 368-370.

Jänicke, W. (1994): Planung der Kuppelproduktion in Sytemen von chemischen Mehrzweckanlagen. *Chem.-Ing.-Tech.* 66:6, 819-824.

McKaskill, T. (1994b): There is a place for expert systems in finite scheduling: Batch process manufacturing. *APICS - The Performance Advantage* 8, 42-44.

McKaskill, T. (1994a): Process planning and scheduling as part of an enterprise wide business system. in: Boucher et al. 1994. 617-626.

Schumann, M., et al. (1994): *Trendreport Rationalisierung - Automobilindustrie, Werkzeugmaschinenbau, Chemische Industrie*. 2. edition. Edition Sigma, Berlin.

Schüpbach, H. (1994): *Prozessregulation in rechnergestützten Fertigungssystemen*. vdf/ Teubner, Zürich/Stuttgart.

Schürbüscher, D., Metzner, W., Lempp, P. (1992): Besondere Anforderungen an die Produktionsplanung und -steuerung in der chemischen und pharmazeutischen Industrie. *Chem.-Ing.-Tech.* 64:4, 334-341.

Wiegershaus, U. (1991): *Entwicklung einer Methode zur Synchronisation heterogener Fertigungsstrukturen*. VDI-Verlag, Düsseldorf.

INTEGRATING SCHEDULING IN LOGISTIC SYSTEMS
FOR BATCH PRODUCTION

Thomas Allweyer, August-Wilhelm Scheer

*Institut für Wirtschaftsinformatik (Institute for Information Systems)
at the University of Saarland, D-66041 Saarbrücken, Germany*

Abstract: For closing the gap between business-oriented information systems and shop floor control systems in the process industries, industry-specific standard software for production planning and scheduling is needed. The compared results of five case studies in different batch-oriented companies show that there are many requirements in process industries which are not met by today's standard software. The paper summarizes the different problems and needs found in each case study. Common findings are compared and requirements for scheduling systems and their integration into logistic systems are developed.

Keywords: Chemical industry, information systems, production control, scheduling algorithms, shop-floor oriented systems.

1 INTRODUCTION

Most of today's systems for scheduling and production management are designed for the needs of discrete manufacturing industries, such as the automotive industry. Only recently software companies began to develop similar systems for process industries, e. g. chemical, pharmaceutical, food, or paper industry.

While for discrete part manufacturers it is possible to integrate all logistic systems (sales, production planning, scheduling, production control) with

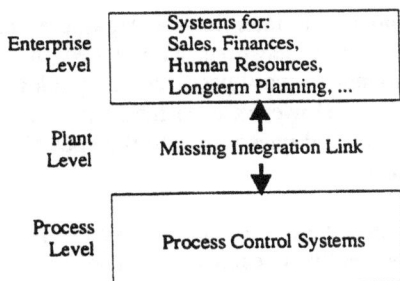

Fig. 1: Missing integration between business systems and process control systems.

standard software (cf. Scheer (1994), pp. 335-351), companies of the process industry are still confronted with a missing link between business systems (e. g. for sales, finances, longterm planning) and the highly automated process control systems (see fig. 1).

If such companies do not want to rely only on manual scheduling and plant management, they need either to develop their own individual solution or to use standard software that has been designed for discrete manufacturing. Individual solutions tend to be expensive, not as powerful and flexible as standard software, and they are difficult to integrate into the overall logistic system. Discrete manufacturing systems, however, have the disadvantage of not meeting the specific requirements of the process industries, since there are many differences between those types of industries.

Specific planning and scheduling requirements of the process industries include topics such as recipe management (cf. Instrument Society of America (1994)), campaign scheduling, hazardous materials, unstable products, by-products, non-linear dependencies, fixed pipelines, etc. Some of these requirements are discussed by Nelson (1983) Hofmann (1992), Schürbüscher et al. (1992), Remme et al. (1994), and Allweyer (1995). Since scheduling of batch production is especially demanding, this paper concentrates mainly on questions related to this production type.

During the last years several approaches have been made to develop scheduling systems especially for the process industry (manufacturing execution systems, MES). An overview about MES has been published by AMR (1994). One of these new systems has been developed in the European research project CAPISCE. The functionality of the resulting system includes scheduling, resource management, inventory management, recipe management, and batch management. The system has been integrated with business information systems, process control systems and laboratory information systems. Loos (1995) gives a detailed description of CAPISCE and the underlying architecture.

The prototype developed in CAPISCE is tested in two plants of the British company Zeneca. However, it was not clear whether all types of process industry-specific requirements can be found within these two plants. To ensure that the software is suitable for a broad range of companies with batch production, it was therefore necessary to get an overview over the situation in different companies and about actually important requirements. This does not only include production structures and technological and organisational constraints, but also the views and demands of the people working in the plant, such as plant managers and operators.

2 DESIGN OF THE STUDY

For getting empirical findings it is eithe possible to make surveys among a large number of companies or to carry out a case study. For large surveys it is necessary to prepare a standardised questionnaire, i. e. the possible answers are already defined by the design of the questionnaire, and it is difficult to get really new information. Case studies, on the other hand, have the advantage of getting much more detailed insight into a specific plant, and it is possible to find facts and correlations that have not been anticipated before. The disadvantage is that each case study covers only one specific example, and it is impossible to tell whether and to which extent this example is representative for other examples, as well.

For the validation of the prototype's requirements definition both detailed information as well as an overview over various enterprises were needed. It was therefore decided to carry out several case studies in different companies, and then to compare the results.

2.1 Scope of the Studies and Selection of Plants

The main objective was to evaluate industry-specific requirements for scheduling systems and their integration into logistic systems, with special emphasis on batch production. Important topics included therefore:
- Production structures within a plant.
- Interconnections with other plants.
- Long term planning on enterprise level.
- Scope and methods of scheduling in the plant.
- Information systems supporting production and scheduling.
- Special problems and requirements concerning production planning and scheduling.

Typical examples of batch-oriented industries can be found in the chemical, the pharmaceutical, and the food industry. From these industries five European companies were selected. In four of them the study included one selected plant, while in the fifth company (a large producer of chemicals and pharmaceuticals) it was possible to get an overview about several plants and departments.

The following kinds of plants and companies took part in the study:
- A dye-producing plant within a large chemical enterprise.
- A large producer of chemicals and pharmaceuticals.
- A tenside plant within a large chemical enterprise.
- A seed production plant.
- A cigarette production plant.

With the seed producer a rather untypical example has been included in the survey, so that it was possible to extend the scope beyond large chemical companies which can be expected to be very similar.

2.2 Interviews and Evaluation

In each plant it was possible to speak with the plant manager or another person responsible for scheduling. In some cases interviews were also carried out with people of the central information systems department.

For making the interviews comparable and to ensure that the same questions were asked in each case, a questionnaire was used as a guideline for the interviews. However, each interview went far beyond the predefined questions into the specific situation of the plant.

The interview protocols were first analysed independently of each other. The results were then presented to the interview partners so that they could be validated or corrected.

In a second step, the protocols of all cases were compared and differences and similarities were evaluated. Finally, the results, i. e. the most important requirements were clustered and common problems were identified.

3 PRODUCTION STRUCTURES IN SELECTED COMPANIES

This paragraph gives a short overview about each company. This includes a description of the company and its products, the production structures and current methods of planning and scheduling, as well as a discussion of specific problems and requirements. A detailed description of the case studies is given by Allweyer et al. (1994).

3.1 Dye Production

Production Structure. The textile dye plant is mainly a supplier of other plants within the company, for most products the plant carries out one production step, but there are also products for which several steps are performed. Since a product may need up to seven production steps, it may be moved through seven different plants until it is finished. Before and after processing, the products are usually stored in a central warehouse outside the plant. There are many strong interconnections with other plants and with suppliers (exchange of products and services, effluent treatment, etc.).

For a certain production the required equipment is usually mounted together to a unit (e.g. a reaction tank with equipment) that is used exclusively for one product at the same time. There are also some products with two or more reaction steps within the plant. For some of them there is a fixed line (e.g. two connected vessels), for others such a line is created when needed.

Planning and Scheduling. Longterm planning is carried out centrally. First the demands for each product are estimated. If the demand causes stocks to sink below a defined level, production orders with amounts and due dates are generated. The planning is based on bills of materials and plant capacities. These capacities are theoretical maximum capacities which cannot be reached in practice, since there are many restrictions and dependencies. The production plan is therefore not feasible, but it can be corrected to some extent by experienced planners. The inaccuracies and the difficulties of coordinating many different plants lead to high (and expensive) stocks of intermediates.

Scheduling is carried out on plant level. It is done manually and consists simply of determining the next time slot in which the required resource combination is available. In many cases it is not possible to meet the due dates. In such cases it has to be decided

whether to interrupt a current production or whether it is possible to change another scheduled job.

Further Requirements. Important requirements include the handling of frequent product changes, shared central resources, changes of resource configurations, a high number of restrictions, strong interdependencies with other plants and inaccurate long-term planning.

3.2 Chemical and pharmaceutical production

The company makes products in the areas pharmaceuticals, dyes, photography, agrochemicals and other chemicals.

Planning and Scheduling. Longterm planning is done regularly, it is based on sales forecasts, from which the net amounts for production are calculated. For most plants infinite capacities are assumed, only for known bottleneck plants the real capacities are considered. This requires much experience. The actual details of each plan have to be discussed with the plants and corrected accordingly. In some divisions there is no central planning department, and the production plans are created by direct co-ordination between the plants. Both ways of long-term planning are not considered to be satisfying.

Scheduling, which is done on plant level, relies heavily on human experience. In most plants it is done manually or with individual software solutions. Some selected plants are currently testing MES and their integration into the company-wide information system. The effects of scheduling decisions on other plants are usually not known exactly. It is tried to avoid problems by having time buffers and enough stocks of intermediates.

Further Requirements. A large number of interconnections between plants has to be handled, the long-term production plan is inaccurate, and it is difficult to get all information for creating a good and feasible schedule. The durations of operations may depend on several factors, like the type of units, or the amount and quality of the material.

3.3 Tensides

Production Structure. This plant mainly makes products from start to finish. There is only a few number of interdependencies with other plants, concerning energy and raw material supply. The raw material requirements of different plants have to be coordinated, since the shared raw material tanks are small, and the actual demands are known only a short time in advance. There are also pipelines that are shared with other plants.

Raw materials are delivered via pipeline, in barrels or railway waggons. They can be stored or immediately be used for production. The same tanks, however, can also be used for storing finished products. For technological reasons there is a minimum size for batches. Due to different equipment requirements, not every product can be made in every reactor. Reactors can also be used as storage tanks. Semi-finished products are mixed and filled into railway waggons. The filling station is a bottleneck. Waggons can also be used as storages. There is a laboratory in the plant, the laboratory tests also have to be scheduled. Each batch is related to a customer order. For technological reasons the produced quantities are usually larger than required.

Planning and Scheduling. Longterm planning is made on division level, it is based on sales forecasts, the actual plan is created in cooperation with the plants. Scheduling is done manually. Scheduling objectives include the feasibility of schedules, low costs, short lead times, high resource utilisation and low cleaning times and costs. Which type of cleaning is required, depends on the sequence of products, it can therefore be influenced by scheduling.

Further Requirements. Important requirements include the handling of by-products, shared central resources, technological restrictions (e. g. minimum batch sizes) and a large number of restrictions. Quality checks and laboratory tests should be regarded for scheduling.

3.4 Seed Production

Production Structure. The plant produces seeds for sugar beets and for sunflowers. The seeds are received from farmers, processed and sold, e. g. to sugar companies. There are about 100 different products made in the plant. The equipment is mainly controlled manually. Raw material supply and customer-demand depend very much on the season. After several production steps there is a large percentage of waste which can be sold as animal feed. Decisions about certain production steps are based on the results of quality tests. Some tests require an interruption of the production for up to ten weeks. There are also processing steps, the duration of which is not known in advance, e. g. polishing until a certain grain size is reached.

Planning and Scheduling. Planning starts with the "crop plan" which determines the amount of each seed type to be grown. The planning is based on forecasts, since it has to be made one year in advance. The main problem is not so much the planning itself, but the quality of the forecasts. Scheduling consists of determining when the seeds should be delivered, and the assignment of batches to resources. The first production steps are planned according to forecasts, the last steps according to customer orders. Scheduling influences the necessary cleaning procedures. The laboratory for quality controls is sometimes a bottleneck for scheduling.

3.5 Cigarette Production

Production Structure. The plant carries out the entire process of cigarette production from raw tobacco to finished cigarettes. There are no interdependencies with other plants during production. Tobacco is produced in batches with certain discrete batch sizes. The lot sizes for cigarette production can be different, they are not limited to certain sizes. There are six to seven production steps. After tobacco production, the production is split into several production lines. The number of possible paths is very high, since there are many possibilities of combining different units and additional equipment.

There are many dependencies between the different production processes. There must be coordination between different types of tobacco-treatment and between cigarette production.

Planning and Scheduling. Production planning is quite complex. There are more than 250 different products, each of them is planned individually for one year, since it was not possible to define non-overlapping product-groups, and there are many complex restrictions.

The schedule is subject to frequent changes. Starting with the final products, all production steps are scheduled backwards. It is tried to achieve a small number of product changes. Scheduling is based mainly on experience. First the standard products with large quantities are scheduled, non-standard products with small quantities are added later. After creating the schedule, its feasibility is checked. If it is not feasible, it is tried to resolve the conflicts and to change the schedule accordingly.

Further Requirements: There are two specific problems that have to be handled. The number of changeovers is very high. It is therefore not possible to maintain changeover matrices for all units and all products. Currently the changeover times are estimated by some "rules of thumb". The second problem is that changeover times on a resource are not only be determined by the two products that are processed before and after the changeover, but in some cases also by the history of production on the unit. For example, some device has been mounted on a unit some time ago which has not been used during the last production, but it is still there and the changevoer time is reduced if the next product needs that device. The changeover time depends therefore on the unit's history or a certain state of the unit.

4 REQUIREMENTS FOR SCHEDULING SYSTEMS

A large number of requirements concerning scheduling methods and -tools, but also production planning and integrated information systems, has been found in the case studies. Some of these requirements are very specific to a certain company, while others have been found in several companies. The most important common requirements are discussed briefly in this chapter.

4.1 Complexity of Production Structures

An important factor for scheduling is the complexity of the production structure. A high complexity results from a high number of possible different resource assignments, many different products, a large number of different paths within a plant and a high degree of interdependencies between plants. An example of a highly complex structure with many product paths is shown on the right side of fig. 2, while on the left side there is a simple case, in which each plant carries out an entire production from start to finish. In the latter case it is still possible that the structure within the plant is quite complex, but at least there are no problems resulting from the dependencies between different plants.

The tensides and the seeds examples have comparedly low complexities, while in the other examples there are higher degrees of complexity, and accordingly there are more demanding requirements concerning scheduling and inter-plant coordination.

4.2 Specific Process-Industry Related Aspects

One of the most important questions of the study was whether there are many process industry-specific requirements which have to be covered by software systems. Table 1 gives an overview over such requirements and in which of the investigated plants they were found.

In many companies, campaigns are scheduled rather than simple batches. Campaigns consist of a sequence of batches that should not be interrupted, since this would lead to increased changeover times and costs. Scheduling systems should provide the possibility to define and schedule campaigns. The reduction of changeover times and costs is an important scheduling objective in some plants, under consideration of industry-specific aspects, such as sequence-dependent cleaning procedures. The use of alternative units can be found in all investigated plants. When using such alternative units, process

Table 1: Scheduling requirements in selected plants.

Aspect	Dyes	Chemicals	Tensides	Seeds	Cigar.
Campaign scheduling	x	x	x		
Changeover times and costs	x				x
Use of alternative units	x	x	x	x	x
Unit-dependent durations		x			
Quantity-dependent durations		x			
Changes of unit-configurations	x	x	x		x
Changeable unit-connections	x	x	x		x
By-products		x		x	
Quality-dependent production	x	x		x	
Scheduling quality-tests	x	x			
Production waits for test results	x	x	x		
Unstable products					x
Shared resources in the plant	x	x	x	x	
Shared central resources	x	x	x	x	

parameters and scheduling restrictions may change. The duration of an operation may depend on the type of unit, or on the actual batch size, usually such dependencies are non-linear.

In many cases it is possible to change unit configurations, e. g. by mounting different type of equipment to a reaction tank. To ensure the feasibility of a job, it is necessary to check the actual configuration, or secondary resources are also scheduled. Not only the resources itself, but also the unit connections may be changeable, e. g. pipes between units. The actual connections restrict the possibilities for scheduling. It may also be an objective for scheduling to reduce the effort for changing such connecting pipelines.

A typical problem of the process industry is the occurence of by-products, which may be used as raw material for other productions. These materials may either be stored, or other processes which consume the by-products have to be scheduled accordingly. In several cases quality-dependent production was found, i. e. decisions are made according to results of quality checks. Since these decisions cannot be predicted, it is more difficult to create a schedule. Sometimes it is also necessary to schedule quality-tests, or the production has to wait until test results are known.

For unstable intermediate products maximum waiting times have to be taken into account. If there are shared resources, their maximum utilisation as well as additional restrictions have to be regarded. There may be also central resources which have to be shared between several plants.

4.3 Integration of Production Planning and Scheduling

None of the investigated companies has an integrated information system including both longterm production planning and scheduling. The coordination between the enterprise level and the plant level is usually based on meetings and telephone calls, and there are a lot of redundancies.

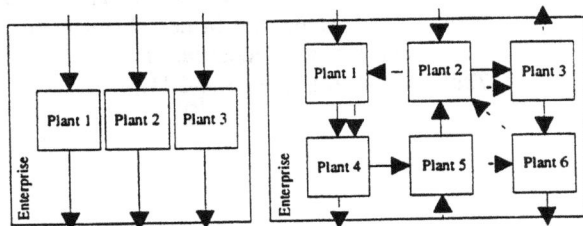

Fig. 2: Simple production structure (left) vs. complex production structure (right).

In each company it was pointed out that they would like to integrate these levels into one system, usually they expect more accurate and more up-to-date information and a better overview over the actual plant situation.

In several companies there is the problem that the capacities that are used for long-term planning are different from the actual resource capacities in the plant. This results from detailed restrictions which lead to reduced capacities under certain circumstances. These restrictions are not known on enterprise-level. Integrated information systems can provide more information about the feasibility of longterm production plans. It would also be possible to design systems which support direct coordination between different plants.

4.4 Requirements for Scheduling Systems

The interview partners were also asked about their requirements for scheduling systems. One of the most important features is a comfortable graphical user interface, including a Gantt-chart. Such an interface is needed to make the current situation and the schedule transparent to the user. In some cases it is also required to get information about which side-effects on other plants can result from a certain decision.

Most interview partners pointed out that it is crucial to have all relevant restrictions represented in the system, and that the system can make a valid check for the feasibility of a schedule. It is also required that the schedule is updated according to process feedback, so that the system can always give an overview over the current situation in the plant, and that it is possible to react quickly to any event.

5 CONCLUSION

The case studies show that powerful and integrated systems for scheduling and production management on plant level are still missing in the process industries. The basic requirements are similar to other industries: The plant managers require accurate, up-to-date information as a basis for their decisions. The integration of production planning, scheduling, process control, and laboratory information systems has to be achieved.

To be useful for the process industry, several industry-specific aspects have to be implemented in a scheduling system, since they are restrictions for scheduling. If these specific aspects are not regarded, the resulting schedules will not be feasible and the system will not be used.

There is not much demand for sophisticated scheduling algorithms, but rather for systems that provide the human planner with the required information and to support him in his decisions.

ACKNOWLEDGEMENTS

This work has been carried out within the CAPISCE project which has been funded by the Commission of the European Union under ESPRIT EP 6168.

REFERENCES

Allweyer, Th. (1995). Produktionsplanung und -steuerung in der Prozeßindustrie. In: *m&c Management und Computer*, **3** (1995) 2 (In preparation).

Allweyer, Th., P. Loos and A.-W. Scheer (1994). An Empirical Study on Scheduling in the Process Industries. In: *Veröffentlichungen des Instituts für Wirtschaftsinformatik* (A.-W. Scheer, Ed.). No. 102, Saarbrücken.

AMR (1994): The SAP Phenomenon Hits MES. In: *AMR Report October/November 1994*, pp. 1-21. Advanced Manufacturing Research, Inc., Boston

Hofmann, M. (1992): PPS - nichts für die chemische Industrie? In: *io Management*, **61** (1992) 1, pp. 30-33.

Instrument Society of America (1994): *ISA-SD88.01. Batch Control. Part 1: Models and Terminology. Draft 11D.* ISA, Research Triangle Park, North Carolina.

Loos, P. (1995): Konzeption und Umsetzung einer Systemarchitektur für die Produktionssteuerung in der Prozeßindustrie. In: *Geschäftsprozeß-optimierung mit SAP-R/3* (Wenzel, P., Ed.). Vieweg, Braunschweig Wiesbaden (In Preparation).

Nelson, N. S. (1983). MRP and Inventory and Production Control in Process Industries. In: *Production and Inventory Management*, **24** (1983) 5, pp. 15-22.

Remme, M., Th. Allweyer and A.-W. Scheer (1994). Implementing Organizational Structures in Process Industry Supported by Tool-Based Reference Models. In: *Proceedings of Rutgers' Conference on Computer Integrated Manufacturing in the Process Industries* (Boucher, T.O. et al., Ed.), pp. 233-247. Rutgers University, East Brunswick, New Jersey.

Scheer, A.-W. (1994). *Business Process Engineering. Reference Models for Industrial Enterprises.* 2nd ed. Springer, Berlin et al.

Schürbüscher, D., W. Metzner and P. Lempp (1992): Besondere Anforderungen an die Produktionsplanung und -steuerung in der chemischen und pharmazeutischen Industrie. In: *Chem.-Ing.-Tech.*, **64** (1992) 4, pp. 334-341.

ADVANCED SOCIO-TECHNICAL SYSTEMS AND PEOPLE EMPOWERMENT: CONCEPTS OF THE QUALIT PROJECT

Federico Butera

University   of Rome 'La Sapienza' and Istituto RSO, Milan, Italy

Abstract: Most of the discussion of the 70s and 80s was concentrated on the social effects of automation, i.e., of new technologies. The conclusion of that discussion was that information technology has little direct effects on QWL and business.

The Qualit project introduces two complementary key concepts: the first concerns the empowerment of people going beyond the previous defensive models, and implying that each individual should also become enabled (i.e., get the power) to actively defend and develop his or her integrity and the quality of life through various ways; this comprise more understanding and knowledge, emotional stability, clear roles, social integration, and to be a person, in order to choose paths and have the freedom for coping with external threats; it also implies that people should control their working processes.

The second is the design of network organization as a structure which encompasses structural flexibility of Joint Design both as a radical rethinking of the organization, and as continuos improvement. Modern socio-technical systems, which are often network organizations, should be built on open professional roles of empowered people.

The qualit project suggests the Change Management Process Framework (CMPF) as a general methodology for re-engineering or continuos improvement of the socio-technical system.

Keywords: Information technology; network organizations; Change Management, People Empowerment, Quality of Working Life, Work Ecology, Socio-Technical System Design,

1. AIMS AND NATURE OF THE QUALIT PROJECT

Qualit is an Esprit Project developed by a Consortium composed by Cap Gemini Innovation, Fiat, Istituto RSO, IPK (Institute for Production Design), SID (Danish Union Confederation), University of Dublin, University of Siena whose aim is to provide practical help to designers, managers, union leaders in introducing Information Technologies achieving at the same time business and technical excellence, as well as improved Quality of Working Life and empowerment of the people involved.

Competitive advantage of organizations and improvement of QWL shall converge in the European private and public organization because new technologies and new organizations require motivated, integral and empowered people. Suggestions and measures for improving QWL and empowerment and positive and manageable examples of good solutions are needed.

On this basis the QUALIT Project aims to identify, design, and spread approaches, practical methodologies of change, tools (software and paper), case studies training packages to help technology designers, interfunctional teams of middle managers, participative employees/management committee when new IT technologies, new organization or new professions are introduced or implemented .

Two main **key missions** of QUALIT are:
• Help *designing socio-technical systems* (processes, technology, organization, professional systems, HR rules) *which provide high technical and economic performances as well as quality of working life and people's personal and professional Empowerment.*
• Help *developing "the workplace within"* people (skills, social capabilities, communication abilities, inner power etc.).

The QUALIT program advocates three **key recommendations:**
• People *should control processes* instead of being controlled: their work should mainly include also communication, co-operation, problem solving and innovation.
• In such a turbulent work environment, manual and clerical workers, professionals, managers should *not carry risks of burn out* and not should *be hurt in their individual integrity* (body, mind, emotional, professional, social, etc.).
• Change programs should be designed in such a way to include from the very beginning *opportunities for training, co-operation and involvement* at any level.

2. THE COMPONENT CONCEPTS OF THE QUALIT PROJECT: THE SOCIO-TECHNICAL SYSTEM AND QUALITY OF WORKING LIFE/EMPOWERMENT

Qualit project is based upon two interrelated set of concepts: the *socio-technical system* and the *Quality of Working life dimensions.*

The socio-technical system

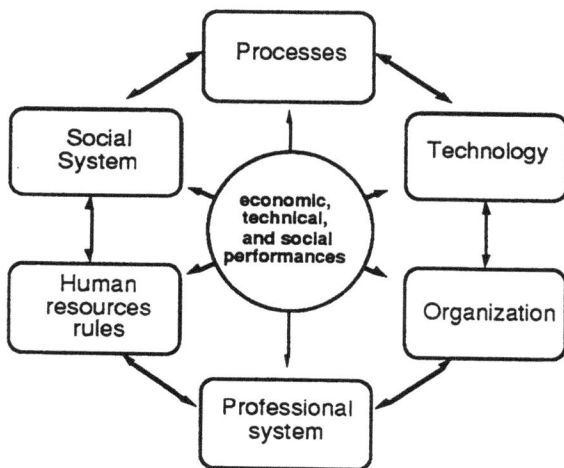

The concept of socio-technical system is based upon the dimensions represented in the following table. According to that, new design, re-engineering and improvement programs should identify and develop

strong and weak connections between business objectives, technical objectives, and quality of working life objectives, creating structural consonance among the various components of the STS system and helping the engagement of people in the accomplishment both of strategies of the organization and improving their own (professional and human) empowerment as real persons.
The dimensions of *quality of working life* have been identified in the following dimensions:
- *Integrity of body* (physical health)
- *Integrity of mind* (cognitive capability and well being)
- *Integrity of psyche* (psychological health and well being)
- *Integrity of self* (identity)
- *Integrity of professional roles* (quality of professional life as variety, complexity, meaning or work: career, social recognition, etc.)
- *Integrity of life roles* (social integration as indicated by public esteem, compatibility of work and family life, etc.).

The QWL concept has been in the past interpreted as a static one (dimensions to be monitored) and defensive one (finding out rules and policies for somebody or something protecting the integrity of people). The Qualit project introduces a complementary key concept based upon the active participation of each individual in protecting his or her own integrity and in the dynamic contribution to the performances of the socio-technical systems: the concept of *empowerment.*

The quality of life model

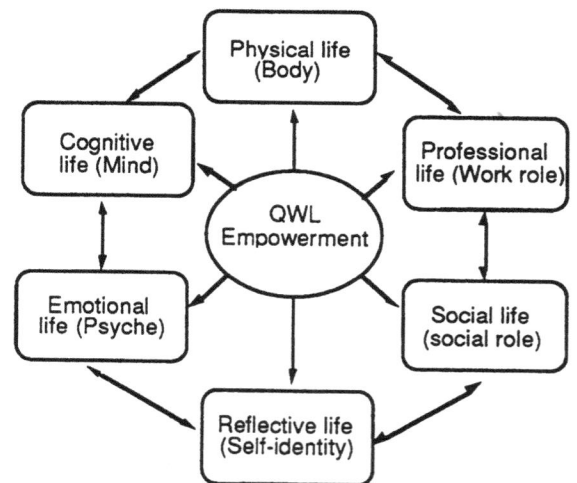

It implies first of all that each individual should not only be protected, but should become enabled (get the power) to actively defend and develop him or her integrity and the quality of life through various means: have more understanding and knowledge, emotional stability, clear roles, social integration and moreover "to be a person", in order to choose paths of freedom and coping with external threats. It also implies that the person should hopefully have control on working processes more than being controlled by the organization and technology. Empowerment is required by modern socio-technical systems that should be also made (and sometimes are) by open professional roles (small firms in the firm") of empowered people, which have in large part the "workplace within", as Hirshorn said. All this is not a "given" but a result process of individual growth.

I define *empowerment as the process through which an individual or a group of individuals improve his or her :*
• individual *and social power to recognize, protect, develop his or her integrity of body, mind, psyche, profession, social integration, self, and ;*
• *the ability to act individually and in co-operation with others in order to control work processes, to positively influence the structures and to improve the performances of a socio-technical system, due to his or her joint physical sanity and strength, level of understanding and competence, emotional stability, professional mastering, social integration, self confidence,*
• *the "being a person able and willingly to properly use the technological tools: it is associated to the degree of the person's maturity, hopefully supported by the inner ability to develop the integrity of body, mind, psyche, profession, social integration, and also to cope with eventual limitations in some of the mentioned sphere.*

3. DESCRIPTIVE, DIAGNOSTIC, ASSESSMENT MODELS FOR DESIGNING AND IMPROVING SOCIO-TECHNICAL SYSTEM ORIENTED TO HIGH QWL AND TO EMPOWERMENT

A *descriptive model* was developed in the past QLIS ESPRIT program, based upon the mentioned components of STS and QWL. It was oriented to clearly identify each of those component dimensions and to measure some values in the considered situations under observation.

MODEL OF RELATIONSHIPS STS/QWL

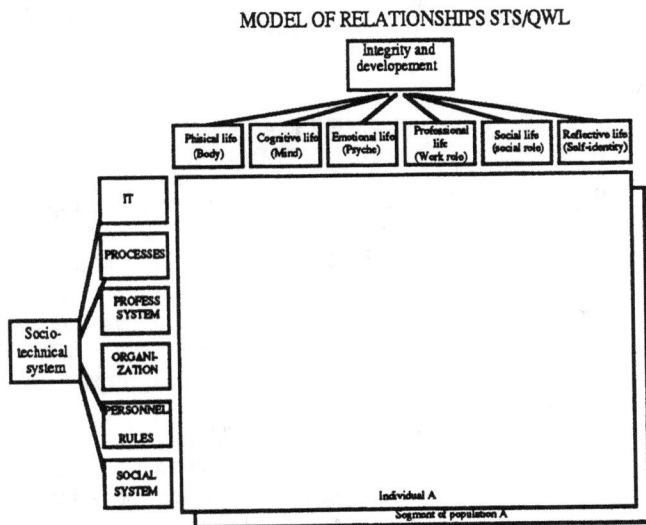

A *diagnostic model* should be developed to measure performances of socio-technical systems and the degree of Quality of Working Life. Measures have to be compared with standards coming from previous (and following) data gathering: this may be possible for some of the selected QWL dimensions (making references to the scientific literature or by developing in the Observatorium a sort of "help-line" with some scientific institutions). For STS dimensions the best path seems to be benchmarking the existing optimum solutions (high performances jointly in technical, economic and QWL) with STS under project.

The existing knowledge and the anyhow limited resources of Qualit Project makes impossible so far to have a *predictive model*: able to predict scientifically the impact of new technology projects, both upon new socio-technical design or upon Quality of Working Life and even less upon the empowerment processes.

Most of the discussion of the 70s and 80s concentrated upon the *social effects of new technologies*. The result was discouraging. We know now that Information Technology has little deterministic effects upon Quality of Working Life for good or for bad: in most cases technology has positive or negative effects upon QWL only when combined together with other dimensions of the socio-technical system. But designing and making improvement taking into account QWL dimensions or trying to improve QWL and empowerment is still a great challenge and it can be done.

4. THE CHANGE MANAGEMENT PROCESS FRAMEWORK (CMPF) APPROACH

The *change management model* includes the previous one, but is mostly oriented to provide tools for supporting groups of experts in factories and offices in carrying on the various steps of a change management process with specific orientation to improve the QWL and the empowerment of people.

The **Change Management Process Framework (CMPF)** is both an approach and a stepwise recursive procedure describing content and process of the QUALIT methodology.

As an approach it leads to the identification of *ad hoc* applications of *solutions* of Socio-Technical Systems design and improvements associated with QWL and Empowerment (as an example: Network organizations; self regulated organization; Professional roles; IT supporting professions and communication; Programs of people Empowerment; etc.). As a procedure, CMPF is focused on design (structuring, re-engineering, etc.) and improvement (implementation phase of design, continuos improvement, evaluation, monitoring etc.) of specific sub-systems or units: design and improvement are interlinked and recursive processes.

Interlinked means involving, in an interconnected form, goals, processes, technology, organization and social system. Recursive means to follow not a water-fall process but a "spiral" model: company plans, design and improvement may start anywhere: pilot design, programs of continuos improvement which may accumulate enough learning in the organization and in people to refurnish higher level company-wide changes.

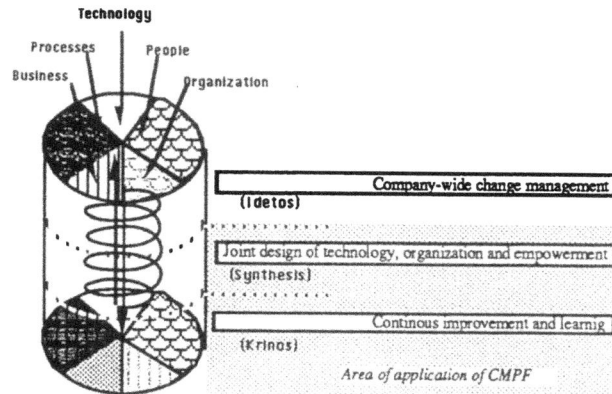

Area of application of CMPF

The CMPF for the Qualit Service Provider and for the final user is also a *guide to manage the process of design* (structuring /re engineering) *or the process of improvement* (implementation phase of design, continuos improvement, evaluation, monitoring, etc.). The CMPF is a *map* which will help the final user to navigate in the QUALIT environment in order to get specific tools that are suitable to his own needs and conditions of application. The CMPF provides the *basic specifications* for the various tools of QUALIT, that are being developed in the project.

CMPF is an ideal process of change including 12 major *steps* now displayed in a sequence for the sake of the understanding of the various views and deepening allowed by QUALIT. In case of very complex projects involving the help of a consultant all steps should be performed. In most projects however some of these steps -as operations- may be skipped, quickly passing through, according to different situations: what remains is the awareness of the different aspects of change, limited as they could be. The sequence of the steps may be changed to a certain extent at anytime. Each step is different from one an other: some are *time consuming* and includes many SW tools (as the block named "Evaluation"), others may be very quick and do not contain sophisticated tools except suggested views, methodological recommendations, and paper or SW check-lists (as the block "Client needs").

There are steps which are more strongly inter-related with each other and should only be performed within a recursive set of action: for example the goal setting and the general diagnosis. Then the notion of *blocks* is introduced and each of the 4 block includes up to interlocked steps of the same nature.

Some steps really deal with the organization and the management of the project: these steps are very important, but they don't concern with information, data, measures, etc. (like the other steps) but rather with promotion, control and co-operation on the project: they have been named *milestones*. **Milestones are not required in case of self-service by the final user**. One of the blocks is really a continuous process ongoing of understanding, learning, participation and growth: it is named *"continuous block.*

The main steps and blocks of CMPF are presented in the following figure. STEPS are included in **BLOCKS** and **MILESTONES**. The **CONTINUOUS BLOCK**, performed along the entire process, is discussed at the end.

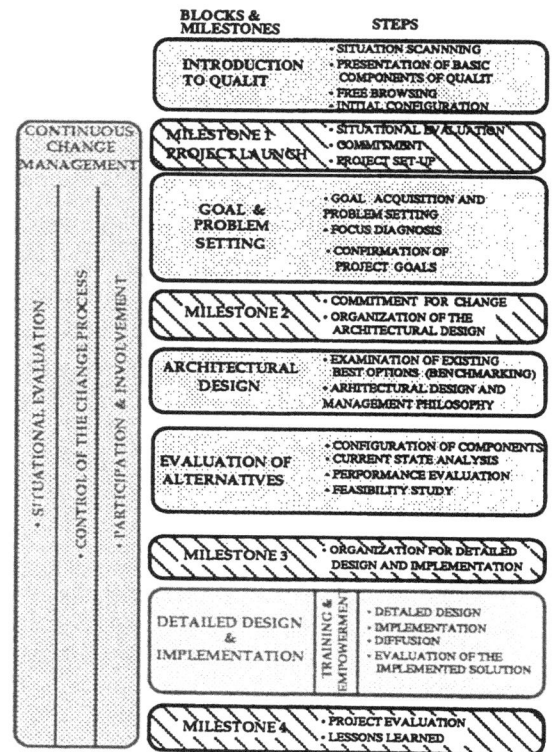

REFERENCES

Bagnara S., (1984) *L'attenzione*, Bologna, Il Mulino

Benjamin, R. I.; Levinson, E., (1993), "A Framework for Managing IT-Enabled Change", *Sloan Management Review*, summer

Bjorn-Andersen (ed) (1979), *The human side of information processing,,* Proceedings of the Copenhagen Conference on computer impact, North Holland Publishing

Blackler,.F.; Brown, C., (1986), "Alternative models to guide the design and introduction of the new information technologies into work organisations", *Journal of Occupational Psychology*, n. 59

Brandt, D. (ed), (1991), *Advanced experience with APS, Technological tools for anthropocentric systems, 30 European Case Studies*, FAST Monitor, v.2-FOP 246, July, 24-25, Commission of the European Communities, DG XIII

Butera, F. and Thurman, J.E., (eds), (1984), *Automation and work design*, New York, North Holland

Butera,F. "Options for the future of work" in Butera, F.,Di Martino,V, Kohler, E (eds), (1989), *Technological Developments and the improvement of working conditions: options for the future*, London Bruxelles and London, European Communities Official Publications and Kogan Page

Butera, F., *Information Technology and the Quality of Working Life*, (1990), 11th IFAC World Congress, Tallinn, Estonia, USSR, August 13-17, 188-194, IFAC, Tallinn '90

Conditions of work, a cumulative digest, (1982-1985), Geneva: International Labour Office

Cooper C.L.; Mumford E.(eds), (1979), *The quality of working life in western and eastern Europe*, Associated Business Press

Davis, L.E., *Organisational Redesign, Concepts, Methods, Processes*

Davis, L.E.; Cherns, A.B., (1975), *The Quality of Working Life*, Free Press, NY

De Michelis, G., *From the analysis of cooperation within work-processes to the design of CSCW Systems*, (1994), Proc. of the 15th Interdisciplinary Workshop on Informatics and Psychology, Schaerding, Austria, May, 24-26

Gallino, L., *Informatica e Qualità del Lavoro*, (1983), G. Einaudi, Torino,

Hackman, J.R.; Oldham, G.R., (1980),*Work Redesign*, Addison-Wesley, Reading, MA

Hall, G.. Rosenthal, J.; Wade, J., (1993), "How to Make Reengineering Really Work", *Harvard Business Review*, November-December,

Hammer, M.; Champy, J., (1993), *Reengineering the Corporation: A Manifesto for Business Revolution*, Harper, NY

Helander, M., (1988), *Handbook of human-computer interaction*, in Martin Helander, State University of New York at Buffalo, North-Holland, Amsterdam, NY, Oxford, Tokyo

Hirshorn, L., (1988), *The workplace within*, Cambridge Mass, The MIT Press

Human factors in organizational design and management-(1986), II: proceedings of the second symposium held in Vancouver, B.C., Canada, 19-21 August, 1986. Amsterdam; New York: North-Holland; New York, N.Y.: Elsevier Science Pub. Co.

Keen, P.G.W., (1991), *Shaping the Future: Business Design through Information Technology*, Harvard Business School Press, Boston, MA

Kern, H.; Schumann, M., (1984), *Das Ende der Arbeitsteilung? Rationalisierung in der industriellen Produktion: Bestandsaufnahme, Trendbestimmung*, C.H. Beck, München

Lawler, Edward E., (1992), *The ultimate advantage: creating the high-involvement organization*, 1st ed. San Francisco, Calif.: Jossey-Bass

Liu, M.; Denis, H.; Kolodny, H.; Stymne, B., (1990), "Organization design for technological change", *Human Relations*, 43, 1

Majchrzak A., (1988), *The human side of factory automation*, Jossey Bass, New York

Markus, M.L.; Robey, D., (May 1988), "Information technology and organizational change: causal structure in theory and research, *Management Science*, Vol.34, n. 5

McIlvaine; Parsons, H., (1985), "Automation and the individual: comprehensive and comparative views", *Human Factors*, 27, 1

Olsavsky, Mary Ann. (1990), *The new work systems network: a compendium of selected work innovation cases*. Washington, DC: The Dept.

Otway, H. J.; Peltu, M. (eds), (1983), *New office technology: human and organizational aspects,*, F. Pinter: London

Orlikowski, Wanda J., (1991), Information technology and the structuring of organizations. *Working Paper, Center for Information Systems Research, Sloan School of Management, Massachusetts Institute of Technology,*

Ovalle, N.K., (1984), "Organizational/Managerial control processes: a reconceptualization of the linkage between technology and performance", *Human Relations*, 37, 12,

Parker, M., *Inside the circle: a human guide to QWL*, Boston MA, South End

Piore, M.J.; Sabel, C.F., (1989), *The second industrial divide*, Basic Books, NY

RasmussenJ; (1986), *Information processing and huma-machine interaction*, New York, North Holland

RSO Institute, (October 1988), Proceedings International Conference *The joint design of technology, organization and people growth*, Venezia

Salvendy, G.; Slauter, S.L., (1987), *Social, ergonomic and stress aspects of work with computers.*", Elsevier Science Publishing, Amsterdam

Schael, T., (1991), "Menschenorientierte CIM-Konzepte für die flexible Fertigung: Praxiserfahrungen in eimen europäischen Industrieprojekt, Mensch und Technik", *Werkstattbericht*, n. 100

Schrage, M., (1990), *Shared Minds: The New Technologies of Collaboration*, Rando House, NY

Scott Morton, M.S., (1991), *The Corporation of the 1990s, Information, Technology, and Organisational Transformation*, Oxford University Press, NY

Shaiken, H., *Work transformed: automation and labour in the computer age*, Holt, Rinehart and Winston. New York

Sheridan, T., *45 years of man-machine interaction*, RSO International Conference, The joint design of technology, organization and people growth, cit.

Sorge, A., (1989), "An essay on technical change: its dimension and social and strategic context", *Organization Studies*, 10, 1

The Experience of work: a compendium and review of 249 measures and their use, London; New York: Academic Press, (1981)

Trist E. and Murray H (eds) *The social engagement of social sciences: a Tavistock anthology*, Philadelfia, The University of Pennsilvanya Press, (Two volumes), 1990

Trist E. and Murray H (eds), (1990), *The social engagement of social sciences: a Tavistock anthology*, Philadelfia, The University of Pennsilvanya Press, (Two volumes)

Volpert, W.; Kotter, W.; Gohde, H.E.; Weber, W.G., (1989), "Psychological evaluation and design of work tasks: two example, *Ergonomics*, 32, 7

Winograd, T., (1987), "A Language/Action Perspective on the Design of Cooperative Work", *Human Computer Interaction*, Vol. 3-1

Wisner, A., (1981), "Organizational stress, cognitive load and mental suffering" in Salvendy G. and Smith J.M *Machine pacing and occupational stress*, London Taylor and Francis

Womack, J.; Jones, D.; Roos, D., (1990), *The Machine That Changed The World*, Rawson, NY

Zuboff, S., (1988), *In the Age of the Smart Machine: the Future of Work and Power*, Basic Books, NY

U.S. Dept. of Labor, Bureau of Labor-Management Relations and Cooperative Programs, (1985), *Perspectives on labor-management cooperation.* Washington, D.C.

COGNITIVE ERGONOMICS AND EMPOWERING ORGANISATIONS

Michele Mariani*
Oronzo Parlangeli*
Sebastiano Bagnara*
Angelo Mc Neive**
Sarah Downing**
Gerard M. Ryan**

University of Siena Multimedia Communication Lab
** *University College of Dublin*

Abstract: empowerment is the new keywords in human resources management. Despite its fast growing popularity empowerment's issues are often disseminated more as a philosophy than as a sound set of knowledge, methods and practical guidelines. The adoption of the point of view of cognitive ergonomics within the QUALIT project appears suitable to give more solid ground to the concept and to get guidelines for the design of an empowering interaction between the organisational artifact and human actors.

Keywords: Ergonomics, quality of work life, organisational factors.

1. INTRODUCTION

Empowerment (Kanter, 1983; Murrell, 1985; Conger and Kanungo 1988; Clement, 1994) is the new 'meme' (Dawkins 1976) in human resources management. Both the empowerment and the cognitive ergonomics (Sperandio, 1980; Norman and Draper, 1986; Bagnara et al. in press) approaches stress the need of improving the design process by carefully considering human factors recommendations and knowledge. A number of areas of overlap are evident from the literature in terms of issues being addressed both in the empowerment and in the cognitive ergonomics fields of work, e.g.: organisational design, employees well being, impact and use of technology, company wide integration, etc. The QUALIT[1] project aims at giving more solid ground to the concept of empowerment and to get methods for the design of an empowering interaction between the organisational artifact and human actors. As a discipline aiming at optimizing the

relation between man and the instruments used, cognitive ergonomics has produced a lot of results which could be proficiently adopted to empower people basing on empirical research.

2. THE POINT OF VIEW OF ERGONOMICS

Ergonomics is concerned with the study of people at work. It can be defined as the discipline that studies which are the principles that optimize the interaction between human being and the tools he/she uses to reach his/her goals. While 'classical ergonomics' has been concerned with the setting up of a body of knowledge adequate for the definition of guidelines for the correct design of the physical interaction between man and tools, 'cognitive ergonomics' has instead focused its attention upon the conditions enhancing the compatibility between the ways in which we understand the world and the cognitive artifacts that we use in these processes. A cognitive artifact can be defined (Norman, 1991) as the artificial device that shows and/or manipulates information and that is able to affect humans' cognitive activity. In this perspective it is possible

---

1 QUALIT is the acronym for the EU funded ESPRIT Project
n° 8069: Quality Assessment of Living with Information
Technology.

to look at organisations as special kinds of cognitive artifacts suited to enable individuals to coordinate their activity toward a performance. The enlargement of the concept of cognitive artifact from computers to the whole work organisation analysis and design is quite recent (Reason, 1988; Hendrick, 1991; Samurcay and Delsart, 1994): until now research in the field of cognitive ergonomics has mostly focused on the aspect of human computer interaction. The reason is that the dominant framework in the discipline has been based on the needs of a single user interacting with a single interface. However, notwithstanding with the recency of this approach, the adoption of the point of view of cognitive ergonomics (that is an interactionist point of view) in the design of organisational artifacts appears suitable to explore the validity of some of the latest claims in the field of human resources management and to give them a sound knowledge base.

Bannon (1991), in a recent paper on the role of psychology in computer systems design, outlined a new approach to understand and conceptualise the relationships between people, work, technology and organisation. He argues that a fundamental shift that is needed is to replace the term 'human factors' (an ergonomic term) with 'human actors'. The reason for this is that the former term has tended to be associated with a 'passive, depersonalised' person while the latter implies an 'active, controlling' one. By changing terms it is suggested that 'emphasis is placed on the person as an autonomous agent that has the capacity to regulate and coordinate his or her behaviour, rather then being a simple passive element in a human-machine system'. Hence, instead of viewing the human as an information processor the 'human actor perspective' focuses on the way people act in real work settings.

The same concern is expressed in one of the last perspectives in Organisational Development: empowerment. Empowerment shifts the human resources management from its orientation towards the search for 'consensus' to the search for 'subjective sense' that each individual must be allowed to prosecute within its own job and with respect to the organisation's constraints.

3. COGNITIVE ERGONOMICS AND EMPOWERING ORGANISATIONS

Ten years ago Lisanne Bainbridge (Bainbridge, 1987) warned about the paradoxes that could stem from a design guided only by the will to adopt a state-of-the-art technology and that does not consider the conditions for the man-machine interaction. The same kind of warning can be raised to those currently involved in the development and implementation of the concept of workers' empowerment.

The new 'meme' of empowerment has reached the cover of many managerial reviews in the last ten years with the merit of raising new interest around issues formerly debated inside the 'democracy of work' movement (see e.g. Bjerknes et al. 1987). However, this reviving of almost forgotten themes has not given them a more solid ground (Burdett 1991). Empowerment issues are often disseminated as a philosophy, a sort of attitude toward 'positive thinking' (Dennis 1991) rather than as a sound set of methods and practical guidelines for the design of organisational features to enable people contributing more to the company's change and development.

Cognitive ergonomics can contribute to fill the gap by researching those conditions for the interaction between human actors and the organisational artifact that will result in an empowering organisation. As Clegg (1994) argues: "cognition is only partly individual: it is also a social enterprise and nested in artifacts (e.g. structures and tools) which support and constraint people's endeavours". It is well known that the introduction of an artifact in an organisation causes not only restricted changes in the task interested by the innovation but on the whole organisation of work. It is essential, therefore, that the design of the new technology (informational and/or organisational) takes into special account the aim of the organisation such as, in the case of empowerment, skill development for employees. For example, the way in which computers are designed and introduced decides if they will contribute to employee empowerment (e.g. enhancing communication, availability of relevant information, monitoring and feedback, etc.) or if they will impoverish people (e.g. authomatizing the easy part of tasks and living the worker with an arbitrary collection of tasks, or overloading an operator with a huge amount of information in which it is hard to navigate).

Thus, cognitive ergonomics will be of help in specifying:

i) the ways in which the concept of empowerment matches the body of proved knowledge about human beings' capacities;

ii) methods to interpret how the world of constraints of a certain organisation can affect cognition, particularly which kind of organisational features can determine or at least support the peculiarities of an empowered workers' population.

4. THE QUALIT APPROACH

Changed market conditions of last years push many Companies toward a change of organisation

(Horizontal Corporations, Networked Enterprises - Butera, 1990). Such transformations tend to focus mainly upon changes in the workflow of activities (see e.g. reengineering approach - Hammer and Champy, 1994). Very few methods, although, deal with human consequences of change and when this theme is faced with it is mainly in terms of the necessity of the participation of people throughout the whole process of change. We claim that wide organisational changes imply a shift to new models of behaviour, culture, and skills that cannot be successfully managed taking only into consideration the processual and technical, objective, aspects of work. A change in organisational artefacts has to be planned taking into account individuals' needs and has to foresee (as much as possible) which will be the effects of such a change upon human workforce integrity and well being.

Within QUALIT it is stated (Butera, 1994) that "through empowerment individuals become nodes within a network of relations and commitments which, by interacting with the other nodes and applying their own knowledge and competencies, contribute more to the evolution of the system than to its stability".

To reach the shape of a network an organisation should undergo a process of change. The levers for such a mutation are searched through a process having some peculiar features:

i) knowledge about human actors is organized in six different dimensions made up of several variables such as: mental workload, feedback availability, interpersonal contact, etc.;

ii) tools for the subjective weighting/evaluation both of given variables and tools for the subjective provision/evaluation of self defined relevant organisational variables are given;

iii) methods for the selection and ordering of a set of guidelines for the new empowering organisation's features design are given.

The QUALIT process for quality of working life assessment and people's empowerment is an open one. Within organisations critical issues are supposed to be situated, that is to be strictly related to the context and to the knowledge of the actors in the process. Thus, variables that can act as levers for an organisational change aiming at workers' empowerment are tested relying on people's subjective evaluation of their working conditions. In addition, relevant variables have both the possibility to be selected out of a set defined according to the competence of experts and to be self-defined by the actors involved in the process. Such a strategy is quite similar to the so called user centred approach along which users of any new system are actively involved all along the design

decision process. Checking workers' opinions at various stages of the new organisation's design is an essential issue that can save resources and time ensuring that unusable or unnecessary procedures and rules are avoided.

5. CONCLUSION

A huge amount of different methods is available for the design of a new 'organisational artifact', however, such methods seldom embed sound ways to find out from the knowledge of the two main actors involved in the process (workers and managers) which can be the levers for organisational change. This happens despite the fact that human agents will be the critical actors for the success or failure of any process of change.

The combined point of view of ergonomics and empowerment can be of help in the research about guidelines for the design of those cognitive artifacts named organisations. Such artifacts (the whole body of rules enabling coordination of the activity of groups of individuals toward a performance) could then directly affect humans' cognitive activity toward empowerment of workers.

The QUALIT process for quality of working life assessment and people's empowerment has been thought to give to the empowerment 'meme' a sound body of knowledge to rely on and a set of methods and tools to interpret how the world of constraints of a certain organisation can affect cognition. In particular to search for which kind of organisational features can determine or at least support an empowered workers' population will be one of the main matters of our contribution to the remainder of the QUALIT project.

REFERENCES

Bagnara S., Rizzo A. and Parlangeli O. (in press). Learning strategies and organisations. In: *Organisational learning and technological change.* (C. Zucchermaglio, S. Stucky and S. Bagnara (eds.). Springer Verlag: Berlin.

Bainbridge, L. (1987). Ironies of Automation. In: *New Technology and Human Error.* (J. Rasmussen, K. Duncan, and J. Leplat (eds.)) Wiley and Sons: New York.

Bannon, L. (1991). From Human Factors to Human Actors: The Role of Psychology and Human-Computer Interaction Studies in System Design. In: *Design at Work: Cooperative Design of Computer Systems*, (J. Greenbaum and M. Kyng (eds.)). Lawrence Erlbaum Associates: Hillsdale, New Jersey.

Bjerknes, G., Ehn, P., and Kyng, M. (1987). *Computers and democracy - A Scandinavian challenge.* Aldershot: Avebury.

Burdett, J.O. (1991). What is empowerment anyway? *Journal of European Industrial training*, 6, pp. 23-30.

Butera, F (1990). Il castello e la rete: *Impresa, organizzazioni e professioni nell'Europa degli anni '90*. Angeli: Milano

Butera, F. (1995). *Quality of Working Life criteria and Empowerment in reengineering and continous improvement of Network Organisations supported by Information Technology*. Working Paper.

Clegg, C. (1994). Psychology and information technology: The study of cognition in organisations. *British Journal of Psychology*, 85, pp. 449-477.

Clement, A. (1994). Computing at work: Empowering action by 'low level users'. *Communications of the ACM*, January, Vol. 37, n° 1, pp. 53-63.

Conger, J. A. and Kanungo, R. N. (1988). The empowerment process: integrating theory and practice. *Academy of Management Review*, 3, pp. 471-482.

Dawkins, R. (1976). *The Selfish Gene*. Oxford University Press: New York.

Dennis, B. J. (1991). The Tao of empowerment. *Performance Improvement Quarterly*, **vol.4**, part 4, pp. 71-80.

Hammer, M. and Champy, J. (1994). *Reengineering the Corporation: A manifesto for business revolution*. Harpers and Collins, New York.

Hendrick, H. W. (1991). Ergonomics in organisational design and management. *Ergonomics*, **vol. 34**, n° 6, pp 743-756.

Kanter, R. M. (1983). *The Change Masters*. Simon and Shuster: New York .

Murrell, K. L. (1985). The development of a theory of empowerment: Rethinking power for organisational development. *Organisational Development Journal*, 2, pp. 34-38.

Norman, D. A. and Draper, S. W. (1986). *User Centered System design*. Lawrence Erlbaum Associates Hillsdale, New Jersey.

Norman, D. A. (1991). Cognitive artifacts. In J. M. Carroll (Ed.), *Designing interaction: Psychology at the human-computer interaction.*. Cambridge University Press: New York.

Reason, J. (1988). An interactionist's view of system pathology. In: *Computer and System Sciences*. NATO ASI series F, **vol. 32**.Springer: Berlin.

Samurcay, R. and Delsart, F. (1994). Collective activities in dynamic environment management: Functioning and efficiency. *Le Travail Humain*, **vol. 57**, n°3.

Sperandio, J. C. (1980). *La psychologie en ergonomie*. Presses Universitaires de France: Paris.

JOINT DESIGN OF ORGANISATON AND TECHNOLOGY - THE INSIDE STORY

K. Mertins*, B. Schallock* and B. Coppola**

Fraunhofer IPK-PLT, Pascalstraße 8-9, D-10587 Berlin
*** Istituto RSO, Via Leopardi 1, I-20123 Milano*

Abstract. The QUALIT team builds an interactive tool to support complex change processes and improve work conditions. The paper concentrates on challenges and obstacles of multidisciplinary tool development by psychologists, sociologists and engineers. The change management framework with a set of tools and a keyword driven case library provides guidance and flexibility and bridges the different requirements of users. Progressing co-operation in this QUALIT team is mirrored with the history of expanding engineering fields that allows a limited prediction of the future consideration for "integrated design" of factory work.

Key Words. Change Management, Multidisciplinarity, Multidimensional Systems, Sociotechnical Design, Quality of Working Life (QWL), People Empowerment, Modelling, Human Factors, Social and behavioural sciences

1. INTRODUCTION

In the last decades the industrial environment has brought a radically new concept of change. Up to the middle of the 80´s the attention of the decision makers has been concentrated much more on the concept of innovation: innovation, mainly technological, has been considered as the strategic weapon to companies´ competitiveness. Change in a wider sense has been considered mainly as a consequence, often an undesired one, of technological innovation: most of the studies concerned impact of technological innovation on organisations and people. Companies' strategies have been quite static and conservative; they could have changed, but such changes have been considered as reaction to major changes in the business and social environment.

Since the mid 80`s the idea is growing that change is no more exceptional and that flexibility is not only important in operations but in strategy itself. The idea of a technology driven innovation has been abandoned, in favour of the idea of change as a demonstration of the ability even to anticipate the dynamic of the environment. Already existing concepts, previously neglected, such as 'learning organisation' and 'change management' have become very popular and even buzzwords.

The increase of systematic connection of the organisation (world-wide competition, diversification proc-esses, diffusion of information and communication technologies), contributes to the major attention paid to change processes. Today, different requirements are given to the management of change. Changes should produce results in the short term and be concluded in not too long time, using few planning staff dedicated and should not be necessarily the result of massive investments. Nevertheless the process has to get people involved in order to ensure success and has to develop the capability of people to get involved and to drive a change process instead of being forced. A change should also show the potential of improving *Quality of Working Life*.

2. TRADITIONAL TOOLS OF SOCIOTECHNICAL DESIGN

Sociotechnical tradition could be considered, in several points of view, the most prepared to such a challenge. The idea of managed global change has been recognised in the 60´s (Emery 1967), while the assumption of the organisation as a complex system of inter-related factors have been discussed by sociotechnical authors (Emery 1981), up to the concept of "joint design of technology, organisation and people growth" (Butera 1988). Sociotechnical tradition has defined several principles, tools and methods to address joint design. Among the others should be here mentioned:

- The principle of compatibility: in order to be compatible and reach its goal, the design should create a system capable of self-modification (Cherns 1973).

- People involvement: "the design team would not design other people's lives" (Davis and Cherns, 1975).

- Wholistic work design: every person should be able to plan, execute and check his work. This offers also learning possibilities and personality growth (Ulich 1992).

- Work Groups and autonomous units: one of the most implemented organisational solutions, being capable to provide social interaction, increase flexibility, improve product quality and support its change, to maximise organisations' profits. The Japanese industry puts this as one of the basis of its restructuring and strength.

- Action Research, which has been used as an instrument to support the change process, and its successful implementation depends on the degree of involvement of both the research team and the client.

All these bases are shared in the QUALIT consortium of the European ESPRIT Project 8162 being discussed in this paper, but under different points of view.

The opportunity is now to integrate more views, more and different actors and more recent research developments in the same tradition. The requirement of a broader and faster change lead to the idea of the change management process framework CMPF (see also Butera in this proceedings).

3. MULTIDISCIPLINARY REQUIREMENTS

To give a support in this wide sense is a big challenge, because it should serve many different companies in different countries and being used by different users from management to planners and shop stewards. Referring to the challenge a QUALIT team was set up that comprises (see fig. 1):

- several countries from Denmark to Italy

- several professions from work psychologists to engineers and computer scientists

- groups of users, of experts and of developers

- experts who cover more the strategic business consulting

- experts who cover IT introduction and

- experts covering organisational change in production environment.

The variety of users, work styles and application areas that should be covered by the QUALIT tools to be built was also represented by the various teams of the partners and individuals contributing. Typical differences that complicated project work are:

Business consultants´ view on *long term strategy* versus engineers´ view on *detailed company structures* versus psychologists´ view on todays *individual*

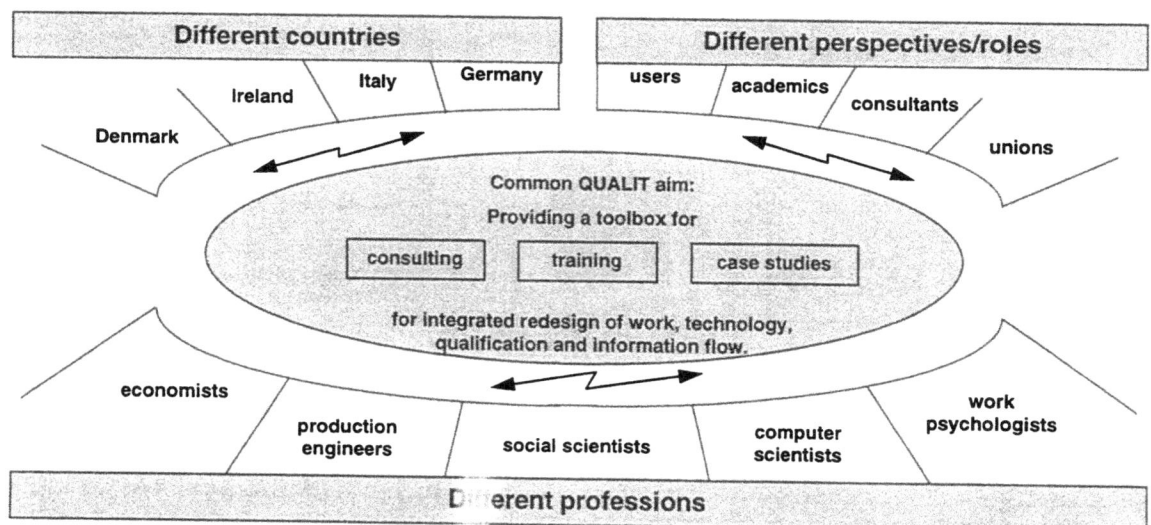

Figure 1: Interprofessional and intercultural work.

114

work situation.

The Danish trade union provided an advanced and *informal co-management practice* versus a *formalised negotiation procedure* by the German union versus a weakened and much less design oriented situation of Italian unions.

Concerning the degree of details in a change management process, managers like workers` representatives are focused on *effects* versus engineers` and planners` focus on the *detailed functions* of a new factory or business process. Obviously managers ask for predicted business results whereas workers representatives ask for planned numbers of employees, changed skill levels and stress factors.

Work habits between scientists and industry (management and unions) differ in that way that scientists express their statements in *abstract and formalised way* whereas managers and unions do much more refer to their last urgent problem and *examples* of the past.

Engineering scientists tend to report in short and structured chapters their *design results* using graphs and figures whereas psychologists discuss phenomonoms in rather lengthy *analytical papers* .

Even if the above given view on the QUALIT team looks worse that daily work really is, it is still an enormous challenge that will be faced increasingly also by other development teams world-wide.

The task to bridge different views and professional traditions is not completely new. An elaborated example of a contribution of one single QUALIT partner should be given from the field of engineering.

4. MULTIDISCIPLINARY ENGINEERING TRADITION

Looking back at the history of the Institute for production systems and design technology (IPK Berlin), the QUALIT project represents an interesting phase between past and future. The reflections should not only help to ease integration within QUALIT but they are transferable to other organisations and teams.

In the beginning of this oldest institute of it's kind in Germany, Prof. Schlesinger covered a remarkable broad field of research and industrial application spanning from machine tools to plant layout, operation management and compensation schemes and psychological hiring tests. This broad scope was lost after he was forced out of Germany in 1936.

The phases of expanding research views beyond machine tools started again in the sixties when computerised controllers became part of a machine tool (Phase I). It was a phase when *solutions* where required by industry. Providing the solution required

then to include the electric/electronic controller as well and required electrical engineers working together with mechanical engineers and material specialists.

This development was followed in the 70´s by designing machine tools as complex systems that also include tool handling and workpiece handling. The additional and expensive devices (e.g. robots) of the new machine tools were not per se accepted by industry and therefor the IPK intensified work on *investment strategy and cost calculation*. This required multi-skilling of engineers and the hiring of economists that introduced different work habits of how to better sell ideas to decision makers within the engineering teams.

This is also the only major case of a related change in profession profile. At that time the academic degree " masters for business and engineering" was created and makes a constant visible percentage of the students meanwhile.

Next to the investment strategy the operation of more complex machining centres (MC) and Flexible Manufacturing systems (FMS) became an additional topic. This involved the design and programming of cell controllers and shop floor control systems (SFC) which required more computer science skills in the teams. It also required more factory design knowledge including industrial engineering knowledge and factory operation knowledge including personnel management rules.

A second phase (*Phase II*) was entered in the 80´s, when *methods* were developed that help companies to develop and implement own solutions. These methods for analysis, simulation, planning, product and process modelling, evaluation and for training mostly began as tools for IPK´s consulting activities and reached the status of products after successful application.

The nature of complex redesign of factories requires the application of several design tools during a consultation. This made IPK enter a third phase of *integrating engineering methods* to reuse results from step to step. Specially organisational and technological development have to be integrated to provide skill enhancing application software for empowered staff in flexible and flat enterprise structures (Mertins et alt. 1992, Schallock 1993). The mean of integration is given by enterprise modelling that was elaborated as a method (IEM) as well as a modelling software (MO²GO) (Mertins et alt. 1995).

Even training skills had to be acquired to provide organisational design and implementation. During such a empowerment training in industrial application, IPK was able to teach social and methodological skills to shop floor workers using the design of their own workplaces as the case for training (Liedtke et alt 1995). Training results and engineering planning results hereby form a synchronised output of an empowerment procedure.

Even if this procedure is already very complex, there is not enough support yet for guided change management (as in QUALIT) and not enough psychological work design skills (as in QUALIT) to provide full & comprehensive service to industry to allow own further development. This enhancement is provided by the current *Phase IV* where QUALIT aims at the integration of engineering, social and business methods in a team beyond the capacities of IPK.

The other partners in the team felt deficits in a similar way, missing the economic or technical side of a consultation for change. The broad QUALIT team and the CMPF give a sufficient platform and a framework of change. This historic consideration can be used to foresee a further development during the next years. The increased complexity of the past will probably be extended in a *Phase V* when the *environment is accepted as one of the resources* to be managed next to machines, material, staff and knowledge. The integration of *biologists and chemists* into factory design teams and business process reengineering teams will be required.

A further future aspect is the *social integration of an enterprise into community life*. Honda gives a known example in Ohio, USA how this integration can be made known and be used for image building. This means that increasingly *political science, sociology and media science* will be part of integrated design teams.

Even if different phases are overlapping, the following attempt of a simplified summarisation of shifted focuses should be given (see fig. 2):

Phase I 1904-1980: Development of solutions (machine tools, application programs, layouts)

Phase II 1980-1992: Development of methods for production planning and operation (simulation methods, planning methods, product and process modelling methods, evaluation methods, training methods)

Phase III 1992-now: Integration of engineering methods (logistics design methods, simulation, business process design, specification methods, control methods, work design and training methods)

Phase IV 1994-now: QUALIT Integration of engineering and social and business methods

Phase V 1998 and beyond: methods for environmental and social integration of enterprises

5. THE BRIDGING APPROACH OF THE CMPF, CSL AND TOOLS

Changed work habits
Changed perspectives and work habits are a prerequisit for a multidisciplinary team. Major changes are performed by *engineers*, who (simplified) have to:

- accept soft facts as e.g. interview results or rumors as facts,

Historic and forseeable development within a large engineering institute of applied research

Phase	I 1904 - 1980	II 1980 - 1992	III 1992 - now	IV QUALIT 1994 - 1998	V 1998 and beyond
Type of result	Development of solutions	Development of methods and related sw-products	Integration of engineering methods	Integration of engineering, social and business methods	Methods for environmental and social integration of enterprises
Typical topics	• machine tools • flexible manufactory systems • factory layout • psychological hiring methods • application programs • robots	• simulation methods • planing methods • modelling methods • evaluation methods • training methods	• strategic planing and enterprise modelling linked to software specifications methods • Training methods for social competence in the simulation game	• participative empowerment • goalsetting related to business goals and human goals • modelling of task, information flow and economic effects • evaluation methods	• green design • industrial and regional interaction • future of paid and unpaid work • knowledge creation • North/South conflict
Team involved	• mechanical engineers • electrical engineers	additional: • economists • computer scientists	additional: • social scientists international technology transfer	additional: • psychologists in a European scale	additional: • biologists • chemists • public administration • cognitive science in an international scale

Figure 2: Scientific development within engineering.

- accept other methods of consensus building, negotiation and empowerment than only management decisions and orders,

- talk about goals, visions and uncertainties and not only about decided structures and procedures

- and coach instead of giving orders and allow a social process instead of a static view on peoples abilities.

Allowing social processes and necotiate is specifically difficult for engineers because the were trained and compensated to deliver ready results and not as facilitators.

Social scientists on the other hand have to:

- change from pure analysist role to the tasks of shaping work with all the resposibility for a wrong advice and have to

- argue for a certain solution and calculate the involved costs instead of only asking for opinions.

A third basic requirement for change is seen with the *unions*, who have to:

- engage themselves for a positive solution instead of only preventing bad solutions.

Even *union members* have to learn to appreciate a co-management attitude of the workers representatives instead of confrontation, because this keeps more jobs on the long run.

The QUALIT team consists of members that had already gone half the way before entering the team but it is clearly visible that not only the project results in further changed work habits of the team but that the QUALIT tools will change work atmosphere in companies as well who apply these tools properly.

CMPF

The CMPF (Change management process framework) is the result of the attempt of the above mentioned inhomogeneous group to join various perspectives and fields of knowledge in a unique framework. The main basis of this difficult work is the consideration given to the *people* within the organisations.

Under the commonly perceived necessity to give relevance to people and follow the sociotechnical design procedures, there are different options how to take into account all the points of view. There is a remaining risk to add approaches more than to integrate them, bringing to the potential dispersion of knowledge more then enrichment. The idea of the consortium is twofold: On one hand it focuses on a *systematic technological and organisational change* as the most relevant process in which different approaches and tools converge. On the other hand it provides a collection of structured, problem and solution based, *multidisciplinary case-studies*.

The different approaches are assembled in the CMPF:

- Consultant´s approaches give relevance to management of change and to conditions for successful projects.

- Psychologist´s approaches give relevance to the involvement of people and to the human requirements both for change and for solutions (stress analysis).

- IT people give relevance to the human-computer interaction and its improvement within the change, also to the performance aspect of information handling.

- Unions give relevance to the risks for the employees associated with change.

The change management process framework is a stepwise structure in which different aspects are relevant with different weight in different phases.

Furthermore, within the CMPF different points of view converge in considering what should be done in order to drive a change in which human actors are empowered. People empowerment within and through the managed change process is a further common issue. It ensures that the change will really happen, avoiding the risk of the 'tutorship effect' of several changes, in which the change seems to take place superficially but is driven only by the intense control and tutorship of the project managing team or external consultants. It quickly collapses when tutorship ends. Following the CMPF gives more value to the individual. It leads to a better level of people well being. It allows, pushes or requires technological improvements.

Selectable Tools

Some of the users are inexperienced and need orientation what to consider and how to proceed, some are experienced and look for detailed help for a particular task within an change process.

The structural approach to that requirement is the collection of selectable analysis and design tools within the framework and the different basic modes of *guided tour* or *free browse* within the CMPF.

Case study library (CSL)

A small remaining risks of the QUALIT approach is to work out a huge tool (the CMPF) that aims to cover all named aspects to some extent but will not be used by larger parts of the inhomogeneous bunch of potential users.

Therefore the consortium gives importance to the availability of a structured, problem and solution based *collection of case-studies* that can be used in a very individual way. The case-studies provide a multidisciplinary way to tell the stories of organisational change, underlining both the problems and the critical success factors in each of the points of view. The result is the idea of the Case Study Library (CSL) (see Schäl in these proceedings) accessible by wide area networks as well.

6. OPEN POINTS

The traditional way of a company to use external support for a restructuring is to use books/video films, to attend workshops, talk to other companies or hire a consultant/university. This means a choice between a cheap, selfguided and risky way and a expensive guided way. The QUALIT Environment with framework, tools and case study library aims at providing a combination of limited consultancy and free use of tools.

It is still not finally clear and proved to which extend and for which applications this QUALIT Environment will be used primarily in the work environment of managers, consultants, planners, workers and worker representatives or public sector and teaching.

Still many managers dont`t like to use tools at all or follow any guidelines.

It is foreseen in long term that for each of the different steps of a change process several appropriate tools for beginners and experts will be offered. Nevertheless in this phase of the project it is not finally decided which additional external tools should be put into the toolbox. Recommendations from other experts world-wide are welcomed.

7 . CONCLUSION

Up to recent years management disciplines did not give much attention to the human factor or better to human actors. Increasingly challenging international business environment forces the companies to deal with continuous innovation in products and services, in technology, in organisation, in competencies and to manage the interaction of all these innovation dimensions. Existing environment and today's ability, behaviour and self esteem of staff makes it difficult to cope with these challenges.

Most of the failures in innovation come from the lack of capability to manage the exploitation of available solutions, the interaction between what has been innovated and internal staff who is often not considered enough as actor of the change. It becomes clearer that market scouting capabilities, self managing, capability to understand one's own strong points in the relation with the market, capability to participate in the evolution of one's individual competence are more and more required at different levels in the company. People empowerment and QWL are no

more a philosophical and ethical concern, but an important management level.

Looking ahead, it is expected that the use of multimedia PCs will still increase and more networks will be used. This makes it more likely that a multimedia tool like the QUALIT Environment will be used. It is expected that more planning skills will be available and the use of advanced tools will be possible for more people.

It is expected that matters of people empowerment, good quality of working life and good workplaces will receive stronger emphasis during the next years and therefore a tool that provides this aspect is more and more desirable.

8. REFERENCES

Butera, F. (1988). Joint design of technology, organisation and people growth. In: *Proceedings of the Joint design of technology, organisation and people growth conference*, Istituto RSO Milano,

Davis, L. and A.B. Cherns (1975). The Quality of Working Life, Vol. 2. The Free Press, New York.

Emery, F.E. (1967) The nine step model. Paper presented at the international meeting on sociotechnical systems. Lincoln, England.

Emery F.E. (Ed.) (1981). Systems Thinking. Penguin Books, Harmondworth.

Emery, F.E. (1993). The nine step Model. In: Trist, E. and H. Murray. *The Social Engagement of Social Science, vol.II: The Socio-Technical Perspective,* University of Pennsylvania Press, Philadelphia.

Liedtke, P.; Roessiger, U.; Spur, G.; Albrecht, R; Heisig, P. (1995). Gestaltung ganzheitlicher Arbeitsabläufe. ZwF 90 Vol. 3, pp104-107.

Mertins, K.; Jochem, R.; Jäkel, F.W. (1995) A Tool for Object-Oriented Modelling and Analysis of Business Processes. Proceedings of the Conference: *"Co-operation in Manufacturing: CIM at Work"*, Eindhoven, Netherlands.

Mertins, K.; Schallock, B; Carbon, M. (1993) Production Management Software Suitable for Group work. In: *Human-Computer Interaction: Applications and Case Studies*; Proceedings of the Fifth International Conference on Human-Computer Interaction, Elsevier, Amsterdam Vol. 1.

Schallock, B. (1992) Skill enhancing shop floor structures. In: *Ergonomics of Hybrid Automated Systems III;* Proceedings of the Third International Conference on Human Aspects of Advanced Manufacturing and Hybrid Automation. Gelsenkirchen, Germany, August 26-28, 1992.

Ulich, E (1992). Arbeitspsychologie. Verlag der Fachvereine. Zürich 1992.

THE QUALIT CASE STUDY LIBRARY TO ENCOURAGE ORGANIZATIONAL CHANGE PROCESSES AND THE IMPROVEMENT OF WORK CONDITIONS

Olivier de Polignac[1], Peter Heisig[2],
Christel Kemke[3], Thomas Schäl[4]

[1] Cap Gemini Innovation, Chemin du Vieux Chene 7, F-38942 Meylan C.
[2] FhG-IPK, Pascalstrasse 8-9, D-10587 Berlin
[3] University College Dublin , Dpt. Computer Science, Belfield, Dublin 4
[4] RSO SpA, Via Leopardi 1, I-20123 Milano
[4] Multimedia Lab, Università di Siena, Via del Giglio 14, I-53100 Siena

Abstract: This paper presents the development of a Case Study Library in Esprit project 8162 Qualit. The scope of the CSL is to support decision makers (e.g., managers, designers, planners, workers representatives) in private and public organisations concerning the management of change processes. The CSL includes cases which illustrate solutions and good practice for improving quality of working life and the empowerment of people in such change processes. According to user requirements, the CSL is very much problem and solution oriented in finding relevant cases for the joint design of organisation, technology and people empowerment.

Keywords: case study, quality of working life (QWL), people empowerment, change management, socio-technical system, Qualit

1. INTRODUCTION

1.1. The Esprit Project 8162 Qualit

Qualit (Quality Assessment of Living with Information Technology) is an Esprit Project to develop a conceptual framework for the examination of Quality of Working Life (QWL) issues in Information Technology introduction and use and the change management processes involved. The aim of the Qualit Project is to help a range of users, such as human resource managers, IT project managers, IT system designers and union representatives in the diffusion and adoption of Information Technology. This support will be developed as a decision support system (DSS) addressing consultancy purposes, and an educational tool addressing training purposes. A library of documented case studies will complete this set of Qualit tools. This paper describes the Qualit Case Study Library (CSL) (Schäl, 1995).

1.2. The Qualit Case Study Library

The CSL is mainly a database of case studies related to people empowerment and the improvement of Quality of Working Life in projects of organisational and technical change. However, this collection of cases could also be seen as a *support system* for policy and decision makers in private and public organisations concerning the understanding of change processes. Typical users are managers, designers, workers representatives, etc. Such users are faced with situations in which they have to find a solution to a problem at hand. Therefore the CSL search modality reflects the user interests in problems, goals and solutions. In this sense the CSL is not a *decision support system* (DSS) for suggesting good decisions, but a collection of experiences for giving heuristic frames of possible actions. Dörner (1987) explains three types of problem solving (analytic, synthetic and dialectic problem solving) which can be applied to the mentioned decision processes for organisational change. The use of the CSL should be seen mainly in the area of dialectic problem solving. Dialectic problems are characterised by the fact that the desired final state is not known. It will become clear only during the problem solving process (Bitzer, 1991). The decision process is thus characterised by the understanding of the own problems (problem setting), the comparison with other realities (cases) and the capacity to link case studies to the own situation (dialectic problem solving). For this reason, the CSL is context, problem, goal and solution oriented in its structure.

The CSL could also be used by the staff of such decision makers to prepare the information needed as an input to the decision process. Direct and indirect users benefit from the CSL for gaining knowledge on the domain of Quality of Working Life (QWL) or to sustain an argument for the benefits of QWL dimensions in design projects, or social development in general. Their needed arguments for sustaining their ideas are supported by the documented examples of case studies and the general framework of Qualit as explained in the CSL.

The CSL includes cases which illustrate solutions and good practice for improving QWL and the empowerment of people. For each case included in the CSL, information is given about the various steps in a process of change according to the *change management process framework* (CMPF) developed in Qualit (Butera, 1995; 1995b; Butera & Coppola., 1995). Problems, objectives, constraints, solutions, change management process, results, evaluation, etc. are retrievable by the CSL users.

The basis of the CSL are projects conducted by organisations of the Qualit consortium and third parties which applied the Qualit model for their in-house projects. This knowledge base of Qualit cases is enriched by cases from the literature explaining successful projects of organisational or technological change which are related to people empowerment and the improvement of QWL.

Unsuccessful cases are rarely reported due to various reasons. Anyway, negative examples for not improving QWL or shortcomings in change management will be included in the CSL as far as reported by attendable sources (personal experience of Qualit partners, published material).

The scope of the CSL is to disseminate the Qualit project and its results, and to promote the idea of Qualit in general as being achieved mainly by the empowerment of people in processes of organisational and technological change.

2. USER REQUIREMENTS AND OTHER EXPERIENCES WITH CASE STUDIES

2.1. The importance of case studies

The uncertainties about outcomes and consequences, which are related to every change process could hinder companies to introduce radical organisational changes. Therefore managers, planners and shop stewards are looking for experiences in other companies comparable to their own. They are actually searching for case studies.

Normally, they are looking for relevant references in literature. First, they will attend conferences and workshops, where companies will present their experiences; they will visit factories to get a concrete idea of the results of changes in organisation, business processes and technology and the related achievements in competitiveness and employment. Their main purpose is to get some knowledge about the procedures and methods applied and the problems solved during the change process, in short: to learn about change management by looking through existing experiences.

2.2. Experiences in the German Program on Work and Technology

In Germany, since the mid 70s, the Federal Government supports the research, development and implementation of modern, efficient and human-oriented forms of work organisation and technology within the program *Humanisation of work*, actually called *Work and Technology*. The approach and the results are documented and published to disseminate the acquired knowledge to other potential user companies. Recently the dissemination of a number of projects is supported by audio-visual material, like video tapes. This approach fits especially the requirements of shop floor members, which are more familiar with this form of presentation. In spite of these huge efforts, the users are not very glad with these cases, because of the time consumed by search and acquisition. In addition, these documents are quite voluminous and oriented more to academics than to decision makers and planners. Nevertheless, there have been some intentions to condense this knowledge , i.e., in the *Handbook of human-centred CIM-Design* (Mertins & Schallock, 1991).

2.3. Case studies on the improvement of living and working conditions

The *European Foundation for the Improvement of Living and Working Conditions* in Dublin is collecting since a long time a variety of case studies about the introduction of new technologies and their effects on economic variables and quality of working life dimensions (e.g., Butera *et al.*, 1990).

The European Foundation for the Improvement of Living and Working Conditions aims at assisting in the effort to improve the working conditions of the citizens of the member states of the European Community, mainly by organising conferences and seminars, the conclusion of research contracts, and the dissemination of information.

In the opening remarks of the Fourth European Ecology of Work Conference (EEWC, 1995), F.-J. Kador stated the importance of case studies: „No general receipt can be given how to design successful work organisation. And so one way has proved to be a good one: to learn by the practical experiences of others to find one's own best practical way."

Over the past 12 years the European Foundation has commissioned several case studies concerning the implications of the introduction of new technology for the Quality of Working Life. The objectives of these case studies were to determine how new technology can be most effectively implemented and used to the best advantage in terms of employees, customers and the public at large. These case studies have been conducted in virtually by all the member states, in a range of working environments - from engineering and other manufacturing through banking and insurance to retailing and the public sector. The case studies also looked at the role of new technology in production, in process operations and in the office. The Foundation's research has looked at the implications of new technology from the viewpoints of managers, workers and customers (Ryan *et al.*, 1994).

2.4. Other European experiences

Several European projects have taken case studies as an important issue for their research agenda, e.g., Discovery, Hermes and FAST Monitor (Brandt, 1990).

Other experience on case studies are reported in Italy. A case study library has been implemented by the Documentation Centres of CEDOC (Regione Toscana), CID (Regione Liguria), SEDI (Regione Emilia Romagna).

2.5. User requirements for using case studies

The requirements on case studies could be summarised according to the different potential users of the Qualit Case Study Library as follows:

Managers want to know whether a certain solution or path of restructuring has already been followed by other companies or whether they are pioneers with a risk. They like to see whether and which external help has been used in other cases. They like to learn about the economic results and the overall duration of projects. They want to become aware of the problems and their solutions other companies encountered while implementing a new technology or restructuring the organisation or business processes. Managers are trained to compare companies within branches.

Planners want to get a detailed view on the solution with specific technical information and costs. Technical specifications consist of technologies, machines, robots, layouts, software (applications, tools). Planners are trained to compare the detailed circumstances of the application of technology. They often like to hear exactly who delivered what particular device at what price to meet the requirements. Often, this information is not publicly available in case descriptions but only by personal visits. Planners should, however, look more at the methods being used than they do today and they should ask less for technical details.

Worker representatives are normally interested in the general changes and in chances and risks that are linked with different options concerning the Quality of Working Life dimensions. This interest is often determined by the legal rights of workers representatives to participate in the decision-making process, by the actual power relations in the company and by the individual capabilities and knowledge of the per-

son. The requirements expressed by workers representatives in Denmark and Germany for the Qualit CSL are reported in de Polignac *et al.* (1995)

3. THEORETICAL BACKGROUND

3.1. Case Study research

Case study research (e.g., Yin, 1993) is an essential form of social science inquiry. The use of case studies is a method of choice when the phenomenon under study is not readily distinguishable from its context. Such a phenomenon is the improvement of quality of working life in the CSL. Sometimes the definition of a change management project may be problematic, as in determining when the project started or ended - an example of a complex interaction between the phenomenon and its (temporal) context.

The inclusion of the context as a major part of a study, however, creates distinctive technical challenges. First, the richness of the context means that the ensuing study will likely have more variables than data points. Second, the richness means that the study cannot rely on a single data collection method, but will likely need multiple sources of evidence.

Business schools have an increasing interest in case studies as a teaching tool. The *Harvard case method* has long been famous for its benefits in exposing students to hypothetical, real-life situations for teaching purposes. Other business schools and departments in other disciplines have similarly used the case study teaching method to good advantage.

Recently, business schools have witnessed a resurgence of interest in the case study, but as a research strategy. Although business faculty and students have always conducted empirical research projects, such projects are becoming increasingly common, and the case study has become valued as one of the possible research strategies. The Harvard Business School has been active in this area. Also in Europe case study based teaching and research is becoming popular, e.g. at the Aarhus School of Business.

3.2. Unit of Analysis

No issue is more important than defining the unit of analysis. 'What is my case' is the question to be posed by doing case studies. Because case studies permit to collect data from many perspectives - and for time periods of undetermined duration - one must clearly define the unit of analysis. The unit of analysis has a critical significance, because the findings of a case study pertain to the specific theoretical propositions about the defined unit of analysis. These propositions are the means for generalising the findings of the case study, e.g., to similar cases focusing on the same unit of analysis. Thus, the entire design of the case study as well as its potential theoretical significance is heavily dominated by the way the unit of analysis is defined.

The CSL requires thus from the case writer to define clearly the unit of analysis. The CSL is concerned about improving the Quality of Working Life through people empowerment. This defines the subject of the CSL. In addition, the CSL wants to document the joint design of technology, organisation and people empowerment. This is reflected in the structure of the CSL. Last, but not least, the boundary of analysis is formalised in the CSL as the area of reorganisation. With these pre-set values, it should be easy for the case writer to define the unit of analysis for his specific case study to be included in the CSL.

3.3. Case study types

Case study research can be based on three main types: exploratory, descriptive and explanatory. Reference to the use of *theory* usually involve the formation of hypothesis of cause-effect relationships. These theories would therefore be considered relevant to explanatory case studies. The CSL is, however, more a collection of descriptive case studies.

Theories, however, can also be important for descriptive case studies. A descriptive theory is not an expression of a cause-effect relationship. Rather, a descriptive theory covers the scope and depth of the object (the case) being described. If the subject is to describe an organisation undergoing change, where should the description of the case study start, and where should it end? What should be included, and what might be excluded? The criteria used to answer these questions would represent the theory of what needs to be described. The more thoughtful the theory, the better the descriptive case study will be. The collection of information about everything does not work, and the case writer without a descriptive theory will soon encounter enormous problems in limiting the scope of the study.

Therefore the CSL gives an outline to the case writer on what should be described in the case study. The descriptive theory is the structure of the CSL. The case writer has to decide the detail of description, but is very much forced on the topics of description. Therefore the CSL has a case writer guideline for writing case studies.

4. CASE-BASED REASONING

Case-based Reasoning (CBR) is a relatively new methodology based on the idea of using prior experiences (cases) in order to solve current problems (Kolodner, 1993).

A CBR-system comprises mainly a case library and a set of access procedures which, for example, search the library for the best stored cases matching a given problem description (e.g., the CSL user's problem in a current project). The basis for this search is an appropriate indexing of stored cases. CBR provides not only a selection of best matching cases for a given problem but furthermore an interpretation of new cases and an adaptation and an evaluation and repair of the selected stored cases to make these suitable for solving the problem at hand.

CBR has aspects of cognitive modelling but at the same time aims at an automated processing by computers. Several prototypical CBR-systems have been developed so far.

In the context of the QUALIT-project CBR is of great interest because it can provide a basis for retrieval, matching, and adaptation of known cases for training purposes as well as for actual use in a consulting context, because cases provide suggestions for solutions to problems as well as a context for understanding and assessing a situation.

5. HOW TO WRITE AND ORGANIZE A CASE STUDY?

5.1. The structure of case studies

According to the proposed Qualit Change Management Process Framework, (Butera & Coppola, 1995) 10 main areas have been identified to structure a case study. The main case study is introduced by an abstract and an introducing situation scanning, being closed by an evaluation. Furthermore, any useful references are given in a final block. Thus the CSL structure foresees 10 blocks:

(1) Abstract gives a short description of the main aspects of the case study.

(2) Situation Scanning for the description of the general background of the case, like the competitive situation, the products involved, the history of the organisation, the cultural settings, in short: all relevant information to understand the subsequent case study. All country specific aspects should be described and made clear in this block.

(3) Needs for change (goal and problem setting) for the presentation of the motivation for change, concrete problem setting and diagnosis of causes, the goals of stakeholders and the agreed project goals by the organisation (business, organisational, technology, social aspects) and the constraints for the organisation.

(4) Architectural design describes the resulting architecture of the socio-technical system and the configuration of its components.

(5) Evaluation of alternatives describes the options which were worked out by the planning groups, their evaluation regarding the project goals and the discussion of these options in respect to the adopted solution.

(6) Detailed Design and implementation describes the concrete solution design and implementation, including activities of training and diffusion of the solution in the organisation.

(7) Change Management describes the organisation of the project work (steering committee, project team), the involvement of different stakeholders and external experts, the training activities carried out for the involved personnel, the industrial relations and the implementation strategy.

(8) Applied Methods presents the methods and tools which were applied during the change management process and the evaluation of the advantages and disadvantages of these tools.

(9) Evaluation documents the implemented solution with the variation introduced during the implementation phases, the achieved results in economic and so-

cial terms as well as a final evaluation and discussion of the case study.

(10) References describe links to persons and information sources (e.g., literature).

The single case studies might not be complete for all 10 building blocks. This counts especially for cases from literature, because they have not been written and documented within the Qualit framework. The CSL has its most beneficial state, if there would be as many significant case studies as possible. Therefore, also incomplete case studies might be included in the CSL. The CSL could be organised for such incomplete or very synthetic cases by more abstract (and short) descriptions. The decision on the inclusion of incomplete case studies is made by the CSL editor

5.2. Contents of the CSL

The CSL is an environment which helps the user to find experiences from others relevant for processes of change in various organisations. The Qualit CSL is not a collection of any case, but only of cases relevant for their implications on people empowerment and Quality of Working Life dimensions. Each case study in the CSL has the following general structure:

- a general one page abstract or figure which makes the main points clear to the reader. This description should be a good level of abstraction for browsing through cases.
- a circa 5000 words detailed case description in 9 blocks
- concepts and keywords associated to single block and to case study on the whole.
- references to other cases, persons or bibliography for further details or additional information

6. HOW TO USE THE CSL?

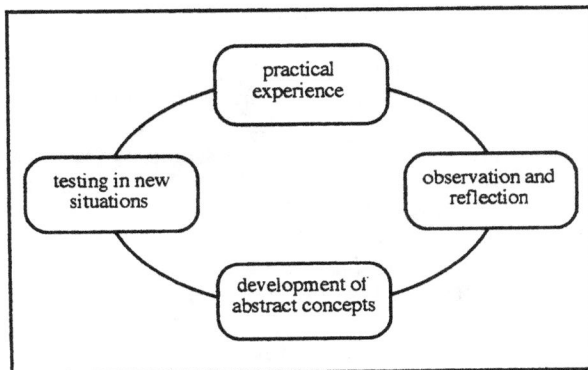

Fig. 1:   The cyclic learning process (Kolb et al., 1974)

The CSL can be seen in the cyclic model of learning as developed by Kolb et al. (1974). The scope is to develop new attitudes and reasoning behaviour as a basis for change processes in organisations. The critical point is to take the experience and knowledge of organisations described in a case study as a starting point to allow for new experiences. The scope of the CSL is to transfer experiences and their abstract concepts into other organisations. This should then lead

in a new project of change to a testing of concepts from the CMPF and other cases in new situations.

The research for a case or a number of cases which matches the interest of the user will be supported by defining the search criteria. These criteria are organised in a hierarchy to allow for bright or narrow queries. In this way, the user defines the area of interest.

In addition to information retrieval mechanisms for finding precise answers to user questions, expert knowledge to guide the user in the exploration of the CSL will be provided, too. This expert knowledge will be implemented by linking the different case studies by defined paths, i.e., in the guided tour.

The CSL has a built-in guidance to help users navigate from their defined problems to the most useful cases (and solutions). These issues are the starting point for specifying the search mechanisms of the CSL.

6.1. Case-based search

Case-based search concentrates on some concrete identifiers of an organisation, e.g., company name, economic sector, product/service class, etc. The user can define a more or less fine filter for finding cases. This search modality is generally used when the user wants to find similar realities to his own or has a certain case in mind. This is a multi-criteria search which means that all specified criteria have to be commonly matched with the case.

6.2. Concept-based search

The tree structured search by concepts is based on the framework developed in Qualit. The tree structure reflects both, the STS and QWL model. In addition, a third model for business factors is developed. The resulting concepts are associated to case studies and the single blocks. The resulting hierarchical reasoning is applied to problem, goal and solution-oriented search mechanisms.

6.3. Keyword search

The user can opt for an unstructured keyword-oriented search on the CSL. The user selects the keywords from a closed list of implemented keywords in the CSL.

6.4. Full-text search

A full-text search facility is implemented on the CSL for searching words or phrases.

6.5. Browsing facilities

Once found a series of case studies or a single case study, the CSL can be consulted on two levels of detail: a synthetic description for browsing through the CSL (e.g., looking only at the abstract) or the study level of detailed and complete cases (e.g., looking at all blocks of the case studies).

6.6 Guided tour through the CSL

The guided tour is a help facility provided by the CSL editor on cases and concepts.

7. TECHNICAL OPTIONS AND TOOLS FOR REALIZING THE CSL

The Qualit consortium evaluated several technical options for the implementation of the CSL. In order to develop the Qualit Case Study Library, three different possible technologies have been identified and analysed:

• the World Wide Web;

• a database management system on a PC;

• Lotus Notes.

These three approaches represent the main streams of what is today available on the market. For each technical possibility, the advantages and disadvantages associated with each of them for building the CSL have been evaluated. Furthermore, techniques for case-based reasoning might be used.

The choice of WorldWideWeb (WWW) for building the first prototype was only partially sufficient for dissemination and visibility. The consortium understood its limit from the fact that the Qualit observatorium has to deal, in addition with research, with potential industrial users who will not use Internet in the near future. Until Internet is not fully accepted and used by industry, these users will have difficulties to know of the CSL and to connect to the CSL. Therefore the consortium decided to develop a PC-oriented CSL as a software tool which can be used possibly in industry as well as in reserach, easily disseminated and up-dated by various technologies.

The choice of a PC-based technology for the CSl allows also for the development of a public released CSL by the *Qualit Observatorium for Quality of Working Life and People Empowerment* which can be integrated with a confidential local CSL. This should be interesting for organisation which want to document their internal projects without necessarily making them public to the observatorium.

8. ACKNOWLEDGEMENTS

Many of our colleagues on the Qualit project and within our institutions have contributed to the ideas and theoretic groundings of this paper, but they are too many to be named individually. They know who they are, however, and we thank them all. We also gratefully acknowledge the assistance given by our institutions and the financial support for this work afforded by the Commission of the European Communities under the Esprit Programme.

REFERENCES

Bitzer, A. (1991). *Beteiligungsqualifizierung zur Gestaltung von technischen und organisatorischen Innovationen.* VDI-Verlag, Düsseldorf

Brandt, D. (1990). *Advanced Experiences - European Case Studies on Anthropocentric Production Systems.* Prepared on behalf of FAST for the International Conference 'Production Technologies, Social Organisation and Competitiveness', Gelsenkirchen, 24-27 September

Butera, F. (1995). *Human Oriented Management of Change - A Conceptual Model.* 6th International Conference on Human-Computer Interaction, 9-14 July, Tokyo

Butera, F. (1995b). *Quality of Working Life and Empowerment in Re-Engineering and Continuous Improvements of Network Organizations Supported by Information Technologies.* Working paper 2/1995, Istituto RSO, Milano

Butera, F.; Coppola, B. (1995). *Architecture of the Change Management Process Framework (CMPF) - Concepts, models and specifications for engineering the Qualit services.* Qualit Technical Working Paper W11A6-0R, Istituto RSO, Milano

Butera, F.; Di Martino, V.; Köhler, E. (1990). *Technological Development and the Improvement of Living and Working Conditions: Options for the Future.* Kogan Page Ltd.

Dörner, D. (1976). *Problemlösen als Informationsverarbeitung.* Kohlhammer, Stuttgart/Berlin/Köln/Mainz

EEWC (1995). *Innovative Work Organisation in Europe and North America.* The Fourth European Ecology of Work Conference, Dublin, 9-12 May 1995.

Kolodner, J. (1993). *Case-based reasoning.* Morgan Kaufmann

Mertins, K.; Schallock, B. (Hrsg.) (1991). *Handbuch der humanen CIM-Gestaltung.* Abschlußbericht des AuT-Projektes "Aufbereitung von HdA-Gestaltungswissen für das Beratungsangebot der CIM-TT-Stellen", FhG-IPK, Berlin

Ryan, G.; Downing, S.; McNeive, A. (1994). *Organisations Concerned with Quality of Working Life.* QUALIT Technical Working Paper T33B1-0, University College Dublin, Dublin

Schäl, T. (1995). *Observatorium Concept: The Qualit Case Study Library.* Qualit Technical Working Paper W16A3-3R, Istituto RSO and Università di Siena, Milano/Roma/Siena

Yin, R.K. (1993). *Applications of Case Study Reserach.* Sage Publications, Newbury Park, California

ORGANIC GROWTH VIA DEVELOPMENT:
THE EARLY LIFE-CYCLE

**Rajko Milovanovic*, Sinclair Stockman*, Mark Norris*,
Danilo Obradovic**, Miroslav Hajdukovic****

* British Telecom, ** University of Novi Sad

Abstract: The paper presents commercially-driven steering of the early lifecycle when developing software-rich systems. The approach is geared towards enhancing the value of the organization's product portfolio, and in parallel towards enhancing its intellectual capital. The implications of exercising the IT architecture during this process are explicitly taken into account.

Keywords: Architectures, Domain analysis, Economics, Product strategy, Requirements analysis, Semantic networks, Software specification, Systems engineering, Systems methodology

1. INTRODUCTION

Significant long-term expansion of a commercial organization's financial results is linked with growth of its portfolio of products and services. Growth occurs in principle along two routes: via mergers with, or acquisitions of other organizations - together with their portfolios; or via organic growth. Organic growth of the portfolio implies either purchasing new additions to the portfolio, for example as licenses, or developing those additions - in house or externally.

This paper concentrates on the early lifecycle for the last option, development. Specific attention is given to developing portfolio-compatible software-rich products and services associated with the information sector (Mandel 1994). Development of additions to the portfolio considered here includes both "pure development" and systems integration.

The paper starts with the economics of the early lifecycle, and then shows how economic implications find its tangible expression in the requirements and design stages of building a service, a system, and a configuration item. Architectural compliance is added as the omnipresent constraint.

2. ECONOMICS OF THE EARLY LIFECYCLE

For software-rich systems, the parallel to micro and macro economics is achieved by considering the *product* level economic impact, and the *company/portfolio* level economic impact. On the product level, if finding and fixing a software problem at requirements/early design stage costs $1, doing the same after delivery costs $100 (Boehm, 1987). Davis et al. (1993) even cite $200 as a cost for the later. Boehm also states that the cost of maintenance outstrips the cost of development by 2:1. Enhancements contained within maintenance should be weighted against setting the requirements out properly in the first instance.

Taking the above considerations into account, we can now tackle a more detailed analysis of the contribution that the early lifecycle (loosely, requirements and architectural design) makes to the overall lifecycle cost of the system. Since terminology as well as software measurement results (or, more frequently, absence of those) vary enormously between different organizations, ratios in Table 1 might be considered as very rough approximations. The reader could find it interesting to insert his/her numbers.

For a s/w product	consists of	with cost distribution	% of cost due to early life-cycle	early lifecycle share in the cost, %
life-cycle	develop ment	1/3		
	maintena nce	2/3		
mainte nance	under-standing	1/2	80	27
	rest	1/2	5	2
develo pment	proper develop ment	1/3	30	3
	rework	1/3	80	9
	quality control, QC	1/3		
QC in dev't	under-standing	1/2	80	4
	rest	1/2	5	0
life-cycle		100%		45%

Table 1 - Contribution of the early lifecycle's to the whole lifecycle cost

Development proper costs only around 11% of the lifecycle. The early lifecycle is just its part, making up less than 7% of the total at best (Boehm 1987). With its bearing on the overall lifecycle cost (45% - see last row of Table 1) being amplified six or more times, impact of the early lifecycle quality on the product's destiny also has very tangible economic consequences.

These consequences are actually even more serious on the portfolio/company level. In the 1995 acquisition of Lotus by IBM, the purchase price was almost double the value of Lotus shares before the IBM move. That value of shares was significantly higher than the value of assets held by Lotus. Accountants will resolve this formally by including a goodwill chunk in IBM's next annual report. But can the pre-purchase nature of "goodwill" of the acquisition targets in soft-ware-rich industry be equated with the nature of goodwill for the downtown subway-exit cigarette kiosk?

Hamel and Prahalad (1994) and - to some degree - Samuelson and Nordhaus (1992) disagree with drawing parallels between the two. The distinction held by organizations in growth, high tech industries is roughly equivalent to the intellectual capital (Hamel and Prahalad 1994, Stewart 1994). This distinction is finding tan-

gible expression in valuations of the business against the economic value added (EVA - see, Tully 1993). The purpose of such valuation is not to sell the business at higher price; rather it is to always know its real, and not just account-ant's value. The difference between these two values is significant, and has a broad catchment: Even for such a "hardware" oriented global group like Siemens (with workforce of 377,000), over 50% of its added value now comes from software engineering.

How does it all start, and how does it prog-ress? Commercial/business optimization tasks for initial links of the value add chain might roughly be presented by Table 2.

	constraint	goal
fundamental research	fixed (time / $ / resource)	deliver new in the area
applied re-search	deliver max possi-ble % of "loose requirements specification"	favourable (time / $ / resource, risk)
development	deliver prototype 100% to require-ments specification	min (time / $ / resource)
software development	deliver product 100% to require-ments specification	min (time / $ / resource)

Table 2 - Nature of optimisation in different stages of R&D

With relatively simple goal functions, at least on this level of abstraction, suboptimal achievement of the goal is very frequently due to poorly defined constraints. The table above clearly shows that the constraints are, except for the fundamental research, primarily spelled out in the requirements specification. This signifi-cance of requirements carries on in the engineer-ing area, Table 3.

	Engineering	Development
Priorities	Business > Technology	Technology > Business
Input	Req.Spec. + Config. Items	Req.Spec + Technology
Feasible solutions	Finite, a few	Infinite
Primarily relies on	"Rules"	"Knowledge"

Table 3 - Differences between engineering and develop-ment

3. REQUIREMENTS

Our industrial experience (Milovanovic et al 1995ap, 1995ag) has confirmed the need for the requirements specification to be fit for purpose (see Davis et al 1993) by being all of:

- **affordable**: in terms of money, time, resource - for development and / or operation
- **correct**: each statement is an actual requirement; statements are noise-free
- **coherent**: no two statements contradict each other; all statements form a unified whole; no statement contradicts with any other approved document
- **complete**: all requirements are written in the requirements specification document(s)
- **verifiable**: in the software product, by an objective finite cost-effective process that does not depend on specific humans
- **unambiguous**: each statement can be interpreted in one and only one way by anyone
- **precise**: numeric quantities used wherever possible; appropriate levels of precision used for all numeric quantities
- **traceable**: it is possible to link each statement with statement(s) of upstream and downstream products of the development life-cycle
- **usable**: all stakeholders can read requirements specification with minimum explanation
- **modifiable**: in order to effect a new / changed / deleted statement of a requirement, the document(s) are easily changed in one or few well defined places
- **achievable**: there can be at least one system that implements the requirements specification
- **concise**: as short as possible without adversely affecting any other requirments specification quality
- **design independent**: there is more than one design and implementation meeting the requirements specification
- **not redundant**: the same requirement is not recorded more than once
- **at the right level of abstraction / detail**: for all stakeholders
- **reusable**: parts of requirements specification can be easily adopted or adapted for use in another requirements specification
- **organized**: readers can easily locate necessary information; logical relationship among
- **adjacent**: correlation of sections is apparent
- **cross-referenced**: to different levels of abstraction for same requirement, or to requirements that depend on the current requirement / current requirement is dependent on

In the case of conflict, the first seven points typically take precedence.

Only requirements like these actually help organization to control the part of intellectual capital (related to software rich systems) which is not in the heads of its employees. Links with the portfolio are primarily achieved through architectural design of the system, and through design constraints based on the IT architecture. We have achieved quite profitable instantiations of these principles by expanding the SRS and associated DIDs of the U.S. MIL-STD-498 (and DOD-STD-2167A earlier), in different application domains, using different development methodologies. Particularly beneficial in these exercises has been the principle of minimal re-

quirements, and the principle of non-logical design.

The principle of minimal requirements relates both the verboseness of the requirements specification (i.e., the more terse the better), and to the seriousness of each individual requirement. The latter can be easily understood if one considers (computer simulation of) the travelling salesman problem. Many an earnest engineer would tend to specify "do all cities" unless explicitly driven to consider cost-benefit trade-offs of the seemingly least attractive 3.5% accuracy (see Table 4).

Cities	100K	100K	1M
Accuracy	0.75%	1.0%	3.5%
Computing time	7 months	2 days	3.5 hours

Table 4 - Travelling salesman - computer simulation

The principle of non-logical design tries to curb the most serious culprit for unsuccessful designs - logical design. With its roots in structured analysis and its derivatives, it has also evolved into "logical" establishment of classes in OO-geared designs. Development of software-rich (and, for that matter, any other) systems involves developers - but it also involves people from marketing, engineering, operations and manufacturing camps. Logical design is typically "logical" only to a subset of developers. Even they admit that it is actually about requirements and not about design; but the highy amplified (by at least 45% + 7%) damage has already been done.

4. SERVICE, SYSTEM, CONFIGURATION ITEM

Requirements actually occur in three different layers of abstraction, as the service requirements specification, as the system requirements specification, and as the configuration item requirements specification - with a common denominator being the domain model (see Fig 1).

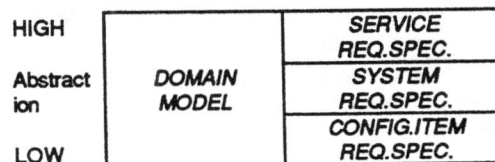

HIGH		SERVICE REQ.SPEC.
Abstraction	DOMAIN MODEL	SYSTEM REQ.SPEC.
		CONFIG.ITEM
LOW		REQ.SPEC.

Figure 1 - Domain model and different requirements specifications

Each stage the of requirements specification takes form of the "product specification" which must be easy reading for any stakeholder. Decoupling "what the product does" from "what is the domain in operates within" has been done by pulling out all domain-specific information into the domain model, whose tip of the iceberg typically looks like Fig 2.

(syntax readable from day one:)

Household *is* Address *and perhaps a* Master *and perhaps a* Mistress *and perhaps* Kids *and perhaps either* Dog *or* Cat *and* FurnitureItems

(syntax typically switched to:)

Household = Address + {Master}1 + {Mistress}1 + {Kid} + {[Dog | Cat]}1 + 0{FurnitureItem}

Figure 2 - Semantic network of the domain model

The nodes of this kind of the semantic network are further annotated, with more detail added as one steps from service to system and on to configuration items. Annotation has two primary aspects - associating all numerical data with concepts defined in the semantic network, and associating all relevant domain-specific algorithms with each concept. Example for the domain specific constraints or algorithm are *birth comes before death* in the public administration domain, or the (1) *comit to interaction* - e.g., lift the handset, (2) *target* - e.g., dial, (3) *interact* - e.g., speak, (4) *terminate interaction* - e.g., hang up sequence for the communications domain. This leaves for the capabilities of the configuration item to spell out only the product/system/application-specific algorithms. Automatic side benefit is clear layering and control of interaction of features of the system/product/service.

(The only remaining class of algorithms are design level algorithms. For example, the requirements might mention "Customer with largest Purchases in past FinancialYear". The design stage would supply a concrete algorithm which defines how the largest one is actually located.)

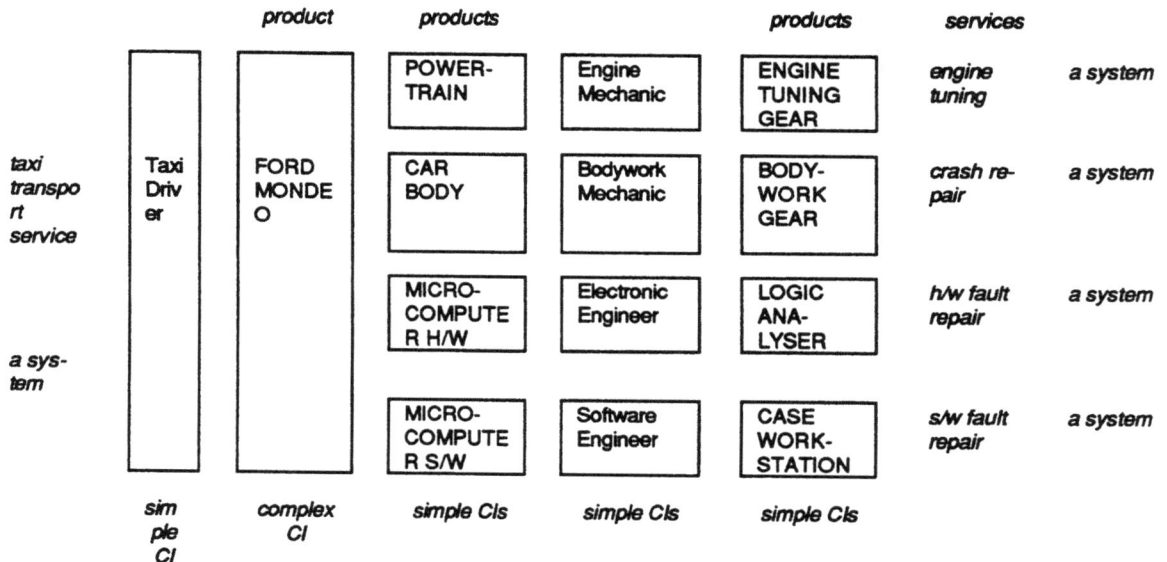

	product	products		products	services	
		POWER-TRAIN	Engine Mechanic	ENGINE TUNING GEAR	*engine tuning*	*a system*
taxi transport service	Taxi Driver	FORD MONDEO	CAR BODY	Bodywork Mechanic	BODY-WORK GEAR	*crash repair*	*a system*
			MICRO-COMPUTER H/W	Electronic Engineer	LOGIC ANA-LYSER	*h/w fault repair*	*a system*
a system			MICRO-COMPUTER S/W	Software Engineer	CASE WORK-STATION	*s/w fault repair*	*a system*
	simple CI	*complex CI*	*simple CIs*	*simple CIs*	*simple CIs*	

Figure 3 - Service, system and configuration items - a car example

A domain model like this immediately strikes the chord with all stakeholders - operations, marketing, engineering, development, manufacturing - for it only uses the words they live with. Its prime advantage over ERD is terseness, which helps dispense with the frequently noisy names of many relations like "owns / is owned by". This basic vocabulary is applicable to any of multitude of services / systems / applications related to that domain; and each one of these can be implemented in a multitude of technologies.

The view on any particular application - or rather, the service and the system whose operation provides the service - is then fairly straightforward (Fig 3).

Further, one can then view the development lifecycle as specifying services, then specifying a configuration for each separate service, then specifying a system, then doing design of the system and its partitioning in the (humans, hardware, software) configuration items, developing (including for reuse) or reusing each hardware or software item and training each

128

human, and progressing towards building the configurations and providing the services (see Fig 4).

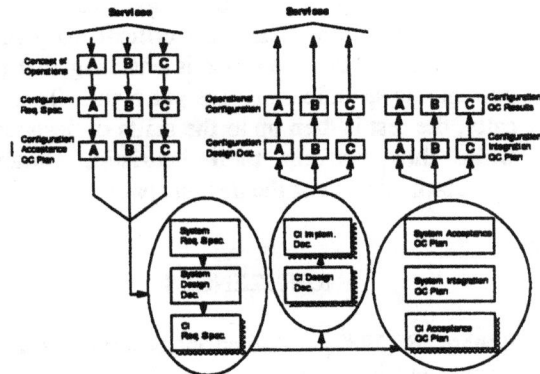

Figure 4 - Service- and configuration item-centered development and quality control

The most appropriate model for any configuration of the system then turns out to be the graph of configuration items and their interfaces.

The key business factor to note is exploitation of the economies of scale by collapsing the configurations into the system: A Ford car, for example, might come in a dozen different configurations, but they all have one of three or four powertrain configuration items. A configuration item is the atom known to non-technical management typically just by name and its commercial value to the business. Constructing a multitude of services upon a small set of these basic building blocks helps shift the customer perception of the external portfolio breadth without undue increase in costly internal portfolio - and with consequential shift of the profit/ROA peak to the right (see Fig 5).

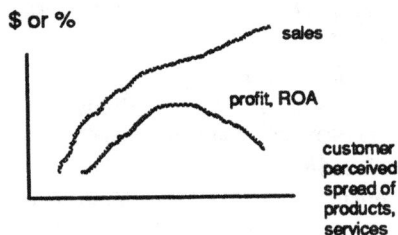

Figure 5 - Maximal sales/variety of products vs. maximal profit or ROA

5. DESIGN

The mechanics of splitting up the system into configuration items are purely business driven: the primary criteria are *business out-of-house* concerns (e.g., portfolio control, options control, marketing, pricing, product/service feature composition, phased introduction to market, contracting); the secondary criteria are *business in-house* concerns (e.g., logistics, configuration management, product evolution, component/product/program reuse, version control, available resource); and tertiary criteria are *engineering* concerns (e.g., modularity, cohesion, coupling).

This partitioning is one of the several design stages. At whatever level the design is done, commercial reality dictates that it is fit for purpose by

- further constraining the feasible (in terms of the requirements specification) set of implementation solutions so that the actual one is closer to satisfying the *product optimisation* criteria
- ensuring satisfaction of organisation's wider-than-product *business goals*
- validation of the *requirements specification*
- speeding up of pinpointing item(s) / aspect(s) to be tackled at each *maintenance* action
- facilitating of the *division of work* among participants of the implementation

6. IT ARCHITECTURE

Compatibility with the portfolio in technical terms essentially boils down to profitably enriching the IT architecture. For organizations whose IT architecture has a large common *platform* denominator (like some of utilities, telecommunications, military, financial sector organizations) this technical compatibility is an obvious constraint. For all other organizations it is a potential lever for achieving the competitive advantage. As far compatibility with the portfolio in marketing terms goes, software-rich systems do not differ significantly from their less hyped-up brethren.

Typical business oriented considerations of IT architecture, differ from those of the mostly-software-engineering camp (Garlan 1995, Perry 1992). While the models of the latter group might well be the reality of the marketplace in several years from now, today's commercial reality views IT architecture principally along these dimensions:

- styles, patterns, interfaces, standards, regulations, (shared) platforms
- common denominators of internal and external portfolio
- expressed as a set of design constraints for a service/product
- embodiment of the technical strategy of the business

Exercised properly, the IT architecture provides numerous benefits for the organization's "front office":

- maintenance cost cut across the homogenous portfolio
- price/cost cut due to reuse of large software lego blocks
- extended current and improved future competitive advantage
- portfolio conformance with other suppliers, customers
- homogenous portfolio of synergistic services/products
- leveraged impact of the portfolio by timing diffusion and lock-in
- explicit mapping of the portfolio - vs. of sum of products - to the market or its segment
- minimal internal portfolio, which can be more profitably combined into the maximal external one
- targeting technology use (see Table 5).

	Offensive Use	Defensive Use
Applied to Customers	VALUE ADD	LOCK-IN
Applied to Competitors	PRE-EMPT	BLOCK

Table 5 - Different purposes of IT architecture

Benefits for the "back office" are of equal relevance:

- right initial definition of a service/product
- longer time horizons of strategic technological evolution
- better coupling of internal technical capabilities with strategic business needs
- larger room for predictable expansion of an investment

The last bullet point is of particular relevance, for it helps steer the commercial assessment of investment in any individual development to explicitly take into account not only the NPV but also the value of options (Brealey and Myers 1991).

Implementing the architecture does not come free. Some of the costs involved are

- establishing, maintaining and evolving the definition of the architecture
- propagating design constraints to specifications of new services/products
- implementing architecture-based design constraints in each service/product
- verifying implementation of architectural constraints
- ensuring reusability of large software lego blocks

7. CONCLUSION

Building technically successful systems is typically a challenge welcomed by engineers.

Building commercially successful systems is almost always a challenge welcomed by businessman. Marriage between the two successes is rather necessary to help the organization compete for the future. The most critical early life-cycle item in business sense is ensuring that requirements specifications are put right. Technically, the rest is then up to the skills of designers while they perform their heuristic, variable granularity search of the design space.

REFERENCES

Boehm, B (1987): "Industrial Software Metrics Top Ten List", *IEEE Software*, Sept, p 84

Brealey R A, Myers S C (1991): PRINCIPLES OF CORPORATE FINANCE, McGraw-Hill

Davis A et al (1993): "Identifying and Measuring Qualiity in a Software Requirements Specification", PROC. 1ST INTERNATIONAL SOFTWARE METRICS SYMPOSIUM, IEEE CS Press

Garlan D, Perry D E (1995): "Introduction to the Special Issue on Software Architecture", *IEEE Transactions on Software Engineering*, 21:4, p 269

Hamel G, Prahalad C K (1994a): COMPETING FOR THE FUTURE, Harvard Business School Press, ISBN 0-87584-416-2

Mandel M J (1994): "The Spawning of a Third Sector: Information", *Business Week*, p 48, Nov 7

Milovanovic R, Stockman S G, Norris M (1995ap): "Standards Foundation for the Top Gun Design College", *Proceedings of the 2nd IEEE ISESS'95 International Software Engineering Standards Symposium, Montreal, Canada, August 21-25*

Milovanovic, R. (1995ag): "GLOBAL CURRENCY", BTTPU 2264, July

Perry D E, Wolf A L (1992): "Foundations for the Study of Sofware Architecture", *ACM SIGSOFT SEN*, 14:4, p 40

Samuelson P A, Nordhaus W D (1992): ECONOMICS, 14th ed., McGraw-Hill, ISBN 0-07-054879-X

Stewart T A (1994): "Your Company's Most Valuable Asset: Intellectual Capital", *Fortune*, p 28, October 3

Tully S (1993): "The Real Key to Creating Wealth", *Fortune*, p 34, September 30

DISCLAIMER

The views presented in this paper are those of its authors, not necessarily of their employers. For additional references send emal to the first author, rajko@ieee.org

Distributed Information Systems and Human Organization

Chikao Imamichi

Mitsubishi Electric Information Network Corporation

Abstract. Decentralized and distributed information systems is replacing traditional centralized information systems. This trend is called as down sizing and the most of concern is focused on architecture, functions and performance of the systems. But the intrinsic changes brought by down sizing reside in the other area; change of systems management, impact to user and operation organization, requirement of computer literacy for users. Analysis of these changes and solutions for both systems and organizations are discussed in this paper. "Skill Oriented System Design " is proposed as a solution.

Key Word. Communication computers applications; Decentralized control; Distributive data processing; Human factors

1. INTRODUCTION

The concept of distributed system appeared earlier in control field rather than in data processing field. DDC (Direct Digital Control) appeared in 1970s had raised the idea of distributed control. Rapid advance of micro processor technology in 1980s accelerated diffusion of distributed control system. In the data processing field, it was PC (Personal Computer) that enhanced distributed information processing combining the idea of end-user computing. After down-sizing started, distributed system means thousands of workstations and PCs linked with LAN (local area network) and WAN (Wide Area Network). Such large distributed system raised the problem of scale that had never appeared in DDC systems.

Such distributed systems require new management system that had not existed in the centralized systems, suitable organization, and skilled person for new functions. Most of the enterprises seem responding to this requirement correctly. But, there are few researches which focused on distributed system from the view point of human skill and organization.

2. A MODEL OF DISTRIBUTED SYSTEM

Down-sizing is general expression of a movement of replacing main frames by workstations or PCs. Data bases, application functions, which run on mainframes once, are allocated on the workstations, PCs, or higher performance small computers called as server.

The model of enterprise-wide distributed system is discussed in this section for further investigation of human skill and organization. The architecture of a distributed computer system has close relationship with the user's organization and its geographical distribution. The model enterprise is assumed to have a headquarters, several branch offices and a large number of local offices.

Centralized information system which had been typical for such enterprise is configured by one mainframe and dumb terminals connected to the mainframe by communication lines. The terminals were placed in every offices. (Fig. 1)

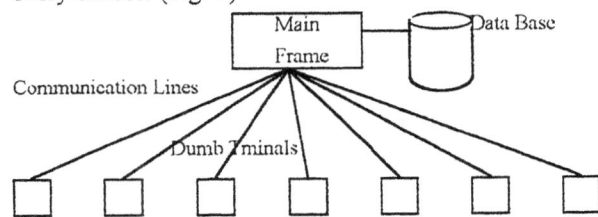

Fig. 1 Centralized Data Processing System

On the other hand, distributed systems has the following structure. Every offices including the headquarters has server / client systems connected by LAN. All these subsystems are connected by WAN which provide enterprise-wide data transfer and

information exchange. (Fig. 2)
There are enterprise-wide data-bases in headquarters, and local data-bases in branch offices. Local offices do not have data-base other than personal one. Application functions are allocated on the servers in the headquarters, branch offices and local offices.
Thus, workstations or PCs are not used as task processors. They are used as man-machine interface for application functions and personal processing like as table calculation / graphical presentation.

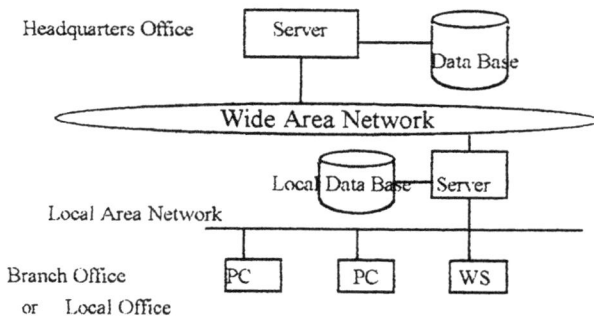

Fig. 2   Distributed   System

3. ANALYSIS OF CHANGES

Distributed systems caused the following three major changes in the aspect of human skills and organization.
Changes of systems management:
Distributed information systems depend heavily on communication network. The performance of the system and availability are affected by network. The management of network become so important as management of computer systems. One of the features of distributed systems is its flexibility. Configuration of the system is easy to change. That demands for real time change management after configuration modification. Distributed systems handle various kind of information as like as graphs, drawings, mails and even videos which demand for more sophisticated management.
Changes of organization:
The centralized system was managed and operated by IM (Information Management) section. When down-sizing has executed, the number of people working in IM section was reduced to half or even one third in parallel with hardware cost reduction. The remained IM section is ordinary posted in headquarters, but there is no such organization in branch or local offices. The

concept of "end user computing" is widely accepted and introduced into new systems. But they encountered the following problems.
1) Who is responsible for improvement of systems ? If each end user must improve the system, that may increase man-hour spending on computer systems than before. More over, ordinary end user does not have enough knowledge about computer, communication, and software systems.
2) Who is responsible for troubles of the systems ? If systems users must solve troubles in widely scattered computers and software in a short time, users do not have such ability.
3) Who is responsible for efficient system operation ? In the centralized system, mainframe was the only one object of improving operation efficiency. But, in the distributed systems, networks, every servers and workstations or PCs are objects of improvement. Reduced IM section does not have power to cover all the systems.
Changes   of   requirement   for   computer   and Communication literacy:
In the centralized computer systems, users need not to know about computer hardware or software, much less communication systems. Only knowledge of key-board operation and application functions are required to the users. IM section covered the special knowledge concerning to computer or communication systems.
But PCs enabled it to users to choose application software packages, data-base software, communication interface software or table calculation software. IM section can not restrict users not to use software package freely because that is the merit of down-sizing. If a user has enough knowledge about PC software, user can build up the PC as like as the user wants. In the other hand, the user who does not have PC software knowledge, operation of PC will become burden. Communication method also has variety of choices. The user must choose proper protocol among X.25, TCP/IP, ISDN, frame-relay, or even ATM. It is more than user's ability to choose proper combination of communication systems and protocol among them.

4. DEMANDED SKILL

Above analysis of changes revealed the unbalance between the system and managing organization and /or user's skill. Traditional IM section was composed of two functional groups. The one is systems design group and the other is operations group. The functions of each group and necessary skill are listed in the Table 1 (Lucas, 1981).

Distributed systems require new management functions other than such functions listed in the table.

1) Geographically distributed systems connected by communication networks made fault tracking difficult. Because so many elements are related to a fault. (e.g. WAN, LAN, router, server, clients)

2) Performance of the system depends on both computers and networks. So the system configuration and performance tuning requires high level skill over both of computers and networks.

3) System administration like as software version control , disk space control, or directory service requires also a combination of networks and computers knowledge.

Table 1. Skill requirement for traditional IM section

	Task Name	Skill level
1. System design and developing group	a. Inception	Analyst
	b. Feasibility study c. System Analysis d. Design	Analyst Analyst System Engineer
	e. Specification	System Engineer
	f. Programming	Programmer
	g. Testing	Programmer
	h. Training of user	System Engineer
	i. Conversion and installation	Programmer
2. Operation Group	a. Controls	Supervisor
	b.Data conversion	Programmer
	c. Output processing	Operator
	d. Maintenance programming	Programmer
	e. Systems programming	Senior Programmer
	f. Data administration	Supervisor

As the result, skill (or knowledge) required for managing the distributed system may be segregated to more specialty than the category shown in the Table 1. Terplan (1995) pointed out the principal management functions of a distributed system as follows:

- Client contact point
- Operations support
- Fault tracking
- Fault monitoring
- Change Control
- Planning and design
- Performance monitoring
- Finance and billing
- Implementation and monitoring
- Security management
- System administration

Concerning to the users, knowledge level required to use the distributed system become higher than the past. Evident difference between the distributed system and the centralized system is freedom of operation. In the centralized system, users do not have any freedom of operation. But, in the distributed system, users are allowed to roll down necessary data from servers and process it by spread sheet software. Or make reports using with the down loaded data on the workstations by themselves. Such interactive operations based on communication functions made users to meet with complex troubles. Mitsubishi Electric Information Network has measured how users are troubled by the distributed systems. As a part of it , help desk data gathered in three weeks is shown in the Table 2.

Table 2. Example of Help Desk Activity

Phenome-non	Necessary Knowledge			
	OS & middle ware	Hard-ware	Applica-tion software	Opera-tion
Terminal start up failure	1	1		
Log-in impossible	2	2		
Mail failure	1	3	1	1
Communication failure	2	4		
Printer failure	2	1		
application failure	2	2		5
Total	10	13	1	6

One thing apparently readable from this data is that users have troubles caused by basic software or hardware four times more than operational problems. More than half of them belongs to communication system like as LAN or WAN.

Based on the above investigation , the skills required for managing and operating the distributed systems must cover both of computer and network systems. Combined knowledge of computer and communication network is required in planning, design, implementation, operation and managing phase . Table 3 shows the skill requirement for distributed systems development, management and operation.

Table 3. Skill Requirement for Distributed System

Phase	Task Name
System design and developing phase	a. Inception b. Feasibility Study c. System Analysis d. Design e. Implementation f. Testing
Operation and management phase	a. Fault tracking b. Fault & Performance Monitoring c. Change Control d. Security Management e. System Administration f. Operation Support

5. DEFINITION OF PROBLEM

Thus, the problems can be defined as follows:
1) How to design organizations and allocate functions matched with distributed system ? Evaluation factors are;
- minimum cost of IM sections and of user hour expenditure on computer and communication systems
- maximum efficiency of IM sections and user themselves
- maximum security of data and systems.
2) What kind of knowledge or skills should each organization have ?
The total knowledge shown in Table 1 and 2 must be covered by new organization and user. Evaluation factors may be;
- minimum education cost of personnel
- minimum length of transient period during

migration from centralized to distributed system
3) What kind of technical support or tools are necessary for distributed systems management ?

6. SOLUTIONS

There is not only one solution of this kind of problems. It depends on the policies of enterprise determined by strategic consideration. The solution presented here is based on the following policies:
1) Concentrate tasks which requires similar kind of high level skill for the purpose of efficient use of qualified human resources.
2) Develop tools which support the above mentioned personnel.
3) Create most suitable organization based on human skill type and tools.
The concept of such design method may be called as "Skill Oriented System Design (SOSD)".
Hammer (1993) warned the risk of functional organization. To cover the weak point of functional organization, management traversing organizations must be prepared.
Types of tools:
The types of tools necessary for distributed system designed by SOSD method may be classified into the following four types.
- Remote system measuring
- Remote system monitoring
- Remote system control
- Remote system administration
The functions of each tool is described below.
1) Remote system measuring:
This tool is used either in the system design / developing phase and operation / management phase. CPU load, disk usage, communication line load, network response time, etc. are the measured items by this tool.
2) Remote system monitoring:
This tool is used in operation and monitoring phase. Failure detection of computers and network equipment or line is the main function of this tool. Operator can track down the failed card or failed line location by this tool.
3) Remote system control:
This tool is used in operation and monitoring phase. Operator can recover failure, tune up system, and support user's operation by this tool.
4) Remote system administration:

Software down load, wiring and floor layout management, equipment and facility management are the functions included in this tool.

By the help of these tools, small number of people will be able to manage and operate distributed systems efficiently.

Organization

From the stand point of skill type, tasks in Table 3 is classified into three group; development, monitoring, and support. Thus, the company organization designed by SOSD principle is as follows:

1) System Development Center : This center is responsible for development of new systems and improvement of existence systems. Inception, feasibility study, system analysis, design, implementation, and testing are the tasks of this center. The term " Development Center " is advocated by Martin (1989). But the scope of tasks in this case is wider than that. Rather it covers whole process of system development. The new or revised programs are down loaded via communication network to servers or terminals wherever they located.

2) Operation and Control Center : This center is responsible for operation of networks and computers. Fault and performance monitoring, fault tracking, change control, security management, and system administration are the task of this center. It furnishes centralized monitoring and control system for WAN, LANs, and computers. Controllers can manipulate computers by remote access tools for the purpose of maintenance and restoration.

3) Desk top service Center : The task of this center is operation support. User access the center with telephone if the user encountered some troubles. The center operator access the user's terminal and read out necessary information from the terminal. If necessary, the operator can directly manipulate the terminal by remote access tools. When the trouble is determined to be occurred by failure or error, necessary information is transferred to operation and control center. Then Operation and control center will follow it.

7. CONCLUSIONS

The impact which distributed computer system gave to the skill requirement / human organization and effect of counter development of remote monitoring and control technology is reviewed in this paper. This example indicate the following lessons :

1) New system must be designed with consideration of human organization which is the user of the system. Or it must be recognized that new technology always forces the change of human organization, so the solution for this change must be prepared with the new technology. Distributed computer system had been seeds oriented but recent development of remote monitoring and control tools seems to be able to offer bridges between the system and human organization.

2) The important point for which attention should be paid, concerning the remote monitoring and control tools, is privacy and responsibility of individuals. As such tools make it possible to enter any workstations or PC s from remote center, no one can not keep privacy of information or data.

3) Design of systems should be paralleled with design of human organization and enhancement of human ability. The fact that computer technology and communication technology is closely related and they should not be considered as separated technology. But, nowadays, most of the engineers are not educated as this way. Does curriculum of information processing in universities properly cover the necessary technologies for still growing distributed systems ? Dose the category of information processing suit for emerging new systems ? (Information Processing Society of Japan, 1985) Those are the questions raised by participants joined in construction of large scale distributed systems.

REFERENCES

Hammer, M., Champy, J.(1993). "Re-engineering the Corporation",

IPS Japan, (1995), *J. IPS Japan* Vol. 36, No. 6, Attachment 12-15

Lucas, Jr., H.,(1981). "The Analysis, Design, and Implementation of Information Systems". McGraw-Hill, Inc. USA

Martin, J., (1989). "Information Engineering, Book 3". Prentice-Hall, Englewood Cliffs, NJ, pp. 499-509

Terplan, K., Huntington-Lee, J. (1995). "Distributed Systems and Netwrok Management", Van Nostrand Reinhold, New York USA

KNOWLEDGE DISTRIBUTION IN ORGANISATIONS

Mārīte Kirikova

Department of System Theory and Design, Riga Technical University
1 Kaļķu, Riga PDP, LV-1658, LATVIA, e-mail: marite@itl.rtu.lv

Abstract: There exists a particular body of knowledge in each organisation, that is shared by employees in performing every-day tasks and planning future activities. The body of knowledge can be visualised and controlled by means of special "tangible" computer aided model of the organisation. The model of the organisation can be interpreted also as a model of knowledge concerning the organisation. Implementation and maintenance of the model provide possibilities to analyse current knowledge distribution in the organisation, stimulate direct communications in sharing knowledge between employees and support individuals continuously with new knowledge with respect to their knowledge needs and organisation's development activities.

Keywords: Enterprise modelling, Methodology, Knowledge acquisition, Knowledge representation, Meta-level knowledge.

1. INTRODUCTION

Development of an organisation is closely related with development of knowledge, that is available for performing different tasks in all the levels of their hierarchy. Naturally knowledge is situated in heads of employees and managers (actors) of an organisation. However, organisation's activities are supported by particular knowledge that is shared between actors not only in performing every-day tasks and planning future activities, but also when new actors are introduced into the organisation. In other words, there exists a particular body of knowledge in each organisation. Without paying special formal attention to it, this body is invisible and can not be controlled. However, it is possible to develop special "tangible" computer aided model of the organisation, that reflects the invisible knowledge body. Implementation and maintenance of this model give the following opportunities:

- current state of knowledge and knowledge distribution in the organisation can be analysed

- knowledge can be shared between employees more effectively by stimulating their direct communications

- new knowledge can be provided continuously with respect to knowledge needs of employees and organisation's development activities

The computer aided model of the organisation is in the same time *a model of the organisation* and *a model of knowledge about the organisation*. The Enterprise Model (EM) introduced by the Swedish Institute of Systems Development (Bubenko, 1993; Bubenko, et al., 1992) is suggested as a basis for knowledge acquisition and representation. The Enterprise Model and Enterprise Modelling methodology are described in more detail in the second section. The modifications of the Enterprise Model with respect to knowledge distribution in the organisations are discussed in the third section. Several variants of usage of knowledge model are demonstrated in the fourth section. Unresolved problems and future work are discussed in the conclusions.

2. THE ENTERPRISE MODEL AND ENTERPRISE MODELLING METHODOLOGY

The Enterprise Modelling methodology with the core of it, the Enterprise Model, originally was introduced as a business modelling environment. The EM is developed in group activities, by employees of an organisation and system analysts, during special knowledge acquisition processes (Bubenko and Kirikova, 1994). Specially equipped room is desirable in the development and usage of the EM.

Knowledge in the EM is organised in three levels. Meta-meta-level consists of several sub-models interrelated by the inter-model links. The sub-models are: Objectives sub-Model (OM), Activities and Usage sub-Model (AUM), Actors sub-Model (AM), Concept sub-Model (CM) and Information System Requirements sub-Model (ISRM). Each sub-Model has its own framework (or meta-structure) and a particular role in the model (Bubenko and Kirikova, 1994). The ISRM is used for reflecting the requirements for information system to be developed and will not be considered in the reminder of the paper. Meta-meta-level of the EM is reflected in the figure 1.

Meta-structure components of the sub-models have been introduced on the basis of practical experience in real life business analysis processes (Bubenko, 1993). They involve the components that have empirically been found to support human thinking and reasoning and are suitable for knowledge elicitation. The modelling technology is computer supported, and permits, to some extent, computer aided checking and analysis of the acquired knowledge. The sub-models of the EM have the following elements in the meta-level of the EM:

- the Objectives sub-Model: *Goal, Problem, Opportunity, Cause, Rule, Development Action,* that can be related by *Motivates* or *Influences* relationships
- the Activities and Usage sub-Model: *Process, Information, Material, External Process,* that can be related by *Has Input* and *Has Output* relationships
- the Actors sub-Model: *Individual, Non-human Recourse, Organisational Unit* and *Role* that

can be related by the *Binary Actor Relationships, Actor Is-a Relationships* or *Actor Part-of Relationships*
- the Concepts sub-Model: *Concept, Attribute* and *Concept Model Component Group,* that can be related by the *Binary Concept Relationships, Is-a Relationships* or *Part-of Relationships*

Basic level of the EM consists of particular knowledge elements reflecting current state of acquired knowledge. For example, knowledge reflected in figure 2 shows that clerk of one of the sub-units of an organisation has to draw and put on the wall the organisation's chart. For performing the task clerk has to know the structure of the organisation. The chart is put on the wall with purpose to make clear for individuals the role of each sub-unit in the whole organisation.

Development of the EM requires special efforts of system analysts and has to be appropriately managed. The Enterprise Modelling methodology does not prescribe particular sequence of steps in acquisition of knowledge. It is flexible and situation sensitive. The only, but not easy to meet, requirement is the quality of the model (Kirikova and Bubenko, 1994). For example one of the consistency requirements is that all the processes in the AUM have to support particular goals in the OM, as well as be related with particular performers in the AM and corresponding concepts in the CM.

The Enterprise Model can be developed continuously by adding or changing its elements if necessary. Several versions of the EM can be maintained. From the point of view of application of Aristotle's explanatory principles to the organisational theory (Dahlbom and Mandahl, 1994) the characteristics of the Enterprise model are comparable with the richest theoretical frameworks for information system development (Sowa and Zachman, 1992; Kirikova, 1995). However, traditional usage and organisation of the EM prescribe amalgamation of knowledge without consideration of information about introducers and owners of knowledge elements represented in the EM. Therefore it does not provide means for explicit reflection of knowledge distribution in organisations.

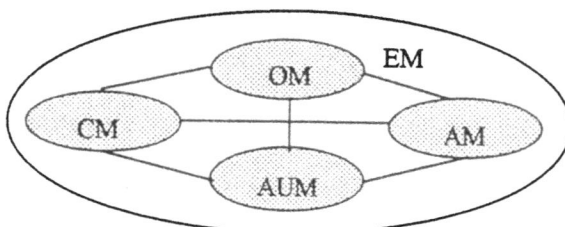

Fig. 1. Meta-meta-level of the EM.

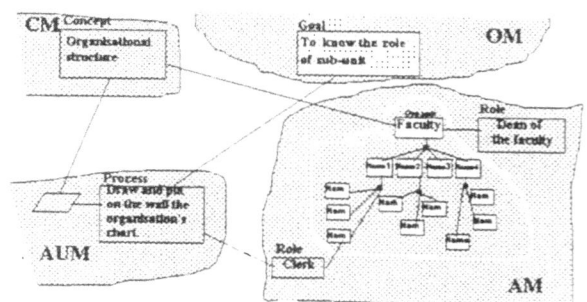

Fig. 2. Basic level of the EM (fragment).

3. REPRESENTATION OF KNOWLEDGE DISTRIBUTION

Knowledge distribution is considered here only with respect to the EM. The traditional organisation of the EM permits to detect answer to the question "Who have to know what?", e.g., clerk in figure 2 has to know the organisation's structure to be able to draw the organisation's chart. However, the Enterprise Model does not provide answer to the questions "Do the clerk know current structure of organisation?" or "Who can clerk get this information from?".

According to the Enterprise Modelling methodology each knowledge element in the EM is introduced by some participant in the group session or interview or elicited from particular documents. Therefore information about origins of the knowledge reflected in the model is available and it is possible to extend the EM by one more model, the Masters sub-Model (Kirikova, 1995). The Masters sub-Model (MM) reflects names of individuals, "masters" of the knowledge amalgamated in the EM (figure 3).

Every element in the EM is linked to some individual in the MM, by "introduced by" or "owned by" arcs. "Introduced by" means that particular person has introduced the knowledge element linked by the arc. "Owned by" means that particular individual is aware about the existence of particular element of knowledge in the model. Individuals in the MM can or can not be actors in the AM.

Knowledge distribution can be considered in the level of elementary units of knowledge in the EM, as well as in the level of different sub-graphs of the EM. Sub-graphs consist of elements that taken together have a particular meaning either in the Concepts sub-Model or with respect to the input information of processes in the AUM. Example of knowledge sub-graph is the organisational structure encircled by the white line in figure 2.

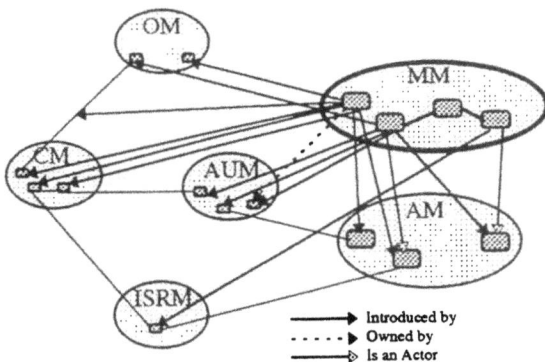

Fig. 4. Distribution of knowledge in the EM (organisation)

Figure 4 illustrates an impact of the MM to the process of Enterprise Modelling. In general relationships to all employees of an organisation can be reflected by introduction of the MM. Knowledge connected with each particular individual in the MM reflects knowledge distribution in the Enterprise Model. In the situations when the EM is developed and maintained adequately, the distribution of knowledge in the EM can be considered as a distribution of knowledge in the organisation.

Distribution of knowledge can be reflected not only by the MM. The alternative representation is actor stamps as an additional field in the description of the nodes and links of the EM (figure 5). Actor stamp reflects introducers and owners of the knowledge element, as well as time of introduction of the element and time when owner has become acquainted with the knowledge element respectively. Time represented in the actors stamp is valid only within the Enterprise Model.

Both forms of representation are mutually transformable. Choice between them depends on background software and implementation possibilities. The Masters Model seems to be preferable when graphical visual computer aided representation of the models is available.

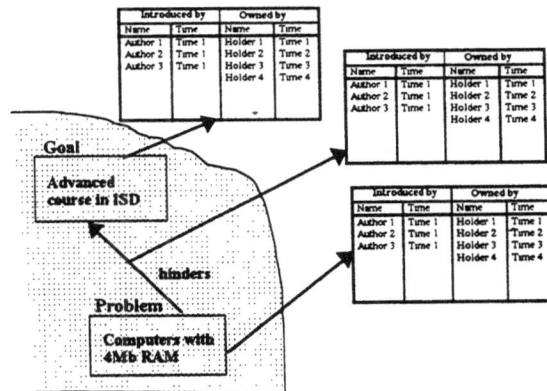

Fig. 3. The Enterprise Model extended by the Masters sub-Model.

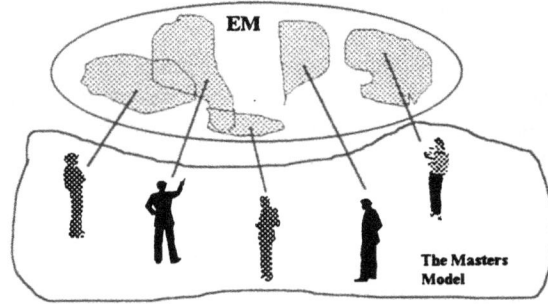

Fig. 5. Actor stamps

4. USING KNOWLEDGE DISTRIBUTION

Visibility of knowledge distribution opens new possibilities in managing the organisation. It can help in every-day activities as well as in planning the development of the organisation.

4.1. Analysis of current state of knowledge

Different related substructures of knowledge in the EM can be considered in analysis of knowledge distribution, for example:

- *Process* in the AUM, *Goal* in the OM and *Individual* (*Role*) in the AM, *Master* in the MM and corresponding relationships. This substructure shows, whether employees are aware about the purpose of the tasks they are performing (figures 2 and 5).
- *Information* and *Process* in the AUM, *Concept* or *Concept Model Component Group* in the CM, *Role* in the AM, *Master* in the MM. Investigation of this substructure and knowledge related to the elements of the CM shows, whether all employees possess knowledge necessary for performing their activities (figure 6).
- *Masters* in the MM and *Roles* in the AM. Considering masters that play roles of heads of sub-units of an organisation, one can see whether management is aware about role of each employee of sub-units managed by them (figure 2).
- etc.

4.2. Supporting direct communications

Recent research work concerning skills and communication in information system development has shown that usage of formal models and direct communications between employees has to be sensitively balanced (Norbjerg, 1994). On the other hand, application of the explanatory principles of Aristotle to organisational theory (Dahlbom and Mandahl, 1994) recovered that recent business models do not support satisfactory the efficient cause, i.e., actual daily activities of organisations. Concerning actual daily activities there is a correlation between them and direct communications, as well as informal organisation of business (Kirikova, 1995).

Modelling in the Enterprise Model extended by the MM is helpful for supporting direct communications and informal relationships between employees. There can be difficulties of understanding formal models reflecting knowledge necessary for performers of particular processes. Therefore it is desirable that individuals receive not only these formal models, but also references to the introducers of the knowledge.

For example, in figure 6 clerk Anita needs knowledge body encircled by white line for drawing the organisation's chart. In current state of knowledge in the EM the encircled body is introduced by dean of the faculty Jon. Thus Anita can receive not only formal model reflecting the organisational structure, but also suggestion to contact Jon, if she has any questions concerning knowledge body. Suggestion is reflected in the MM by corresponding knowledge elements.

4.3. Supporting active tasks

Existence of "tangible" knowledge model can provide additional means also for supporting active tasks. Generally this is unusual feature of business models (Dahlbom and Mandahl, 1994). Possibilities of the extended Enterprise Model are limited, too. However, when automatic knowledge procedures are available, the EM, can be helpful in supporting employees with up-to-date knowledge, whenever changes in corresponding knowledge structures are introduced. An example of the supporting possibilities is illustrated in figures 6 to 9.

State 1 of knowledge concerning organisational structure is reflected in figure 6 by encircled knowledge body. State 2 of knowledge is reflected in figure 7. The only change to compare with figure

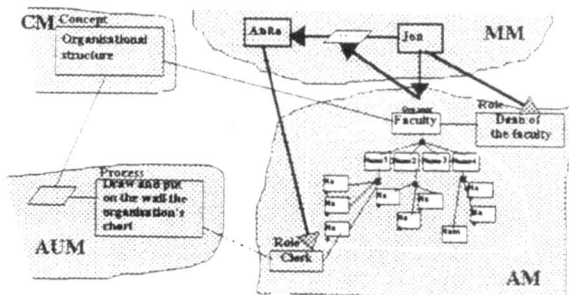

Fig. 6. Support for knowledge sharing (state 1)

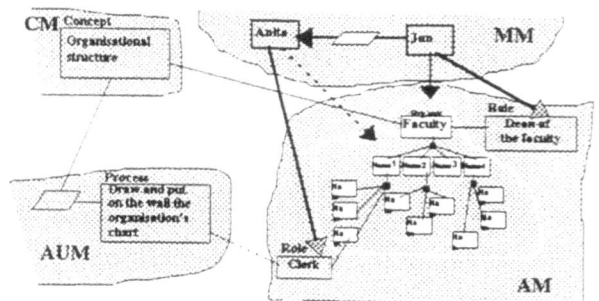

Fig. 7. New owner of knowledge (state 2)

6 here is that Anita has become an owner of the encircled knowledge body. Figure 8 says that new structural unit is introduced by Ralph. Suppose, Jon has signed a paper concerning adding a new sub unit to the organisation's structure and has become owner of the knowledge introduced by Ralph (state of knowledge 3). There is information in the system of knowledge that Anita is preparing task, based of knowledge introduced by Jon. After state 3 there is an inconsistency between Anita's and Jon's knowledge (figure 9). Therefore Anita has to receive corrections concerning organisational structure immediately after they had been introduced in the EM.

As seen from examples, the Enterprise Model extended by the Masters sub-Model is helpful in analysis of knowledge distribution, sharing knowledge between individuals or groups and supporting active tasks with up-to-date knowledge. Development and maintenance of the Enterprise Model are relatively time consuming and complex task. Deep system analysis in the organisation is necessary to obtain basic sub-models of the EM. The most convenient point of time for starting enterprise modelling activities is the time, when new information system is introduced into the organisation. Development of information system naturally includes process of requirements engineering. The EM including the Information

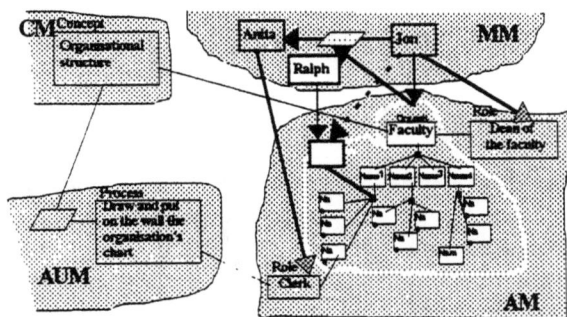

Figure 8. Adding a structural unit of an organisation (knowledge state 3)

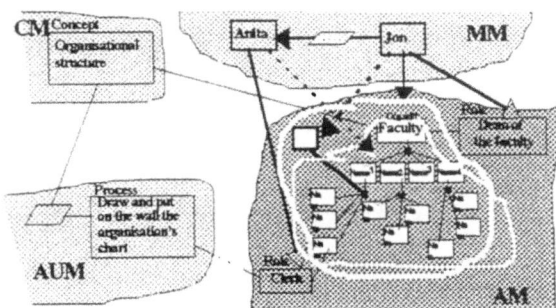

Fig. 9. Inconsistency between shared models of knowledge

Systems Requirements sub-Model (section 2) can be used for this purpose. Later the EM can be maintained continuously with respect to the needs of the organisation.

Benefits from the usage of computer aided knowledge body in the organisation depend on the skills of system analysts, degree of automatisation of knowledge analysis procedures and management support for knowledge acquisition and maintenance activities.

5. CONCLUSIONS

The computer aided model of the organisation is in the same time *a model of the organisation* and *a model of knowledge about the organisation*. There are several features of the model essential for successful use of knowledge amalgamated in it. They are:

- possibility to develop the model continuously
- good understandability and adequacy of the model
- possibility to model the organisation from different viewpoints

The first feature requires special tools for representation, maintenance and changing of the model. Computer aided procedures of checking consistency and other quality characteristics of knowledge are desirable. Current implementation of the Enterprise Model is UNIX based and therefore not always available in smaller organisations.

The second feature means, that model has to be organised in a way, that not only facts, but also explanations of the situations are available. Computer maintained model of knowledge, that consists of mutually related Objectives, Activities, Actors and Concepts, sub-Models, i.e., the Enterprise Model, introduced by the Swedish Institute of Systems Development, and Masters sub-Model, is suggested for this purpose. The combination of these sub-models quite appropriately satisfies four explanatory principles of Aristotle, proved to be essential in understanding of organisations (Dahlbom and Mandahl, 1994).

Possibilities to reflect organisation from different viewpoints depend on the structure of the Actors and Masters sub-Models and their relationships with other sub-Models. Representation of different views in the model and algorithms for detection and meeting knowledge needs of individuals and groups are the matter of further investigation. This representation shows different aspects of distribution of knowledge in the organisation, and can be helpful in organising every-day's activities as well as in planning of development of the organisation and educating new employees.

ACKNOWLEDGEMENTS

This paper is based on the work in the research project "Knowledge Distribution During Requirements Engineering", sponsored by the Danish Research Councils. Dr.sc.ing. Jānis Tenteris from Riga Information Technology Institute (RITI) is acknowledged for providing valuable comments on a draft of the paper.

REFERENCES

Bubenko, J.A. (1993). Extending the Scope of Information Modelling. *Fourth International Workshop on the Deductive Approach to Information Systems and Databases.* Floret - Costa Brava: Universitat Politecnica de Catalunya.

Bubenko, J.A. and M. Kirikova (1994). "Worlds" in Requirements Acquisition and Modelling. *Proc. of the 4th European - Japanese Seminar on Information Modelling and Knowledge Bases.* Stockholm, 1994.

Bubenko, J.A., Ch. Nellborn and W. Song (1992). Computer Support for Enterprise Modelling and Requirements Acquisition. *From Fuzzy to Formal,* Esprit III Project 6612: Deliverable 3-1-3-R1 Part B.

Dahlbom, B. and M. Mandahl (1994). A Theory Of Information Technology Use. *Proc. of the 17th IRIS,* P.Kerola, A.Juustila, J.Jarvinen (Eds.), Department of Computer Science, University of Oulu, P. 66-77.

Kirikova, M. (1995). Principles of Aristotle in Requirements Engineering. *Proc. of the 5th European - Japanese Seminar on Information Modelling and Knowledge Bases,* Sapporo, Japan.

Kirikova, M. and J.A. Bubenko jr (1994). Enterprise Modelling: Improving the Quality of Requirements Specifications. *Proc. of the 17th IRIS,* P. Kerola, J. Jarvinen (Eds.), Department of Information Processing Science, University of Oulu, Finland.

Norbjerg, J. (1994). *Skill and Cooperation in Systems Development,* Ph.D. thesis, Department of Computer Science, University of Copenhagen, Copenhagen, Denmark, DIKU-report No. 94/28, 1994Dissertation. Thesis Publishers Amsterdam.

Sowa J.F. and J.A. Zachman (1992). Extending and Formalising the Framework for Information Systems Architecture. *IBM Systems Journal,* Vol. 31, No. 3.

BEHAVIOURAL CLONING:
PHENOMENA, RESULTS AND PROBLEMS

Ivan Bratko[1,2], Tanja Urbančič[1], Claude Sammut[3]

[1] *Jožef Stefan Institute, Ljubljana, Slovenia*
[2] *Faculty of Electrical Eng. and Computer Sc., University of Ljubljana*
[3] *University of New South Wales, Sydney, Australia*

Abstract: Controlling complex dynamic systems requires skills that operators often
cannot completely describe but can demonstrate. Behavioural cloning is the process
of reconstructing a skill from an operator's behavioural traces by means of Machine
Learning techniques. In this paper we analyse various phenomena and problems observed in experiments in behavioural cloning in several domains: piloting, driving a
container crane, production scheduling and pole-balancing. The analysis includes the
"clean-up" effect and the time delay between state and action. We derive from this
analysis some elements of an emerging methodology for behavioural cloning.

Keywords: Machine learning, control system synthesis, human-centered design

1. INTRODUCTION

Controlling a complex dynamic system, such as
an aircraft or a crane, requires a skilled operator who has acquired the skill through experience. The skill is subcognitive and the operator is usually only capable of describing it incom-
pletely and approximately. Such descriptions can
be used as basic guidelines for constructing automatic controllers but, as discussed for example in
(Urbančič and Bratko, 1994a) the operator's descriptions are not operational in the sense of being
directly translatable into an automatic controller.

If we are interested in designing an automatic controller based on an operator's skill, we have the
following situation. An operational description of
the skill is not available, but the manifestation
of the skill is available as traces of the operator's
actions. One idea, explored in several projects,
is to use these traces as examples and extract
operational descriptions of the skill by Machine
Learning techniques. Extracting symbolic models of a real-time skill from traces of the operator's behaviour was termed *behavioural cloning*

by Donald Michie (1993). To our knowledge, behavioural cloning has been applied in the following dynamic domains: piloting a simulated aircraft (Sammut *et al.*, 1992; Michie and Camacho,
1993), operating a crane (Urbančič and Bratko,
1994b), pole-balancing (Michie *et al.*, 1990) and
production scheduling (Kibira, 1993).

Behavioural cloning is normally performed by
applying standard machine learning (ML) techniques. Of course, the cloning problem needs
a formulation and representation that fits these
techniques. The usual ML-based approach to behavioural cloning, employed in all the domains
above, has been as follows. The "behaviour trace",
that is, a sequence of the states of the controlled system and the operator's control actions
is viewed as a set of examples of correct control decisions. Each example consists of a pair
($State, Action$) where $State$ is an attribute-value
vector and $Action$ is a "class value" for a learning program. In such a formulation, attribute-
based learning techniques can be applied to induce
a functional relation

$$Action = f(State)$$

Both the attributes of *State* and the class value *Action* can be discrete or continuous. The learned function f then represents an artificial controller, or behavioural clone, which is supposed to mimic the original operator.

Sometimes a time delay is considered between *State* and *Action*. The reason is that the current *Action* is not viewed as the operator's response to the current *State*, but to some *previous* state. This delay between the state and the action is assumed to be due to the time needed for state recognition and physical manipulation of the controls by the operator. The functional relation to be learned in such a case is

$$Action(Time) = f(State(Time - Delay))$$

In some domains there are several control variables. In piloting, for example, the throttle, flaps, ailerons etc. can be manipulated simultaneously. In such domains the learning problem is partitioned into several sub-problems, each of them dealing with one control variable. These may be dependent, so the decision regarding one control variable may be used as an attribute for a decision regarding another control variable.

A way of structuring the learning problem, applied typically in piloting, is the division of the behaviour traces into phases. For example, a flight may be divided into take-off, straight-and-level flight, turn to a specified heading, etc. Separate controllers are induced from the traces for each phase of the flight. To carry out the control task, these controllers are invoked according to the particular phase of the flight plan. The phases, the plan, and the phase recognition conditions are hand crafted and not learned automatically.

In this paper we give a comparative analysis of some of the phenomena observed in behavioural cloning, taking into account the results in all the domains mentioned above. Among other aspects we discuss the "clean-up effect", time delay between state and action, robustness of induced clones and representation for inducing human-like control strategies. Another important issue is the comprehensibility of induced clones. This was addressed in (Urbančič and Bratko, 1994b).

2. PROBLEM DOMAINS

Before comparing and analysing the phenomena, successes and problems experienced in behavioural cloning, we present the main characteristics of the problem domains considered in this paper:

- *Pole balancing* (Michie *et al.*, 1990)
  The well-known problem of balancing a pole on a cart moving on a track of limited length has often been used to demonstrate new approaches in control synthesis, e.g. in (Michie and Chambers, 1968; Barto *et al.*, 1983; Anderson, 1987; Varšek *et al.*, 1993). For cloning, a "line-crossing" variant was used by Michie and coworkers (Michie *et al.*, 1990). Here, subjects are asked to make as many crossings of the mid-line of the track as possible, within the test period, without crashing.

- *Piloting* (Sammut *et al.*, 1992; Michie and Camacho, 1993)
  Two studies have been reported: piloting a Cessna (Sammut *et al.*, 1992) and piloting an F-16 (Michie and Camacho, 1993). In both cases, the task is to fly according to predefined flight plan, including take-off, climbing to a specified altitude, straight and level flight, turning and landing.

- *Operating cranes* (Urbančič and Bratko, 1994b)
  The task is to transport a container from an initial position to a target position. Performance requirements include basic safety constraints, stop-gap accuracy and as high capacity as possible. Here, capacity is measured as the total load transported within the allotted time.

- *Production line scheduling* (Kibira, 1993)
  The task is to schedule and control a serial manufacturing system. The problem is to determine an optimum or near optimum allocation of labor for a period of time on a production line at any time during a shift.

All the domains, except pole-balancing, are of interest for potential applications. On the other hand, pole balancing is useful as experimental domain because it allows clearer study of separate phenomena in cloning. Production line scheduling is atypical in that it requires control decisions at time points separated by several hours. Therefore, this domain does not have many of the characteristics common to most dynamic control problems and will be discussed in less detail than other domains.

In all the experiments, simulators of the dynamic systems were used. The flight simulator used in (Sammut *et al.*, 1992) was provided by Silicon Graphics Incorporated. In (Michie and Camacho, 1993), the authors used the ACM public-domain simulation of an F-16 combat aircraft. For pole balancing and crane control, the simulators were developed specifically for these experiments. The crane simulator was assessed by a specialist crane designer as very realistic. It runs on an IBM compatible PC and provides real-time performance on a 33 MHz 386 or faster.

In Table 1 we compare the characteristics of the problem domains, along with the parameters of human and machine learning. The table gives some idea of the complexity of the tasks. However, some qualifying comments are required. One of informative characteristic for a domain is the time a human operator requires to master the task. Since individual differences can be very large (see e.g. (Urbančič and Bratko, 1994b), this should be treated only as a rough approximation. For pole-balancing, human learning time was around one hour. For operating the crane, approximately ten hours of training were needed. The F-16 piloting problem was more demanding than the Cessna variant. Also the meaning of event should be clarified. In the flying domains, an event corresponds to change in control action, while in the crane problem, events are actually snapshots, recorded at regular time intervals.

3. CLEAN-UP EFFECT

Michie and Camacho (1993) described the *clean-up effect* as follows. "When induction-extracted rules were installed in the computer as an 'autopilot', performance on the task was similar to that of the trained human who had generated the original behaviour trace, but more dependable ..." They continue with an explanation of this effect. "A trained human skill, ..., is obliged to execute via an error-prone sensory-motor system. Inconsistency and moments of inattention would then be stripped away by the averaging effect implicit in inductive generalisation, thus restoring to the experimenters a cleaned-up version of the original production rules."

Michie, Bain and Hayes-Michie (1990) were the first to report the clean-up effect which they observed in the pole-balancing domain. They also gave a quantitative assessment of clean-up. When a clone induced from an operator's traces is used as a *predictor* of the operator's actions, the prediction error rate often exceeds 20 %. Michie and Camacho interpret this error rate simply as indicative of the cumulative sum of human perceptual and execution errors. These errors are presumably filtered out by the induction program. (They used Quinlan's C4.5 (Quinlan, 1987). As a result, the clone's behaviour is much smoother

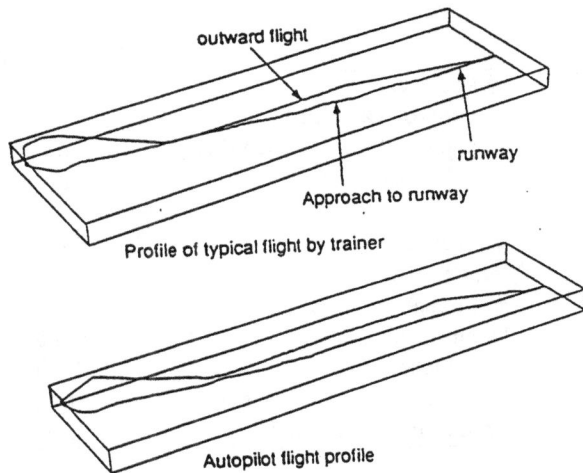

Figure 1: Human performance (upper trajectory) and the clone's execution of the same flight plan (lower trajectory).

than that of the human operator. Michie and colleagues (1990) measured this *performance* (not prediction!) clean-up in terms of ranges visited by the four system variables: position, angle, and their velocities. In general, better control results in smaller ranges. The ranges achieved by the clone were much tighter than those achieved by the human trainer. They were reduced by the clone to between 17 % and 45 % of the original human's ranges (depending on the system variable). This result is very illustrative although the simple measure of clean-up is debatable as it considers each of the state variables separately. The *fitness* function introduced in (Varšek et al., 1993) is probably a better measure of performance and is more in the spirit of traditional control engineering.

Michie and Camacho (1993) also report on clean-up in flying the simulated F-16 aircraft. They considered the error in straight and level flight measured as the plane's deviation from a straight line. The clone's deviation was only about 15 % of the human's deviation.

In learning to fly the Cessna (Sammut et al., 1992) the clean-up effect was also observed. Figure 1 shows the trajectories in three dimensions of two flights, one by a human operator and the other one induced from the same operator's traces. Both flights accomplish the same flight plan. Clean-up is easily noticeable in the approach to the runway.

Table 1: Characteristics of the domains and parameters of learning

	Pole and cart	Cessna	F-16	Crane
# state variables	4	15	15	6
# control variables	1	4	9	2
# types of control variables	boolean	real, integer	real, integer	integer
# subjects	10	3	1	6
# traces	1	90	20	450
# events in data set	3500	90.000	25.000	450.000
length of a trace	5 min	5 min	18 min	1 - 3 min
# phases	1	7	8	1
learning program	C4.5	C4.5	C4.5	Retis, M5
delay [seconds]	0.4 - 0.5	1 - 3	1	0 - 0.1
preprocessing of data needed	no	yes	yes	no
predefined plan needed	no	yes	yes	no

Similarly, clean-up was also observed in the crane domain. In the container crane, there are six system variables: position of the trolley and its velocity, rope length and its velocity, and rope inclination angle and its velocity. The task is to move the load from a start position to a given goal position. When the goal position is reached, the state variables must be kept sufficiently close to the goal values for some minimum time interval. Figure 2 shows two time behaviours. One is by a human operator. This contains precisely those events that were used to induce the clone that produced the other trajectory. The clone was induced by Quinlan's M5 (Quinlan, 1993) which generates regression trees that are by default drastically pruned. As in the previous cases, it can be seen in Figure 2 that the clone carries out the task in a style very similar to the original. The clean-up here is reflected in that the clone is noticeably more successful than the original with respect to the time required to complete the task (75 seconds for the clone compared to 90 seconds for the original).

In the production scheduling domain (Kibira 1993) the problem is to allocate production resources to tasks in the assembly line. The goal there is that at the end of an 8.5 hour shift, the queue sizes of currently available subassemblies at various stages of the line are as close as possible to specified goal levels (500 in Kibira's experiments). The queue sizes at the start of the shift deviated grossly from the target levels. Kibira (1993) gives the time behaviour of the queue sizes at various points in the assembly line for both human expert scheduler and the clone. The clean-up is here reflected in the fact that the clone's final level (at 8.5 hours) is always closer to 500 than the human's final level.

4. OTHER ISSUES

4.1 Reaction time delay

Some researchers strongly believe that there should be a time delay between the system's state and the control action. Such a delay seems necessary because an operator cannot react to a stimulus (that is system state) instantaneously. Sammut et al. (1992) in their experiments used a delay that varied between 1 and 3 seconds. Their choice of the delay was pragmatic and determined experimentally. The choice was not made in a principled way although they paid considerable attention to this question in their discussion. They believed that the delay was important, but did not have a firm theoretical basis for determining appropriate delays. Although a delay appeared to be critical, slightly altering the length of the delay for recording a subject's behaviour did not critically affect the clone's performance. The clone also had to implement a delay in order to accurately mimic the human subject. The length of this delay was critical since it had to match the recording delay.

Experiments in the crane domain were also carried out with various delays (Urbančič and Bratko, 1994b). It seems that in this domain zero delay does not produce inferior results compared to other delays. Some discussions lead to the belief that the operator's delay in fact varies and depends on the situation. There are quick, purely reactive decisions, and there are also strategic, longer term decisions that have to do with setting new intermediate goals, for example start accelerating until a goal speed is attained. Also it seems that a skilled operator is capable of compensating for delays in reaction time by predicting

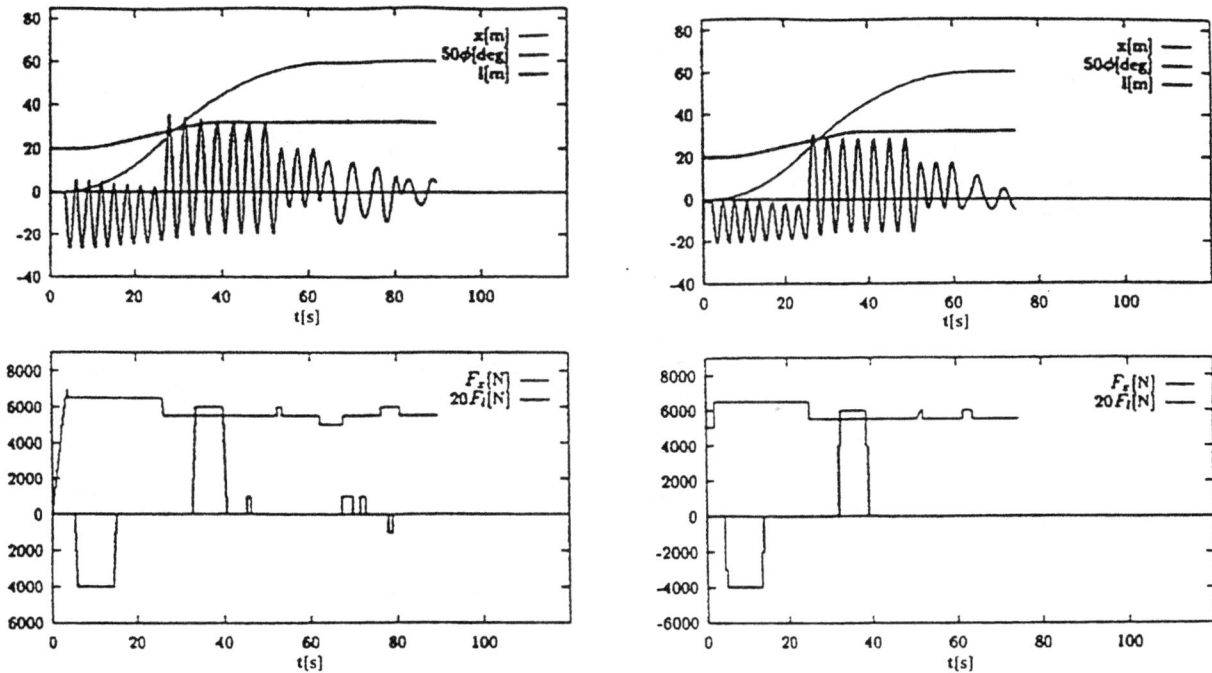

Figure 2: Diagrams on the left: one of the most "tutorial" traces by a human controlling the crane; on the right: the trace of a clone induced from the human's trace.

short term future states of the system. In extreme case this would even indicate that a "negative delay" would make sense. So straight-forward adoption of reaction time delays to unpredicted events, known from psychology, does not seem to be appropriate. To conclude, there seems to be no clear indication, either theoretical or experimental, of what would be an appropriate delay.

4.2 Choice of example traces for learning

One question that must be answered when designing a behvaioural cloning system is which traces to choose for learning among all the available example traces. The style and control strategies obviously vary significantly from operator to operator. or example in the crane domain, some operators tended towards fast and less reliable operation, others were slower, more conservative, and more reliable. Some operators were avoiding large angular accelerations at the expense of time. Such strategies produce reliable, but slow performance. This is in contrast with some operators strategies that tend to achieve faster times, but require higher accelerations of the trolley which causes large angles and requires very delicate balancing of the load at the end of the trace. There were

also differences between the operators in the order of attaining the subgoals. Similar differences between individuals were observed in the learning to fly experiments (Sammut et al., 1992).

To avoid mixing individual styles, the commonly agreed practice in behavioural cloning has been to combine training examples from the same subject only. However, even the example trajectories of the same subject may vary considerably. For example in the crane domain, the same subject using the same control style will produce trajectories whose finishing times are quite different. According to this, when trying to induce the most "tutorial" and "unadventurous" clones, the most conservative example trajectories have been found to be by far the most useful.

4.3 Brittleness

Experiments have shown that the generation of behavioural clones is feasible and that the clean-up effect can be observed in all the domains investigated so far. Successful clones perform similarly to the human subjects, although the clones' trajectories are of course not literal reproductions of the original trajectories. However, in the more complex domains, such as flying and crane driving, the original experiments produced clones that were usually very brittle with respect to changes and were only successful within a fixed plan. They were sensitive to small changes in the task or the parameters of the problem domain.

These problems can be cured by training in a noisy environment. For example, the first piloting clones were built using a flight simulator that did not include turbulence or wind disturbances. Therefore, when flying straight and level, it was sufficient to leave the controls alone and the aircraft would continue along its original altitude and heading. A trace of such a flight provides no examples of what to do when the aircraft either begins off its desired course or what to do if it is pushed off course. This can be corrected by introducing turbulence and wind drift. The human pilot must now generate examples of corrections which can be used to train a more robust clone. Arentz (1994) performed just such experiments and found that clones can be constructed that are quite robust to substantial disturbances. Clearly, if the disturbances encountered by the clone are greater than those encountered by the trainer, we can expect loss of control since circumstances have been created that are outside the clone's range of experience.

4.4 Inducing human-like strategies

In all the work until now the clones have taken the form of decision or regression trees or rule sets. These clones are purely reactive and inadequately structured as conceptualisations of the human skill (unless embedded in a hand-crafted fixed plan). They lack the conceptual structure typical in human control strategies: goals and subgoals, phases and causality. The simple form of the clones as mappings from system states to actions does not suffice to express such a conceptual structure. This conceptual difference between the clones and the humans' own descriptions of their skill is analysed in (Urbančič and Bratko, 1994b).

Here we note some requirements for the representation of human-like controllers. First, such a controller should have some internal memory to maintain the current goals and phase of task. Furthermore, to enable the learning program to discover conceptual structure in a behavioural trace, the program should have access to some background knowledge about the domain.

In his work on improving yield in process control, Leech (1986) developed a two-stage method in which variables critical to the yield were first identified by induction. A further inductive step was used to construct control rules to achieve desired values for the critical variables. Donald Michie has suggested that an analogous scheme might be used in behavioural cloning. This approach is currently under investigation in the piloting domain. Initial results suggest that it may, indeed, be possible to construct more goal-oriented clones. However, there is still much work to be done.

Regardless of the representation, we believe that the study of human skill should take into account the constraints that humans must live with. One constraint is that the human can only look at a small number of state variables a time. Here are comments from one of our crane operators: "At this stage I only look at x very little; I never look at $\dot{\theta}$. [Later:] Here I never look at $\dot{x}$; if I do I get very confused". This suggests that at any given time the operator's decision only depends on a very small number of attributes. An important part of human strategy is to know what instruments to look at at various stages of the task.

5. CONCLUSIONS

Experience in behavioural cloning described in this paper indicates some elements of an emerging methodology which we summarise in the following paragraphs.

1. *Choice of example traces for learning.* Style and control strategies vary significantly from operator to operator. To avoid mixing individual styles, the commonly agreed practice in behavioural cloning has been to use training examples from the same subject only. However, even the example trajectories of the same subject may vary considerably. When trying to induce the most reliable and "unadventurous" clones, the most conservative example trajectories were found to be by far the most useful.

2. *Time delay between state and action.* Human response times for sudden stimuli do not necessarily give any indication of an appropriate delay for behavioural cloning. A reasonable method is to try first with zero delay and increase the delay gradually, looking for the best performance.

3. When *designing the representation*, i.e. choosing attributes, it is useful to take into account the operator's verbal description of his/her skill. The introduction of such a descriptions into the "cloning cycle" is discussed in (Urbančič and Bratko, 1994b).

These observations are relevant mainly when the goal of cloning is to achieve good performance. However, in using clones as *conceptualisations* of human skill, there is much to be done. Earlier, we noted the large conceptual difference between the clones, represented by decision or regression trees, and the humans' own descriptions of their skill. In (Urbančič and Bratko, 1994b) it was possible to establish only a partial correspondence between the operator's instructions and the clone. ML-based analysis of the operator's instructions helped to reveal that the operators occasionally were not doing what they believed they were doing. In part, this is due to the fact that subcognitive skills are not available to introspection and therefore, the operator's instructions are *post hoc* justifications for his/her behaviour. However, a significant part of the difference between the clone and the human's conceptualisation appears to be due to the limited representational capabilities of pure attribute-based learning systems. Structured induction techniques of the sort described in section 4.4 may be more appropriate. Inductive Logic Programming (ILP) techniques also hold promise for behavioural cloning because of their ability to accommodate background knowledge more flexibly than other learning methods. However, as most of the variables in control systems are continuous, further research is required to allow ILP to handle numeric data more effectively.

ACKNOWLEDGEMENTS

We thank Donald Michie for the many helpful discussions and suggestions that have contributed to our experiments.

REFERENCES

Anderson, C.W. (1987) Strategy Learning with Multilayer Connectionist Representations. *Proceedings of the 4th International Workshop on Machine Learning*, Morgan Kaufmann, pp. 103–114.

Arentz, D. (1994) Experiments in learning to fly. Computer Engineering Thesis, School of Computer Science and Engineering, University of New South Wales.

Barto, A.G., Sutton, R.S., Anderson, C.W. (1983) Neuronlike Adaptive Elements That Can Solve Difficult Learning Control Problems. *IEEE Transactions on Systems, Man and Cybernetics*, Vol. SMC-13, No.5, 834–846.

Kibira, A. (1993) Developing an expert controller of a black box simulation of a telephone line using machine induction. Unpublished technical report, University of New South Wales, AI Laboratory.

Leech, W.J. (1986) A rule-based process control method with feedback. In: *Proceedings of the ISA/86 International Conference and Exhibit*, Houston, Texas.

Michie, D., Chambers, R.A. (1968) BOXES: An experiment in adaptive control. In: Dale, E., Michie, D. (eds.) *Machine Intelligence 2*, Edinburgh University Press, pp. 137–152.

Michie, D. (1993) Knowledge, learning and machine intelligence. In: L.S.Sterling (ed.) *Intelligent Systems*, Plenum Press, New York.

Michie, D., Bain, M., Hayes-Michie, J. (1990) Cognitive models from subcognitive skills. In: Grimble, M., McGhee, J., Mowforth, P. (eds.) *Knowledge-Based Systems in Industrial Control*, Stevenage: Peter Peregrinus, pp. 71–99.

Michie, D., Camacho, R. (1994) Building symbolic representations of intuitive real-time skills from performance data. In: K.Furukawa, S.Muggleton (eds.) *Machine Intelligence and Inductive Learning*, Oxford: Oxford University Press.

Quinlan, R. (1987) Simplifying decision trees. *International Journal of Man-Machine Studies*, Vol. 27, No. 3, 221–234.

Quinlan, R. (1993) Combining instance-based and model-based learning. *Proceedings of the 10th International Conference on Machine Learning*, Morgan Kaufmann, 236–243.

Sammut, C., Hurst, S., Kedzier, D., Michie, D. (1992) Learning to Fly. Sleeman, D., Edwards, P. (eds.) *Proceedings of the Ninth International Workshop on Machine Learning*, Morgan Kaufmann, pp. 385–393.

Urbančič, T., Bratko, I. (1994a) Learning to Control Dynamic Systems. In: D.Michie, D.Spiegelhalter, C.Taylor (eds.) *Machine Learning, Neural and Statistical Classification*, Ellis Horwood, pp. 246–261.

Urbančič, T., Bratko, I. (1994b) Reconstructing Human Skill with Machine Learning. In: A.Cohn (ed.) *Proceedings of the 11th European Conference on Artificial Intelligence*, John Wiley & Sons, Ltd., 498–502.

Varšek, A., Urbančič, T., Filipič, B. (1993) Genetic Algorithms in Controller Design and Tuning. *IEEE Transactions on Systems, Man, and Cybernetics*, SMC-23(6):1330–1339.

A NEW PATH TO SELF-SUSTAINING MARKETS FOR PHOTOVOLTAIC

Anne Kreutzmann, Dipl. Ing. Wolf von Fabeck, Philippe Welter

Solarenergie-Förderverein e.V. · Herzogstr. 6 · 52070 Aachen · Germany

Abstract: A new form of market development for photovoltaics is growing in Germany. "Rate-based incentives" allow the public to install PV systems and recover their investment over time through a per kilowatt hour payment for clean energy generation. The payback is funded through a low surcharge on electric utility bills. The important aspect: there is no need for subsidies from the state, local authority or public utility with this system. The advanced electrical measurement and control techniques are the prerequisites for this development. Some years ago grid-connected PV-systems were very unusual. This paper will provide the specifics behind the rate-based concept and its advantages for PV market development.

Keywords: Renewable energy systems, Ecology, Economic systems

1. COVERING THE COSTS FOR ELECTRICITY FROM RENEWABLE SOURCES

It's true that there are relatively few privately owned solar facilities in operation today. The main reason is that despite incentive programs, those who produce solar electricity still pay a lot more for their power than those who don't. This situation could be changed if the operators of solar facilities were paid for the electricity they fed into the grid at a price that would cover their costs just as at any other power plant.

The owners of solar facilities put all solar electricity they produced into the electric mains.

Thus there would be a slight increase in electricity rates for the regular customer.

1.1 Rate-Based Incentives means:

Individuals or businesses who invest in photovoltaics are paid for every kilowatt hour of solar energy the PV system feeds back to the grid. PV investors in Germany receive two deutsche marks per kilowatt hour (U.S. $ 1.34/kWh) so that they can fully recover their cost of purchasing and installing the PV system.
Cost recovery is funded through a surcharge on electric utility bills paid by the public "rate-base" of the utility.

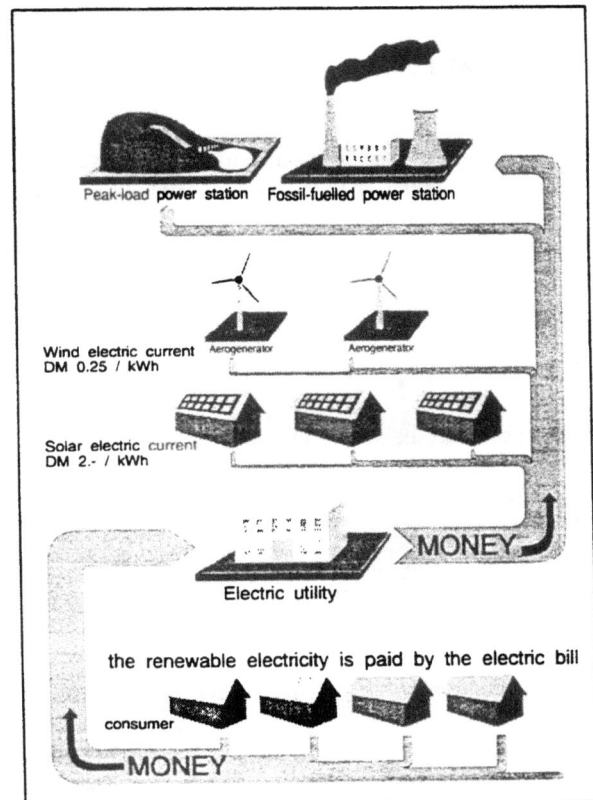

Fig. 1.   Rate-based financing of renewable energies

The limit of this surcharge is still in discussion and varies from state to state between 0.6 (Bavaria) and 3%. Mostly 1 % is accepted by the ministers of economy. In Northrhine-Westfalia 1 % is the accepted limit even though an official report ordered by the ministry proposed a limit of 5 %. The limit in Badenwürttemberg is 3 %.

1.2 The Aachen Model

Rate-based incentives were first proposed in Germany in the city of Aachen near the Belgian border. The 2 DM/kWh payback rate is paid for total system installations and is intended to pay for both the installation and the cost organizing the capital. The 2 DM/kWh are paid for twenty years, the average life of a PV-system. After a certain time, for example after one year, the price of one kilowatt hour must be determined for new installed systems again. It is expected to be less than 2 DM. People installing a PV-system now will get the new price, but are also guaranteed payments over twenty years to recover their investment. This helps to avoid penalizing "solar pioneers" in the payback rate when the cost of solar electricity goes down.

The Aachen Model was proposed in 1992, and was implemented in Aachen as of September 1, 1994. The electric utility bill surcharge is limited to 1%. This will be sufficient to install 1.5 MWp of PV and about 7 MWp of wind. However, the installation limit is determined by the surcharge of the electric utility bills.

2. PRIMARY ADVANTAGES OF RATE-BASED INCENTIVES:

A comparison of rate-based incentives with government subsidies shows that the rate-based model offers several advantages over past government programs for PV market development. In fact the government programs are no real **market** development but research programs.

2.1 Stable funding source

The public rate-base provides a constant funding source that is directed specifically to long-term PV market development. By contrast, government programs are funded by the federal tax-base and are subject to unstable political cycles and budget constrains that can start and stop PV market development abruptly.

2.2 Stable markets help reduce system costs

Since the rate-based model provides an uninterrupted fundig source, market development is consistent over time. Rate-based incentives therefore provide a more stable market opportunity than subsidies given by the government. This allows manufacturers to confidently invest in increased production and process automation, which reduces the cost of PV systems.

2.3 Demand based on public choice

Under the government model, market demand is based strictly on the cost of PV systems. Under the rate-based model, market incentives are provided for PV system cost recovery to increase public demand.

2.4 Competitive pricing for consumers

The rate-based model encourages price competition among distributors and system integrators in selling PV systems to the private sector. If utilities and commercial power providers install PV systems, they will shop for cost-effective systems to potentially make a profit from the 2 DM/kWh payback rate.

2.5 Emphasis on total kilowatt hours leads to better system design and inverter efficiency

The payback rate is based on total kilowatt hours produced by the system. Past government and utility programs have emphasized PV system output as it relates to peak demand hours for the utility. The peak demand emphasis led to wide variations in system output in the German 1.000 Roofs program, where PV systems of similar size and insolation levels varied as much as 2 to 1 in power generation. Rate-based incentives focus consumer emphasis on total system power production, which redirects priorities toward improvements in inverter efficiency and system design.

2.6 Rate-based incentives can attract commercial interest in PV

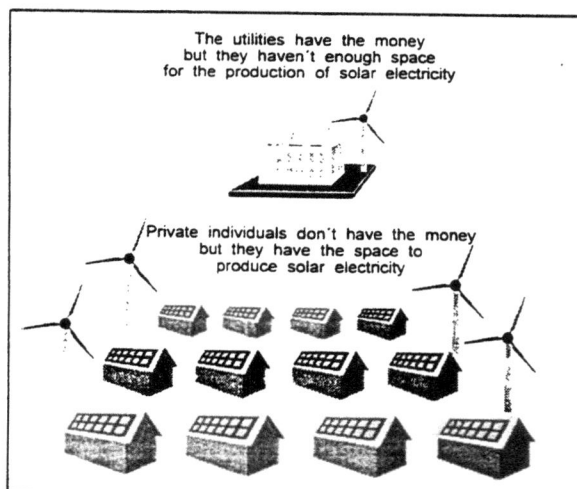

Fig. 2. The use of solar-electricity is no question of space

The 2 DM/kWh rate can attract commercial power providers and utilities to invest in PV and makes photovoltaics a more cost-effective option. The model can also present a potential source of utility business development and service diversification, by offering systems to homeowners who could not afford to purchase PV systems on their own. The commercial involvement in PV may expose banks to the technology and spread the availability of low-interest loans to individuals and small businesses to purchase PV systems.

3. QUESTIONS ABOUT RATE-BASED INCENTIVES

3.1 What impact does this program have on monthly electric utility bills?

The surcharge to pay for the program is less than 1% per kilowatt hour. If 1 MWp were installed in a city of 250.000 people, the average monthly bill would increase from DM 30,00 to DM 30.30 (U.S. $22.40 to $22.60). Utility ratepayers would essentially be asked to contribute less than half a dollar per month to support clean energy generation and improve the environment. If a community wanted to protect industry, customers with special rates (industry), could continue to pay old prices. However, the question of whether industrial customers also have to pay higher electric bills to support solar energy has not yet been solved. A state surveyor has declared that both ways (to include the industrial customers or to exclude them) are feasible.

3.2 How is the program monitored?

The system power production is fed directly into the utility grid and read by a meter. Abuse of the system is nearly impossible: electricity from sources other than solar would be quickly noticed because a given system can produce only a certain number of kilowatt hours per year.
Another positive effect: a defective system wouldn't pay - its feed-in meter would read „zero". Thus every photovoltaic owner would have a strong interest in buying good quality equipment and keeping the installation running.

3.3 How is the payback rate of 2 DM/kWh calculated?

The Federal Rates Regulation for Electricity calls for an investigation of electricity costs in regard to economic and cost-effective operation. This means that the analysis of power-producing costs must apply to systems installed at optimal conditions (southern exposure, appr. 35° tilt angle, proper ventilation behind anels etc. ...)

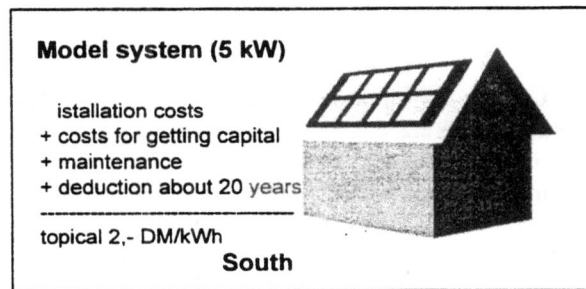

Fig. 3. The calculation of the price per kWh solar electricity

Solar arrays on rooftops are nearly maintenance-free and work fully automatically. For this reason, operation costs are limited to the investment cost for the equipment, financing costs of meters, insurance and a sum to cover the required yearly revision only. This leads to a present calculated cost for electricity production at DM 2,- per kWh.

3.4 How long does the program last?

Program specifics vary by city and country in terms of the payback rate and the length of the time paid for the solar or wind electricity. As a general rule, an installation ceiling of 1 MWp for 250.000 inhabitants has been established for PV, with programs running from as low as ten years (then 2 DM/kWh are not really cost-covering) up to 20 years. Once the installation ceiling has been reached, cities can review the acceptance and impact of the program.

3.5 How high are the administrative costs?

The rate-based incentives are free of red tape.
For example, the "1.000 Roofs program" gave rise to costly and time-consuming test procedures. In 1991 in Northrhine-westfalia the administration of a single application cost about DM 3500,- and lasted about a year.

3.6 Does the rate-based program touch the utility economically?

No. it doesn't. The program is a straight pass-through cost that's supported and funded by the public.

3.7 Will this rate-based model impede other measures to reduce CO_2 levels?

There is no rivalry because these programs are paid by different sources. One is paid by taxes, with the government deciding how the money should be spent. Another is paid by a utility bill surcharge. where the money is mandated to be spent on PV system output.

3.8 What legal basis is there for rate-based incentives?

1) German law on power fed back to the grid (Strom-einspeise-Gesetz, § 2, 2nd sentence) explicitly refers to a minimum sell-back rate for electricity from renewable sources and thus provides leeway for higher rates.
2) The Federal Electricity Rates Regulation also allows for compensation going beyond the long-term costs if certain conditions are met (§ 2, 1st sentence). One condition demands, for example, rational and cost-effective operation in electricity production (e.g. no systems facing north).
3) The admissibility of passing on extra costs to the electric utility bills has been confirmed in writing by the Federal Economics ministry, the Northrhine-Westfalia rates board and the Bavarian state government.

3.9 Will the rate-based incentives make solar electricity cheaper?

The rate-based incentives will be based on a model installation implemented by the city council or the local utility. It is also possible to find out the price for a kilowatthour by a market investigation.
Anyone with a higher investment compared to the model will operate at a loss; anyone producing solar energy at a lower cost will turn a profit. This competition will bring prices down.
The model will also lead to a sustained and long-term increase in demand for solar systems. This is a pre requisite for mass production and thus for lower costs. Resulting large-scale manufacture could bring electricity generation costs down to 0.50 DM (study by RWE, Siemens, Bayernwerk), perhaps to 0.23 - 0.30 DM per kWh (Commission of the German Bundestag).
Production costs of photovoltaic equipment would begin to decrease at a production of about 100 MW/year.

3.10 What chances does the Aachen Model have of becoming a national program?

It's very hard to convince each city of the effectiveness of this model. However, to reach a volume of PV-production that will clearly decrease costs, the support of more than a few towns is necessary. Thus a national program is necessary. On the other hand the current government is opposed to the rate-based concept. Perhaps the German industry will advocate the rate-based model to stay competitive with other countries. The interest of a lot of cities is very great (see next page).

4. MARKETS

4.1 Rate-based Incentives in Germany

The program has been implemented in cities where the public and the politicians support the idea of a utility bill surcharge to encourage the installation of photovoltaics. the next page will give a survey on the status of rate-based incentives in cities which have implemented the program and also look at those cities which are considering the concept. Other cities known by the SFV to be actively studying the rate-based model, but it is not necessarily a complete list: Creilsheim, Göttingen, Hannover, Karlsruhe, Kassel, Krefeld, Magdeburg, Münster, Stolberg.

4.2 Rate-based Incentives worldwide:

In the USA the cities Arcata and Davis (CA) are considering Rate-Based Incentives. Following the lead of several European communities, a program is outlined in which Davis can support private investment in solar photovoltaic power through a 1 % residential electricity rate surcharge fund (about $8/ year per home). Based on city residents' present electricity consumption, the fund would be sufficient to pay for the annual electricity generated by 100 to 200 household PV systems. It is anticipated that with a payment rate of 50 cents per kilowatt-hour (kWh) delivered to Pacific Gas & Electric Company (the utility), residents will be able to pay for their initial investment in about 12 years.

In Brasilia the electric utility of Sao Paulo wants to adopt the model.

5. SOLARENERGIE-FÖRDERVEREIN

The Solarenergie-Förderverein publishes the topical state of the Aachen Model four times a year, as part of its coverage of solar energy market developments in Germany, in the magazine **Solarbrief**. To receive the magazine or further information, please contact SFV President Wolf von Fabeck, Philippe Welter or Anne Kreutzmann at:
Solarenergie-Förderverein, Bundesgeschäftsstelle, Herzogstr. 6, 52070 Aachen, Germany
Phone: +49 241 511616   Fax: +49 241 535786

Table 1 German Cities practising Rate-Based Incentives or a high reimbursement (2nd part of the table):

City	Population	Adopted	Rate (DM/kWh)	Rate ($/kWh)	Prog.-Ceiling	Rate paid for (years)
Aachen	250.000	1992	2.00	1.34	1 % surch.*	20
Hammelburg	13.000	1993	2.00	1.34	15 kWp	20
Remscheid	125.000	1995	2.00	1.34	100 kWp	20
Soest	?	1995	2.00	1.34	125 kWp	20
Elmshorn	47.000	1994	2.00	1.34	150 kWp	10
Freising	42.000	1993	2.00	1.34	100 kWp	10
Hamburg	1.6 Mio	1995	2.20 **	1.34	300 systems	15
Marburg	50.000	1994	0.99 ***	1.34	1 % surch.	20
Raisdorf	7.200	1993	2.00	1.34	20 systems	20

*surch. = surcharge on electric utility bill
** Hamburg pays differnet rates depending on the size of the PV-system. 1-5 kWp: 1,80 DM/kWh; 5-10 kWp: 1,60 DM/kWh; 10-50 kWp: 1,50 DM/kWh. In addition Hamburg will pay for the most effectiv systems a bonus of 0,20 DM/kWh and furthermore a bonus of 0,20 DM/kWh for every system producing more than 50 % of the owners electric consumption.
** Marburg: the state Hessen pays a subsidy of 40% of installation costs in addition to the 0,99 DM/kWh.

Table 2: German Cities considering Rate-Based Incentives:

City	Population	Adopted	Rate (DM/kWh)	Rate ($/kWh)	Prog.-Ceiling	Rate paid for (years)
Bonn	300.000	1995	2.00	1.34	1 % surch.	20
Eschweiler	55.400	1993	2.00	1.34	1 % surch.	20
Gießen	74.000	1994	2.00	1.34	1 % surch.	20
Gütersloh	90.000	1994	2.00	1.34	1 % surch.	20
Herford	65.000	1995	2.00	1.34	1 % surch.	20
Krefeld	249.000	1995	2.00	1.34	not defined	20
Lüneburg	64.000	1993	2.00	1.34	1 % surch.	20
Neuwied	64.000	1995	ca.2.00	ca.1.34	?	?
Passau	51.000	1995	not defined	not defined	not defined	not defined
Pleinfeld	ca.10.000	1995	2.00	1.34	1 % surch.	20
Schwäbisch Hall	40.000	1994	2.00	1.34	100 kWp	20
Wuppertal	388.000	1993	2.00	1.34	1 % surch.	20

Table 3: German Cities with a decision for Rate-Based Incentives but no practising: The program is approved by the city, but not yet implemented by the local utility.

City	Population	Current Status
Berlin	3.2 Mio	Aachen Model unanimously approved by Berlin Parliament (11/24/94). The Berlin Senate is requested to discuss adoption of the model by the electric utility BEWAG. The Senator for Economy wants a limit of 5 MWp for all PV-systems installed.
Darmstadt	141.000	Decission by town council on 06/27/95 expected.
Düsseldorf	578.000	Town council asked (!) the utility to pay r.-b.i.(05/18/95).
Frankfurt	664.000	The party Die GRÜNEN made an application for rate-based incentives which is still in discussion.
Fürstenfeldbruck	?	Decission within a pre-gremium of the town council on 07/19/95.
München	1.3 Mio	Active discussion of model in local politics; first decision of the "Energiekommission" expected on 07/04/95.
Rosenheim	?	strong activities for r.-b. i. from the local solar-energy-association
Saarbrücken	192.000	The "Bezirksrat Mitte" is discussing the model.
Solingen	167.000	1993 first application for r.-b. i.; but no decision untill today
Ulm	114.000	application from the Social Democrats (SPD)

THE ROLE OF INFORMATION NETWORKS IN PROMOTING SUSTAINABLE DEVELOPMENT

M.A. Hersh

Centre for Systems and Control & Department of Electronics and Electrical Engineering,
University of Glasgow, Glasgow G122 8LT, Scotland.
Tel: +141 330 4906   Fax: +141 330 4907   E-mail: M.Hersh@elec.gla.ac.uk

Abstract: This article discusses the role of information technology and networks in promoting sustainable development which harmonies the conflicting demands for environmental protection and economic growth, particularly in the 'developing' countries.

Keywords: Sustainable development, information technology, 'developing' countries, ecology, appropriate technology, information networks

1. INTRODUCTION

Increasing concern about the destructive impact of human society on the environment has led to wide acceptance of the need for 'sustainable development' which meets the developmental needs of humanity, without damaging the natural environment or compromising the ability of future generations to meet their own needs (Our, 1987).   Technology has also often been implicated in present environmental problems, giving rising to hostility towards technology and suspicion of those who develop and work with it.   However, it is not technology itself, but the ways in which it has been applied, that has caused environmental degradation.   There is thus a need for better scientific and technical education and a two-way communication process between scientists and engineers and the 'lay' population (Hersh, 1993).

Particularly in the last few years there has been a great expansion of information technology, leading to the so-called 'information superhighway'.   While information of all types is becoming more widely and easily available, this proliferation of information may increase inequalities of access, both between countries at different stages of 'development' and different socio-economic groups within a given country.   The spread of information and IT is also likely to have a dramatic effect on society as a whole.

Thus the theme of this paper is IT and information networking and their role in promoting sustainable development, with a focus on information sharing between engineers and scientists.   Technological transfer between 'developed' and 'developing' countries is also discussed.   The layout of the paper is as follows: information networking is discussed in section 2 and information technology in section 3; technology transfer is considered in section 4 and conclusions are presented in section 5.

2.   INFORMATION SHARING BETWEEN EXPERTS AND OTHERS

Information is required by planners and decision makers in order to formulate appropriate policy and support projects which promote sustainable development and by individuals to make changes in their life styles, so as to use resources more wisely.   Scientific and technical information has an important role to play in promoting development (Lim, 1985).   However information in itself is not sufficient.   To be of use it must be presented in a comprehensible form.   IT has a role to play in presenting information in an appropriate format, as in the use of geographical information systems to improve understanding of the changing relationship

157

between the environment and economy and draw attention to potential conflicts over resource use (Manning, 1990).

Since it is, at least in part, the effect of modern technology on the highly complex interconnected systems composing the planet which is challenging sustainability, much of this information is, of necessity, highly technical. However the general level of scientific and technological education of the majority of the population in most countries is very poor. Similarly to other professional groups, scientists and engineers have developed their own 'tech speak', incomprehensible to outsiders, and cultivated an air of mystery about their work, colluded with by the general population. Training in communication skills has rarely been a priority in the past in the scientific and engineering communities, though this is slowly changing. Thus the lay population is in general intimidated by scientific argument and technology and reliant on the advice of 'experts' (who may have their own agendas) in many important areas of life.

However individuals and non-government organisations (NGOs) require access to and a certain level of understanding of scientific and technical information in order to participate in important policy debates on a more equal footing and to give extra weight to the pressure they exert on governments (which are unlikely to do so otherwise) to take positive action for sustainable development.

Particularly in the industrialised countries little validity is given to knowledge and information which does not originate from 'experts' with recognised qualifications. However, for instance, indigenous people, particularly women, have much greater knowledge about certain aspects of the natural environment than the established 'experts'. The importance of information sharing and joint work by scientists and indigenous people has been officially recognised (Agenda 21, 1992).

There are also often inequities in access to information with a given society. For instance, women generally have much poorer scientific and technical educations than men, as well as being underrepresented in the scientific and technical communities for a variety of reasons, including the way they are socialised, the perception of some jobs as unsuitable for women and unequal access to education and training for women in many countries. Breaking down of these barriers is essential to allow full social expertise to be drawn on and to prevent this lack of facility with science and technology being transmitted to future generations.

Information sharing and interaction between scientists and environmental groups and non-

governmental organisations has in the past been responsible for focusing attention on a number of environmental problems (Juma and Sagoff, 1992). For example a combination of pressure from environmentalists and scientific evidence was responsible for drawing attention to the destruction of the ozone layer by CFC's (Grove-White et al., 1992).

Since only a very small minority of politicians have an adequate technical education and even fewer were originally trained as scientists or engineers, the majority of politicians and other decision makers are themselves suspicious of science and technology. There is a tendency for government to try to control science and technology, including using financial constraints to put pressure on scientists to follow particular lines of research in order to obtain funding (Boehmer-Christiansen, 1992). Since scientific theories evolve over time and different scientists often disagree, governments can generally find a body of scientists which backs up their policies.

Lack of information or disagreement between scientists on the extent of environmental threats has often been used by governments as a reason not to take action on, for instance, global warming. However this is purely a pretext, since recent official recognition of the causal link between emission of greenhouse gases and global warming has not led to a flurry of government action. Scientific reports have also been used to reassure the population about the effects of pollution incidents such as the Chernobyl disaster. However independence of the scientific and engineering communities from political control is vital for public confidence.

3. INFORMATION TECHNOLOGY: A MIXED BLESSING?

Since many of the advantages of IT are well known, this section will concentrate on the disadvantages. These can be categorised loosely into environmental, cultural, employment-related, equity and democracy-related issues. Despite talk of the so-called 'paperless office', IT has not reduced paper consumption, presumably because of the facility with which documents can be produced, altered and duplicated. Much of this paper is still neither produced from recycled material or collected for recycling when no longer required.

Present computer design and the continuous development of software which requires ever-increasing amounts of memory has led to machines becoming 'obsolescent' after about three years. When this happens some computers are down-graded to less demanding applications within the institution, supplied to schools or shipped out to 'developing' countries, but the remainder are presumably just dumped, since facilities for recycling computers are very limited. This problem of the appropriate disposal of 'outdated' machines, is further exacerbated by the replacement of main-frame computers, for which there are few alternative uses, by networked personal computers or supercomputers. Despite the large use of plastics in computer production, with associated consumption of scarce petroleum supplies, few computer manufacturers use recycled plastics.

Working with terminals can lead to a number of health problems, including headaches, eyestrain and repetitive strain injury, though some of these problems could be alleviated by improved workplace layout and frequent breaks.

The sheer volume and variety of information now available are themselves a potential source of serious problems, unless carefully managed (Morris-Suzuki, 1988). In addition to the increasing divisions between those with access to IT, there are differences in access to training and facility with technology. In particular, at least in the short term, use of IT will probably sharpen the cultural gap between generations, as younger people gain easy familiarity with IT, but older ones experience difficulties. This lack of proficiency may make it even more difficult for older people to find employment.

In much of Asia and Africa, there is a strong oral cultural tradition (Lim, 1985). Analogously to the move from oral to written culture caused by the development of the printing press, the spread of IT is likely to lead to a devaluation and loss of oral traditions (Lundu, 1989; Weyers, 1990). Although some of the important information about, for instance, local fauna and flora, embodied in this tradition will be preserved, cultural barriers between western trained information specialists and local people will cause loss of much valuable information. However other aspects of local cultures will be totally lost, leading to a serious loss of cultural diversity.

In most 'developing' countries communication is based on personal contact, which has the advantages of promoting group and neighbourhood cohesion (Chasia, 1978), whereas the pattern in 'developed' countries is for more impersonal contact using telecommunications and information technology. This lack of social cohesion is probably a contributory factor to prevalent feelings of alienation, often leading to anti-social behaviour, in many urban areas. Organisations also suffer from the loss of friendly atmosphere and group identification and the small time-savings associated with electronic communications rarely outweigh this

loss when personal meetings are replaced by electronic communication media.

The labour saving potential of IT could cause severe problems in 'developing' countries with their large pools of unskilled labour (Tiamiyu, 1993) and 'developed' countries with their growing unemployment rates. Some countries, such as Japan have avoided associated job losses by increased product diversification, including the production of a wide range of electronic gadgetry (Morris-Suzuki, 1988), but at the cost of unnecessary resource consumption.

IT is also changing the nature of many jobs and, for instance, in Japan is leading to increasing similarity between office and factory work (Morris-Suzuki 1988). In some cases IT use has led to deskilling, but more often to a shift in skills required from high level manual skills to a range of low or mid-level intellectual skills. IT could also make many jobs more interesting by integrating a whole range of operations in the hands of each worker, but this is being resisted by management policies of divide and rule or infected by ingrained, culturally determined prejudices, whether unable to imagine factory workers writing computer programs in Britain or women repairing machines in Japan (Morris-Suzuki, 1988). However the potential of IT to make jobs more challenging and interesting will only be realised if appropriate training is supplied (McEniry, 1989). IT has, in general, made administrative activities more efficient and a whole new range of administrative activities possible, but this may have contributed to the growth of unnecessary bureaucracy.

Advances in electronic communications have also allowed businesses, particularly in the United Stated, to move out of central business districts to take advantage of cheaper office space in the suburbs. The pace of this development has led to considerable traffic congestion which it will be difficult to find solutions for (Orsaki, 1987).

Developments in IT have important implications for democracy. On the one hand IT gives governments unprecedented powers of surveillance and data storage, particularly about political opponents and grass roots activists. Many countries do not have data protection acts or agencies or freedom of information acts to offer even a minimal legal protection to individuals. On the other hand IT gives individuals and non-government organisations (NGOs) access to information from a wide range of sources, including information their own governments would rather suppress, and facilitates the distribution of information on, for instance, human rights violations and environmentally damaging projects round the world, allowing organisations across the globe to network and co-ordinate action.

IT poses a number of hard questions about the ownership of information and puts a whole new dimension on the question of copyright (Massil, 1985). There is an increasing trend to electronic publishing which could eventually lead to hard copy publishing being superseded in some areas. Whether, in the long term, IT will lead to books and journals being almost totally replaced is still an open question. On the other hand IT has made much smaller print runs feasible. IT has also made it possible for small NGOs to produce and disseminate professionally presented information, thereby helping to counterbalance the enormous disparity of resources between NGOs and governments and multi-national companies.

In the short term 'developing' countries will be disadvantaged whether or not they acquire IT. The short term need to import both components and scientific and technical information will increase dependency (Anafulu, 1985), whereas non-acquisition of IT may block development. Despite the fact that hardware is becoming cheaper, high foreign exchange costs may be prohibitive. Importing technology is also out of line with the policy of countries such as Nigeria, which are trying to become less dependent on imports (Tiamiyu, 1993) and has environmental costs in terms of the fuel requirements and pollution associated with long distance transportation. Acquisition of IT in many parts of the world is based on outside finance, often accompanied by a preference for prestige projects rather than meeting real needs (Eldin, 1988; Ezekoye, 1984).

The need for a new world information order with equitable distribution and availability of scientific and technical information has been recognised. The United Nations has ratified a resolution to set up a global information network for the benefit of the 'developing' countries, but no practical measures have been taken to do this (Anafulu, 1985).

In general 'developing' countries have a largely rural population engaged in agriculture, whereas in 'developed' countries population is mainly urban and production industrial (Eldin, 1988; Subramanyon, 1983). Although there is clearly a need for modern and appropriate technology to be transferred to rural areas, urbanisation of rural areas in 'developing' countries is not the answer (Rao, 1986). The tendency to site IT facilities in urban areas in no way makes them accessible to the majority rural population, particularly in view of inadequate and unreliable telecommunications systems in many 'developing' countries. Since databases in 'developed' countries have little

information of relevance to 'developing' countries (Subramanyon, 1983), unless care is taken the spread of IT will only widen these divisions, both between and within countries.

4. TECHNOLOGY TRANSFER

Since there is already a body of experience in the development and use of IT in 'developed' countries, acquisition of IT by 'developing' countries involves at least an element of technology transfer. This then gives rises to the question of what (information) technology is best suited to the needs and conditions of a given 'developing' country. This is part of the extensive debate about 'appropriate technology'. The term was first used in United Nations documents in 1971 and has come to mean an overall approach which takes into account local socio-cultural and political as well as economic conditions, all aspects of community needs and results in a continuous improvement of the quality of life and community development (Vaid-Paizada, 1981). Appropriateness also implies that equipment can be used and maintained locally and that spare parts are available (Robredo et al., 1991). One of the pioneers in appropriate technology is the Appropriate Technology Development Association in Lucknow, India (Unambowe, 1979).

With regards to appropriate technology, to date experience with CD-ROM has been particularly positive and it is, for example, extensively used in many parts of Africa (White, 1992). It is a robust technology and not sensitive to heat, humidity, dust and other harsh environmental conditions found in many 'developing' countries. High impact can be obtained with limited resources. Costs are known and fixed in advance, unlike those for on-line retrieval systems and decrease per query (Weyers, 1990). Since there are no cost benefits associated with minimising search time, skilled searching is not required, which is of benefit in countries where there are few information specialists, as well as giving the power of search to the user (White, 1992), thereby promoting self esteem. Owing to its combination of relative operating simplicity and sophisticated search capacity, CD-ROM can be used to promote local computer literacy. It can also be used for regional resource sharing and communication or distribution of, for instance, pan-African databases by pooling the costs involved. A complete information retrieval and dissemination system could be obtained by combination with related IT, such as microfiche.

In terms of computer hardware, microcomputers are preferable to large machines due to the availability of software, increased tolerance of harsh environments and relative cheapness, making them more accessible to small institutions (Weyers, 1990), though costs still seem prohibitive in some countries. Despite their slightly higher costs, laptops are preferable to micros on account of their increased robustness and ability to cope with temporary failure of air conditioning and power supply deficiencies, such as surges and failures, frequently occurring in many 'developing' countries. For data transmission pocket radio is preferable to PSS services, since it is cheaper and more reliable and many 'developing' countries have radio communications (Weyers, 1990).

With regards to technology transfer, Dearing (1993) considers than the main barrier to success is difference and, therefore, that the solution is one of difference reduction. Many of the differences which complicate technology transfer are cultural. These include differences in working patterns and attitudes to work (Mehkati, 1986). However the frequently held view that progress requires 'developing' countries to adopt the cultural attitudes of the 'developed' countries is itself a barrier to development and, if followed, would result in an undesirable loss of cultural diversity. It is preferable that visiting experts learn about conditions in the host country and try to adapt themselves to them. The pattern of industrialisation and development followed by western countries has led to unsustainable resource use. as well as causing destruction of many ecosystems and pollution of the environment, in addition to (historically) very bad working conditions. It is therefore not an example to be emulated.

Cultural attitudes in the west are not the 'norm' and those in 'developing' countries a deviation from this norm. For instance, a Chinese survey (Bond et al, 1987) has identified uncertainty avoidance as a uniquely western dimension of culture. Therefore many of the applications of IT in the west concentrate on the production of timely, accurate and reliable information which can be used to reduce uncertainty in decision making. It may seem self-evident to readers with a western cultural background that this is an appropriate role for IT. However uncertainty reduction is not so valuable in countries which accept uncertainty and therefore, successful implementation of IT in these countries requires concentration on its other facilities. It should also be noted that the differences between the 'developing' countries are even greater than those between the 'developed' ones (Subramanyan, 1983) and therefore an approach that works for one country will not necessarily work for others.

It has been shown that there are no differences between the cognitive capacities of different ethnic groups (Winser, 1985), though there are differences in task performances, largely based on differences in

training, as well as differences in manual dexterity and body shape (Meshkati, 1986). However, although some ethnic or national groupings may, on average, acquire keyboard skill faster than others, owing to particular manual dexterity, IT can in general be used by all ethnic groups without adaptation.

Other important preconditions for successful transfer of technology include harmony of political priorities of recipients and donors (Robredo, 1991), or at least similar expectations of the transfer. This is often not ·the case. Developing countries also require adequate information in order to make appropriate technology choices (Unambowe, 1979). However information about alternatives and their performance is often unavailable and foreign consultants may encourage the purchase of expensive and unsuitable equipment from the donor country (Weyers, 1990).

There are different measures of success of technology transfer projects and it is important that aims and time scales, including those for training, are realistic (Johnson, 1992). It is inappropriate to measure success purely in terms of number of machines transferred, as done by some donor agencies (Weyers, 1990). Unless appropriate training and transfer of 'know-how' occurs, expensive hardware will probably remain unused (Ezekoye, 1984; Weyers, 1990). For understandable commercial reasons, donor organisations rarely transfer full information. However transfer of understanding of how a system works is more valuable than purely hand-over of hardware. Successful technology transfer is a two-way process. The donor should learn from the recipient as well as vice versa (Dearing, 1993).

5. CONCLUSIONS

Information sharing has an important role to play in promoting sustainable development by allowing a more realistic appraisal of threats to the environment and the likely consequences of particular policies and projects. It is important that it is a genuinely two or multi-way process and information from 'developing' countries and 'lay' people is not devalued.

The spread of IT is inevitable and should therefore be carried out in such a way as to reduce rather than increase existing inequalities and to empower 'developing' countries. Like all technologies, its dissemination is likely to have both positive and negative consequences. On the one hand it has the potential for the development of a technological police superstate, by giving governments unprecedented powers of surveillance. On the other it has the potential to promote democracy and reduce human rights violations by allowing NGOs and individuals to access and distribute information globally.

In the medium to longer term it can be used to aid the development of the 'developing' countries, but in the shorter term it will increase dependency and indebtedness. Spread of IT will almost certainly be accompanied by a devaluation of oral cultural traditions and some loss of cultural diversity. However successful transfer of IT requires a recognition of the cultural values and context of the recipient country.

IT is likely to change employment patterns and skills and will probably cause some job losses, but has the potential to make many jobs more interesting. Use of IT has the potential to save energy and other resources. However production of IT itself consumes resources and there is a need for governments to promote the production of environmentally friendly computers which use recycled materials and can be easily upgraded or recycled, through financial and other incentives.

REFERENCES

Agenda for action in the 21st century (1992), produced by the Rio summit.

Anafulu, J.C. (1985). Using automated systems. In: *University libraries in developing countries*, IFLA.

Boehmer-Christiansen, S. (1992). How much 'science' does environmental performance really need, *SPRU Report*, University of Sussex.

Bond, M.H. (1987). Chinese values and the search for culture-free dimensions of culture, *J. of cross-cultural pyschology*, **18**, 143-164.

Chasia, H. (1976. Choice of technology for rural telecommunication in developing countries, *IEEE transactions on communications*, **COM24**, 732-6.

Dearing, J.W. (1993). Rethinking technology transfer, *Int. j. technology management*, **8**, 479-85.

Eldin, H.K. (1988). Problems of technology transfer to developing countries, *technology management publication*.

Ezekoye, A.N. (1984). Before you phase out, think of technology transfer, *Proc. Int congress on technology and technology transfer*, 331-334.

Grove-White, R. S.P. Kapitsa and V. Shiva (1992). Public awareness, science and the environment. In *An agenda of science for environment and development in the 21st century*, Cambridge Univ.

Hersh, M. (1993). The role of scientists and engineers in promoting sustainable development, *Science, technology & development*, **11**, 272-290.

Johnson, J.S. (1992). Computerizing information systems in developing countries, *Quarterly*

bulletin of int. association of agricultural librarians or documentalists.

Juma, C. and M. Sagoff (1992). Politics for technology. In: *An agenda of science for environment and development in the 21st century*, Cambridge University Press.

Lundu, M.C. (1989). The information gap: reflections on its origins and implications, *Info. development*, **5**, 223-227.

Mc-Eniry, M. (1989). Appropriate technology transfer to the developing world, *Proc of the human factors society*, **33**, 757-760.

Manning, E.W. (1990), Geographic information systems and sustainable development, *Government info. quarterly*, **7**, 329-342

Massil, S. (1985). New information technologies available in the industrialised world. In: *University libraries in developing countries*, IFLA

Meshkati, N. (1986). Major human factors considerations in technology transfer to industrially developing countries, *Human factors in organizational design & management*, 351-367.

Morris-Suzuki, T. (1988). *Beyond computopia*, Kegan-Paul Int.

Orski, C.K. (1987). Managing suburban traffic congestion, *Transportation quarterly*, **41**, 457-476

Our common future (1987). World commission on envirnment and development, Oxford Univ. Press.

Rao, A.J. (1986), Role of universitites in rural development and technology transfer in India, *World conference on continuing engineering education*, 391-394.

Robredo, J., T.M. Robredo and M.B. da Cunha (1991). Some problems involved in the installation of advanced information systems in developing countries, *Resource sharing and info. networks*, **6(2)**, 81-95.

Subramanyan, K. (1983). Online searching in less-developed countries, *Nat online meeting*, 539-549.

Tiamiyu, M.A. (1993). The realities of developing modern information resource management systems in government organisations in developing countries with particular reference to Nigeria, *J. of info. sci.*, 19, 189-198

Unamboowoe, I. (1979), A unified information system for appropriate technology, *UNESCO science policy study documents*, **45**, 117-122

Vaid-Paizada, V.K. (1981), Is computing technology appropriate for developing countries?, *Western education computation conference*, 105-110.

Weyers, Y. (1990), Friend or foe? the microcomputer in developing countries, *Microcomputers for info management*, 7, 217-226.

White, W.D. (1992). CD-ROM in developing countries, *CD professional*, 32-35.

Wisner, A. (1985). Ergonomics in industrially developing countries, *Ergonomics*, **28**, 1213-1224.

COMMUNICATIVE SYSTEM DEVELOPMENT
ABOUT THE CHARACTER OF COMPLEX SOCIO–TECHNICAL SYSTEMS AND IMPLICATIONS FOR SYSTEM DEVELOPMENT METHODS

K. Münker, H. Jansen-Dittmer, M. Dahm, D. Meyer-Ebrecht

Institute for Measurement Technology, Aachen University of Technology, D-52056 Aachen, Germany, e-mail: Münker@RWTH-Aachen.De

Abstract: The relevance of the social-organizational context for the development of a computer workplace-system is underlined and illustrated by empirical results of two workstation-development processes in the medical field. The consequences for the system development process are pointed out and the novel method of Communicative System Development is introduced. This method is distinguished by its appropriateness for complex work environments. The proposed method includes instruments for the analysis of the interaction between organization, task and experience in the field of work and ensures the participation of the later user in the decisive phases of the project. Under the consideration that computer technology is "organization technology", organizational goals are worked out in cooperation with the later user. During analysis and realization phase all stages of experience are regarded concerning the handling of information, so that parameters relevant for the development of experience can be transformed into appropriate representations.

Keywords: Communication Systems, Information Technology, Socio-technical system design, Medical Applications, Organizational factors, User interfaces

1. INTRODUCTION

Increasingly, computer systems are developed to support users in highly complex applications like medical diagnostics or process control. In these areas, fixed work flows that can be modelled and transferred directly into a sequence of technical procedures give way to experience-based work. Experts working in their environment are frequently confronted with unknown situations where decisions have to be made on the base of varying sets of information. For an external consultant, who wishes to obtain indications for technical support of this work, individual problem solving strategies applied by the expert are hardly recognizable, because the expert himself is often not able to describe the knowledge used to solve a problem. New development methods have to be found that meet the requirements of the special characteristics of work. Additionally, the linkage of computer workplaces, forming computer systems, and the emerging need for support of cooperative work (CSCW) demands the acquirement of far-reaching and adequate structuring concepts and system architectures.

The object of work, the historic development of the work environment, the dissipation of power and hierarchy in a work field and the resulting organization, the various forms and stages of experience are unique and interdependent "parameters" in a certain working field. They form unique circumstances for an introduction of a new technology, especially a workplace network. The less regulated the work is, the more it is characterised by a complex structure of cooperation dependent on an intensive exchange of information and transfer of experience from experts to novices. Informal processes form an essential part of that complex structure and have to be considered in the analysis and the design phase to prevent a "computer-supported" loss of transparency and reliability in that organization.

Development methods that are currently in use show either a tendency to a pure technical treatment or an emphasis on psychological aspects. Even the combination of both in a development project is insufficient as long as sociological and psychological analysis on one side and technical concepts and implementations on the other side are strictly segregated.

163

The results from the analysis are only transferable into a socio-technical realization if both disciplines engage in the concerned work environment and if visions for a future work situation are developed in close cooperation with the user-to-be. Recognizing the necessity for a method that combines engineering and social and psychological sciences with the expertise of the users-to-be to a systemic approach that covers the complexity of this problem led to the project DIBA (Digital Image Workplace for the Radiology). The goal of the project was to develop a method how to design workplace systems that meet the requirements in complex work environments and to use the method in a representative work field. In the DIBA project a digital workplace as a part of picture archiving and communication system (PACS) for medical radiologists had been developed on the base of the novel method.

The article is structured as follows: In the second chapter the character and fundamental conditions for the acquisition, use and transfer of experience are described. In the third chapter, the structuring effect of technology is stated and enlivened by an example of PACS applications. The necessity of new approaches that have an emphasis on organization is strengthened. The fourth chapter draws the consequences for new development methods that have to involve the user in strategic decisions of the use of technology. The fifth chapter gives a summary on the method of Communicative System Development and the sixth chapter illustrates the use of the method for the development of an image workplace for the radiologist. The seventh chapter introduces the current project of the development of a workplace for the support of hospital care and is followed by the conclusion.

2. CHARACTERISTICS OF WORK BASED ON EXPERIENCE AND THE RELEVANCE FOR A SYSTEM DESIGN

In recent years experience, formerly widely ignored in the context of mechanization of the working world, is slowly regarded as an outstanding human ability to cope with large amounts of information. Experienced work characterized by an uncommon ability of perception and decision-making. Experts in their environments are able to handle and to evaluate a complex set of information in an instant. Thus, beside the cognitive-analytical, law-governed and rational expert knowledge, there exists a practical, procedural, intuitive and holistic knowledge, that is based on individual experience. As an example, the radiologist gains his diagnostic experience in long years of practice. The phenomena to be discriminated on the radiological images are very often discrete and unobstrusive features or feature combinations where an expert may often spontaneously and within a few seconds recog-

nize complex medical facts (Dahm, *et al.*,1993). The more competent experts become in their domain, the less able they are to describe the knowledge they use to solve problems. For a transfer of experience from experts to novices the institutionalization of cooperative diagnosis in small teams plays an important role.

The circumstances that promote acquisition, use and transfer of experience have been summarized by Witt (Witt, 1994): Experience plays a major role in complex situations with a high diversity of situations, where simple rules are not applicable. Mulitmodal sensory perception of processes and the feedback after interventions promote the acquisition of experience. A scope that offers a spectrum of alternatives for interventions as well as the emotional relation to the own task are unrenounceable conditions for the constitution of experience. Communication with colleagues in cooperative work situations allows the transfer of experience-based knowledge. Thus, the relevance of experience in work environments is not only determined by the actual task, the available tools and the circumstances in the close surroundings, but reaches to organizational conditions that determine factors like competence or the organization of work flow in cooperation with other colleagues. For a workstation design and also for the entire system architecture it is essential to gain insight into the organizational and technical conditions that hinder the use of experience and the conditions that promote the use of experience. This knowledge is then to be used in the workplace development process to promote the development of the worker's competence (Rose and Jansen, 1981) preventing that experience-based work is undeliberately reduced to strictly regulated work.

3. ORGANIZATION AS ELEMTENTARY OBJECT OF STRUCTURING

In most applications information technology and related computer-mediated communication are supposed to increase the availability of information for all members of the organization, thus integrating locally separated departments. Data flow and formal information exchange are then the principle guidelines for the conception of digital workplaces within the organization. Due to the inherent organizational meaning of data flow and information exchange, this results in a technical copy of existing structures and therefore a fixation of organizational structures. Also, the inherent interdependency among information and power is widely ignored. Studies show, that computer-mediated communication is usually designed precisely to support ongoing hierarchical relations (Child and Loveridge, 1990).

For example, the development of PACS had been triggered by processes outside the clinical work. Technological progresses allowed the handling of large data volumes concerning their storage, display, transfer and processing, and the medical field was found to be a challenging and prestigious field of application. The relevance of the PACS-technology is thus measured on the base of technical arguments like digital storage of images as a solution against the overflow of conventional archives, improvement of medical care by an increased information transfer and reduction of radiation dose by means of digital imaging. These arguments, however, lose in importance if the entire socio-technical system is considered. Incompatibilities between work processes and implemented network structures or even system functions can easily wreck the technological achievements. After years of research and implementation work side-effects came up that have their origin mainly in the organizational domain. It is reported in (Gel, 1994) that the application of a PACS leads to conflicts among departments and that the role of a radiology department as a central unit is put into question. Indeed, the availability of information for the physicians in the ward could have been improved by the network. Instead, this had been impeded by the radiologists. Experiences like this underline the results of studies of other work groups (Mantovani, 1994; Child and Loveridge, 1990). The PACS literature illustrates that PACS concepts have been established neglecting operational and organizational aspects of the underlying work and their interrelation. The design of the reporting workplace itself showed a purely technical treatment. Structuring effects of technology had not been considered. Today, radiologists reputed in the area of PACS research state that the complexity of the process of work in the radiology had been underestimated. The false estimations lead to the so called "banana products", supposed to mature at the customer, leaving the customer alone with the modification and completion of an otherwise useless system (Peters and Imhof, 1994).

The example demonstrates, that the introduction of technology leads to involuntary and thus unpredictable structural outcomes with corresponding problems for the individuals of that organization, if organization is not an object of design in advance. An information technology, offering maximal functionality and leaving the configuration to the user, pretends an illusionary adaptability to differing organizations. Such a retreat of project teams from the responsibility concerning the structuring effects of technology in organizations leads to an unpredictability of the structural outcome. Computer technology is primarily organizational technology, see for example (Hirsch-Kreinsen, 1984). But still, this is seldomly taken into consideration. Essential problems in modern settings of tasks, like CSCW, can hardly be solved adequately, if the organizational components are not elementary parts of the analysis and if they are not considered prior to the design of the tasks of work. Although the importance of organizational factors for a system design has been

stated in several publications in the last time (Volpert, 1993; Kötter and Volpert, 1993), substantial consequences for the development process are not visible. Instead, instruments for the analysis related to the individual workplace are proposed which are a contradiction to the demand for a design process "covering the whole area of the socio-technical system" (Kötter and Volpert, 1993).

4. CONSEQUENCES FOR THE SYSTEM DEVELOPMENT

Today, the necessity of a participation of the future user is out of question. But as the design process is focused on the design of the technology, the user-to-be is only involved in the design of displays and dialogues. The essential aspect, the motivation for the use of technology implying the concept for the system architecture and design-metaphor, is usually already defined by intentions of key actors and by attributes of the technology involved.

The maxim, that a balanced development process is only achievable if the users-to-be have a considerable influence on the motivation and the goal of the use of technology and not only on its realization was central for the development of CSD in the DIBA-project. This demand implies, that the method has to include instruments that allow the participation of a large amount of users from various levels and departments of the organization in an early phase of the project. The users with their varying backgrounds and fields of interest have to find a consensus on how the technology is used and which form of organization is chosen to be actively supported. This requires an expansive application of instruments of analysis over a large spectrum of users and additional design sessions in work groups. Here, the user is not the passive subject of analysis with the right to change details of implementation, but is seen as the user with the competence to reflect and evaluate the own work environment and to form visions for structural improvements with respect to a future use of technology. The analysis phase serves as an amplifier for this process.

5. COMMUNICATIVE SYSTEM DEVELOPMENT

The name-giving feature of the Communicative System Development (CSD) method (Dahm, et al., 1992; Jansen-Dittmer, et al., 1994; Dahm, et al., 1993) is the involvement of the actual user-to-be in the decision making process for the future socio-technical structure. This implies, that the design goal has to be open before the development process starts - a precondition that appears as a self-evident fact, but that is usually not the case due to analytical instruments that incorporate hypotheses to be investigated. The CSD method initiates with a situation analysis during which the current situation of the affected work environment and organizational conditions are investigated. For the development of prospections the analysis of currently existing structures, their comparison and rating is a vital precondition. The stages of the development method are embedded in a "version concept". Prototype versions are designed, implemented and tested within the work environment in various iterations. Fig. 1 shows the phases of the cyclic development process that is iterated for the development of the prototype versions.

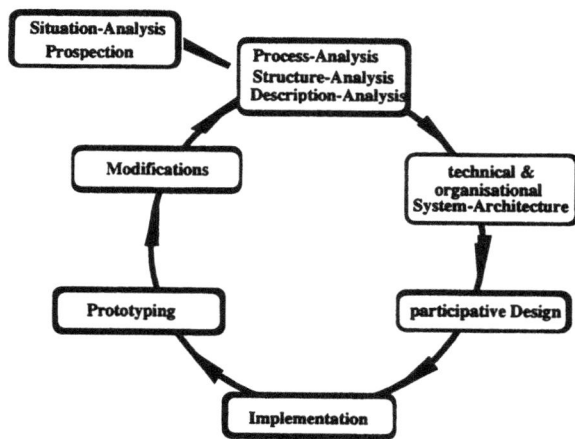

Fig. 1:   Version Concept of the Communicative System Development

During the situation analysis phase several user work groups are formed where existing working- and structural conditions are discussed and compared with possible new solutions. In the phase of the process analysis the organizational structures, forms of interaction, strain and conditions that hinder an improvement of working conditions are identified and analysed. The instruments applied permit the analysis of organization and tasks and their interaction and the seizing of qualitative aspects of work, for example the users' experience gained in the work environment. A mixture of qualitative and quantitative instruments is applied to gain the interdependencies between formal and informal parts and to ensure a synergetic orientation that corresponds with the dynamic character of the system. Alternative analytical instruments are selected in cooperation with the user (Dahm, 1992).

Prototyping and analysis are closely interrelated by the approach that demands the systemic view of tasks and their organizational embedding. It is mentioned above that the experience gained in a domain is dependant on a large amount of factors. Correspondingly, the possibility for the user to evaluate a technical approach is only given, if a task is done within its organizational context. Thus, the prototype has to cover integral tasks that may be reduced in their complexity.

It is not the detail of a function that is evaluable but the consecutive stream of working steps that form a flow of actions.

During the analysis as well as during the prototyping all stages of experience are taken into consideration. The successive enlargement of experience, that is initiated by the experienced situations is included in the working analysis. Relevant parameters for acquisition, use and transfer of experience are then transformable to efficient and usable representations that go further than the so-called "problem oriented views". In case of a new organizational setting the prototype will form a provocative technical approach that initiates discussions about the existing working situation in comparison to potential new work situations. The prototype serves as an initial proposal and a platform for modifications by the users-to-be thus offering clues for the technology to be developed. As already mentioned, the concrete field of application and the realization of technology is open with the begin of the analysis phase. It is emphasized here, that the realization of an "open" prototype is only possible, if known approaches like standard interface metaphors and network topologies are not seen as self-evident technical means with the begin of the development process.

6. THE REALIZATION OF THE DIGITAL IMAGE WORKPLACE FOR THE RADIOLOGY

During the situation analysis in the DIBA-project a prospection for the use of a reporting workstation had been worked out in cooperation with radiologists from various radiology departments. After several interviews and observations of radiologists during their reporting work it became apparent that an optional future technology would only be adequate if it is usable in the daily routine. Experienced radiologists were critical concerning the use of image processing. The gained experience in evaluating images is closely related to the image formation and image carrier and thus to the image characteristics. At the begin of the analysis phase it was obvious, that the utility value of an image workstation stands or falls with a beneficial image handling comprising image selection, image positioning on the display, switching between images and their removal from the screen. Thus the image handling being the crucial factor for the success of this technique therefore formed the central working field in the first iteration of the development process. The actual reporting task, the handling of large-scaled images where object of analysis. However, this task had to be seen in the context of work flow and organization so that work steps resulting from organizational conditions are separable from work steps resulting directly from the task itself.

For this, radiology departments of five hospitals of different medical direction and therefore differing organizational structures had been analysed. As expected, the organizational settings of the departments varied. There exists a continuum with the two extreme values describable as "clinical radiology" and "control radiology" (Dahm et al., 1992). A radiologist working in a "clinical radiology" usually diagnoses under consultation of clinical information, a radiologist from the "control radiology" deliberately excludes any additional information and wishes to evaluate images "unprejudiced". Thus, the work flow concerning the entire diagnostical process showed considerable differences, even the task itself was affected. If more information is consulted and team work and presentations form an essential part of the work process, tools are required, that support this style of work.

The major organizational aspect that had been considered for the workstation design was the support of the "clinical radiology". Concerning the task itself the analysis uncovered that the individual experience of the radiologist is directly connected to individual strategies of image-handling, thus leading to the design guideline that flexible and fast handling of images are required if the diagnostical experience can furthermore come into action at the image-workplace. These and other aspects had been taken into consideration for the design of the user-interface. Here, the questions arising in context with the design and evaluation of the interface could not be isolated and then analysed individually because they are strongly related to each other and have to be evaluated in their context. The interface had to facilitate the complete evaluation process with all its diagnostical process phases.

The limited display area of one or two monitors in comparison to the nearly unlimited surface display area of a lightbox or an autoalternator resulted in the need of a novel image handling concept. The close interrelation of experience with the image-handling and the individual strategies in image viewing demanded an image handling concept, that conserves this interrelation. By using monitors instead of large lightboxes, a parallel viewing strategy is replaced by a sequential viewing modus. This seriously interferes with the requirement of a simultaneous image comparison needed in the evaluation of the progress of an illness. A list of demands was set together in cooperation with users, that form unrenounceable conditions for a successive replacement of the lightbox by a digital workplace without reducing the quality and efficiency of the diagnostic process: First, a high switching rate between images had to be supported to guarantee acceptable working conditions. Secondly, the possibility of several images be viewed simultaneously offering an overview had to be given. And third image-viewing had to be kept undisturbed by interactions with the system by an intuitive interaction concept.

In a first design approach two displays where chosen, one for image-viewing and the other one for the organization of work. The "organization display", equipped with a touch-screen, was the interaction medium to obtain access to the patients data corresponding to the actual phase of the diagnostical process. The second high-resolution monitor for image-viewing, called "reporting-monitor", was positioned upright above the organizational display. With the idea of scaling down the images to small icons it was possible to display several of those miniature images on this "organization display". They could be shifted over the screen to give users the possibility of organizing individual image landscapes. Touching an icon resulted in the display of the large-scale image on the "reporting monitor". Thus, overview and detail was integrated in this configuration. The "reporting monitor" was kept free of text-information or menus to avoid distraction during image viewing. A fast image display and thus the possibility to exchange the images on the reporting monitor with at a high frequency had been realized by a special hardware configuration that had been developed in a parallel project (Winkler 1990).

To evaluate the quality of this workstation-approach in relation to the questions mentioned above, an experimental prototyping was found to be the only practical method. For a simulation of the diagnostic process, complete cases with several images including the clinical question had to be collected and stored on disc. In about 50 prototyping sessions radiologists diagnosed these set of images. The interaction via the interface was recorded in a logfile. Besides the direct communication with the user these logfiles offered an important tool to examine the behaviour of the user during a session. The results from the logfiles manifested the result from the analysis, that radiologists use individual, thus completely different image-viewing strategies during the reporting work. The configuration of the prototype allowed the continuation of the individual strategies. With the realisation of these elements the basics were laid for a reporting workstation unit that can withstand the demands for an application in the medical routine.

7. THE DEVELOPMENT OF A WORKPLACE FOR THE SUPPORT OF HOSPITAL CARE

In the PLUS-project (Multimedia Information and Communication Technologies for the Support of Hospital Care, BMBF 01HK152/3) a digital workplace for the support of the care process in an interaction-oriented nursing concept is being developed. CSD is the applied method. Here, the interrelation of organization and task becomes even more evident than in the DIBA project and there is a wider spectrum of users to be involved. Although the project is still in the analysis- and design phase, the concept of participation has already presented as offering promising innovations. The design approach is based on an organizational model that has been determined in close cooperation with the future user. Thus, structural changes that have been defined after negotiations among user groups are to be supported by the appropriate technology.

The development approach is focused on the ward and intends the integration of diagnostic and therapeutic data in the ward system thus promoting closer cooperation (but not computer-mediated) between the various disciplines and a patient oriented care. This approach offers new possibilities to reduce the strain in the nurses' daily routine. The PLUS-concept covers the entire care process including information retrieval, identification of resources and problems of the patient, definition of medical and nursing goals, planning of nursing operations and the assessment of their results (Jansen-Dittmer et al., 1995). The method of CSD allows the direct involvement of nurses in the structuring of their future work by the development of a future nursing concept. The nursing concept and the resulting structure then define the technology that is intended to support the care process.

8. CONCLUSION

There can only exist a framework for a development process that implies, that the goal of the use of the technology and its organizational embedding is not defined in its essentials in advance. Thus, the method applied has to be very flexible and has to be adapted to the work field with its unique character. Otherwise, the applied method "forms" the field of application, which is not the intention in using a method. The Communicative System Development offers one framework that offers a guideline for the course of process steps. Unrenounceable precondition for the successive application of this method is the openness of the development goal and the development process that incorporates the welcome influence by the user. Due to the openness of this process the project team has to engage in the changes resulting from the users influence. This requires flexibility of applied instruments and finally the flexibility of the project team itself.

9. ACKNOWLEDGEMENT

The DIBA project was funded by the German Ministry of Science and Technology, (Nr. 01HK 577 3). The PLUS project (BMBF 01HK152/3) is currently promoted. Only the authors are responsible for the content of this paper.

We would like to thank the participants of both projects; i.e. nursing service managers, nurses, medical technicians and radiologists from the following hospitals who contribute considerably to the results of the DIBA project and the ongoing work in the PLUS-project: Allgemeines Krankenhaus Ochsenzoll, Allgemeines Krankenhaus Eilbek, Wilhelm Augusta Krankenhaus, Clinic of the RWTH Aachen, University clinic of Münster.

REFERENCES

Rose H, Jansen H. (1981). Behinderung statt Entwicklung der Arbeitnehmerpersönlichkeit durch Computertechnologien?. In: *Zeitschrift für Arbeitswissenschaft* 1981/4 35(7 NF)

Child J., Loveridge R. (1990). *Information Technology in European services - towards a microelectronic future.* Oxford: Blackwell

Dahm M., Glaser K.H. Jansen-Dittmer H., Meyer-Ebrecht D. Münker K., Rudolf H.(1992). DIBA-Digitaler Bild-Arbeitsplatz. Arbeitswissenschaftliche Analyse als Gestaltungsgrundlage für einen radiologischen Befundungsarbeitsplatz. In: *Zeitschrift für Arbeitswissenschaft* 1992/2 46(18 NF)

Dahm M., Münker M., Jansen-Dittmer H., Meyer-Ebrecht D. Münker K (1993). DIBA - Digitaler Bild-Arbeitsplatz. *Abschlußbericht über das DIBA-Projekt* (Förderung BMFT, Knz. 01 HK 577 A)

Gell G. (1994). PACS-2000. In: *Radiologe*, Springer-Verlag 34:286-290

Hirsch-Kreinsen H. (1984). *Organisation mit EDV - Bedingungen und arbeitsorganisatorische Folgen des Einsatzes von Systemen der Fertigungssteuerung in Maschinenbaubetrieben.* Fischer R. G. Frankfurt

Jansen-Dittmer H., Münker K., Dahm M., Meyer-Ebrecht D. (1994). Prototyping als Element Kommunikativer Systementwicklung am Beipiel der Entwicklung eines digitalen radiologischen Befundungsarbeitsplatzes (DIBA). In: *Zeitschrift für Arbeitswissenschaft* 1994/1 48(20 NF)

Jansen-Dittmer H., M. Dahm, R. Schuber, Münker K., D. Meyer-Ebrecht, W. Hargens, L. Kleine-Besten, E. Zehn (1995). PLUS - Technikentwicklung unter Gesichtspunkten struktureller Veränderungen im Krankenhaus. In: *Informationssysteme im Unternehmen Krankenhaus.* Krämer, Cotta (Hrsg.), GeSi Mannheim

Kötter W., Volpert W.(1993). Arbeitsgestaltung als Arbeitsaufgabe - ein arbeitspsychologischer Beitrag zu einer Theorie der Gestaltung von Arbeit und Technik. In: *Zeitschrift für Arbeitswissenschaft.* 1993/3, S. 129-140

Mantovani G. (1994). Is Computer-Mediated Communication Intrinsically Apt to Enhance Democracy in Organizations? In: *Human Relations*, Vol. 47, No. 1

Peters P.E., Imhof H. (1994). PACS oder die schleichende Revolution. In: *Radiologe* (1994) 34:285 Springer-Verlag

Winkler, W. (1989). I.D.E.A. - ein Datenflußkonzept zur Verarbeitung, Speicherung und Darstellung digitaler Bilder. *Dissertation at Lehrstuhl für Meßtechnik RWTH Aachen.*

Witt H. (1994) Erfahrungsgeleitete Arbeit - Ein empirisch begündetes Handlungskonzept. Reader of the work group "Berufliche Erfahrung und Technikentwicklung". 39. Kongreß der Deutschen Gesellschaft für Psychologie.

Volpert W. (1993). Von der Software-Ergonomie zur Arbeitsinformatik. In: Rödiger K.H. (Hrsg); *Software-Ergonomie '93 - Von der Benutzungsoberflläche zur Arbeitsgestaltung.* Bericht des German Chapter of the ACM. Teubner: Stuttgart 1993, S. 51-66

MACHINE TOOL DESIGN FOR SKILLED WORKERS: INFORMATION TO EMPOWER THE MACHINE OPERATOR

R. Daude, H. Schulze-Lauen, M. Weck, C. Wenk

*Laboratory for Machine Tools and Production Engineering (WZL)
at the Aachen University of Technology (RWTH), Aachen, Germany*

Abstract: The development of man-machine systems for the control of manufacturing plants is the subject of several research projects. This paper shows concepts and realisation examples for the design of such systems for milling technology with CNC controlled machine tools, as they were developed at the Laboratory for Machine Tools and Production Engineering at the RWTH Aachen. The spectrum of covered examples reaches from compact universal machines to complex five-axis manufacturing centres for coon areas.

Key Words: Man-machine systems; CNC; Machines; Software development; Information.

1. INTRODUCTION

For a few years now, a tendency toward reincorporating the human into a central role on CNC-controlled machines can be observed during the design of machine tools. The strengths of the human being are seen as his flexibility and creativity, which are shown amongst other things in his reaction to unexpected occurrences. Over and above this, the specialist craftsman has a special extensive knowledge that offers great potential regarding technological parameters as well as planning activities [Carus, Schulze 1995]. These are only insufficiently utilised on current machine tools and controls, and, if the human were to be pushed out of this manufacturing process, would be lost. There is therefore, during the design of new machine tools, the tendency to combine the concepts of conventional and computer-aided manufacturing possibilities, in order to combat this problem. The aim is to combine the advantages of automation and control technology with the strengths of the specialist worker. High accuracy and manufacturing speeds in connection with high flexibility and the inclusion of the valuable specialist knowledge deliver advantages in form of higher productivity and more attractive work to the human and the company.

In the following paper, two concepts for the efficient support of the specialist worker during production on numerically-controlled machine tools will be intro-duced as examples, that have this symbiosis of man and technology as their aim. Both concepts, developed within the framework of research projects at the WZL, are solutions in the milling machine area of technology: Chapter 2 introduces the design of the user interfaces of a shop floor universal milling machine, that combines the intelligent CNC technology with ergonomic specialist worker support under consideration of his knowledge base and the possibility of an extended information accessibility. The third chapter presents a control concept for five-axis milling, which allows the worker to include his experience in complex applications such as coon shapes.

2. SUPPORTIVE ACTION FOR THE WORKER ON THREE-AXIS MILLING MACHINES

The completely manual operation of the machine by the worker is characteristic for conventional machine tools. The actions carried out here are largely based on experience gained in daily work due to information attained from the manufacturing process [Rose 1991]. Due to the design of the machine, the resultant "closeness" of the specialist worker to the manufacturing process and the complete character of the work, this knowledge base can be continually increased.

The acquisition and application of this valuable experience is impeded when working on numerically controlled machine tools. The necessary encapsulation of the machine, limited possibilities for process control due to purely electronic operating elements without feedback mechanisms and the NC programming in pre-production carried out in many manufacturing plants do not only create a spatial separation of process and worker; a lot of important information necessary for the acquisition and utilisation of experience is lost to the specialist worker due to the splitting of integral tasks.

The constantly changing market situation demands that the companies react in a flexible and quick manner [Weck, et al. 1990]. This increase in flexibility is only possible by the increased inclusion of the specialist worker in the production process, especially in single and small series production. The necessary support for the worker, as realised in the concept for the user interface of a milling machine introduced in the following chapter, is offered in two areas, which will be described in two sub-chapters. On the one hand the subject is improved possibilities for process control, on the other hand the extension of the information access to all data, that concern the manufacturing task.

2.1 Multimodal In-/Output media

Investigations of manufacturing practice show that specialist workers have a need for improved possibilities for manual process control of NC-controlled machine tools, especially during set-up as well as for exercising simple machining operations [Mertens, Rose, Ligner 1993]. This need is given credit in the example of the design of a shop floor-capable three-axis milling machine introduced here, in as far as, on the one hand, suitable control methods are integrated in the form of handwheels and a joystick for the simultaneous feed movements of up to three axes of movement. On the other hand, these control elements are utilised as force feedback media, that give the operator suitable feedback on the occurring process forces.

Various possibilities are available for the evaluation of the process forces. Force measurement platforms based on piezoelectronics, that are installed on the worktool side in the force flow between the machine table and the clamping fixture are suitable for the direct assessment of the axis-relevant force components during milling operations. However, high measuring accuracy is offset by high costs and limited practicability. For this reason machine-integrated measuring facilities that permit the evaluation of the machining forces via an appropriate sensor system on the main spindle or the feed drives and the deformation of these components are more suitable. A more elegant solution is offered by evaluating the current flow in the drive motors. These values can be used better for digital drives than for analogue drives due to the higher resolution,

and for the machine type used here correlate well with the occurring forces and moments. The information on the size of the drive current is available for this NC at the appropriate interfaces, making them easy to read and process. An additional sensor system for the evaluation of the process forces is therefore unnecessary. However, the evaluation of the axis-drives *and* the main spindle is necessary. Exclusive observation of the axis drives leads to problems during climb milling, where the cutting forces occur in the direction of the feed movement. On the other hand, the information of the main spindle current is not always sufficient in order to determine the direction of the cutting forces.

Mechanical and control-relevant solutions were found for the technical components for process control mentioned before - joystick and handwheel. Both components have an active restoring mechanism which, in comparison with a purely braking, passive system, has the advantage that it can break down occurring process forces and ensure cutting free of the tool even when there is no input from the user. This adds to an increased machine safety.

A step motor is used for the restoring mechanism of the handwheel, which is shown in Fig. 1, that, as opposed to other motor variants, is smaller and lighter and delivers high torque at low speeds. Its low price is a further main advantage. The disadvantageous locking due to the step movement is suppressed by a suitable control of the rotating field. The regulated moment of resistance occurring in dependence on the process forces is created by a controlled twisting of the rotating field against the stator field and by pulse width modulation of the motor current.

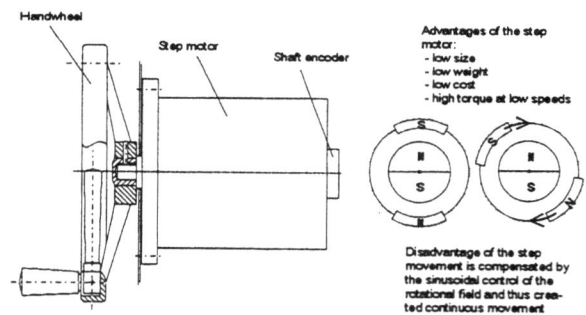

Fig. 1. Handwheel with integrated force feedback

The decisive advantages of the joystick against the other instruments for movement control of the machine axes are found in the possibility of definite allocation of movement between the joystick- and the NC-axes and the possibility of being able to position up to three axes simultaneously. Over and above this it is possible to transform the deflection of the joystick axes into a speed-proportional input value for the NC-machine. This results in an extremely comfortable input of the manually set control values for the user [Mertens 1994]. The integration of the control elements in a control-relevant concept is shown in Fig. 2 for the example using the joystick. The configuration possibilities of the variable force and speed characteristic curves can also be seen in this figure. The left half of the picture shows the control circuit for the non-feedback operation, whilst the circuit is extended by the elements shown in the right side of the picture for the feedback mode. The application of feedback capable operating elements on machine tools significantly improves the information access to the machining process, especially for the specialist worker.

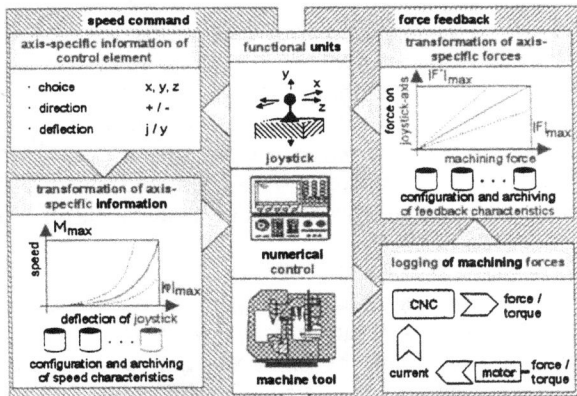

Fig. 2. System integration of operating elements for manual process control

2.2 Shop floor information system

The development of the man-machine system for machine tools still offers potential for improvement. Today, the specialist worker only partially profits the manufacturing and production data in the company. The experience of the specialist worker is still not sufficiently utilised [Fleig, Rundel, Schneider, 1993]. The creation and dissemination of feedback or remarks concerning manufacturing processes or manufacturing documents in pre- and post-production areas is still insufficiently supported.

As a solution possibility, the WZL of the RWTH Aachen has developed the concept of a shop floor information system called „InfotiF" (Fig. 3).

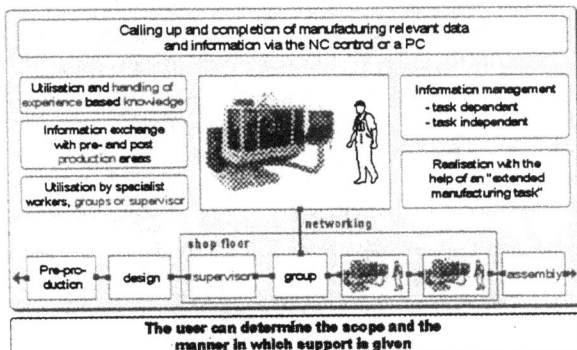

Fig. 3. Outline InfotiF

As a software information system for the shop floor, InfotiF is aimed at offering the possibility of calling up, completing and transmitting manufacturing-related information to any networked shop floor workplace.

As the basis for this functionality, appropriate areas such as pre-production, design, NC programming, shop floor, assembly etc. must be networking into a company information net. On the basis of this network, the shop floor information system has access to any manufacturing-related information that have be made available for this reason.

Via a PC-terminal or a PC-based NC control at his place of work, the specialist worker has access to all information relevant to his production tasks such as NC-program, work plan, tooling sheet, part drawings, complete drawings, documentation concerning clamping situations etc. and, using these or copies, can comment or edit them, or initiate feedback.

Comments in text or vocal form regarding any accessible information or general information - e.g. for the description of machine fault - can be input for the utilisation of experience of specialist workers, groups or other user groups. This information can be stored either related to certain projects (task dependant) or as general information (task independent). The user can make his information available to other users in a controlled fashion, as long as it is of general interest. Remarks that are also to be available for similar tasks, other machines or other workgroup can then be stored or managed in an indepedant manner by the shop floor information system.

The functionality of the shop floor information system can also be used by workgroups or the supervisor in the shop floor area.

Information entered into the system is allocated to the appropriate production task via a task status file (Fig. 4). The task status file contains, amongst other things, the combination of production information with the creator of this information as well as the location where the information is stored.

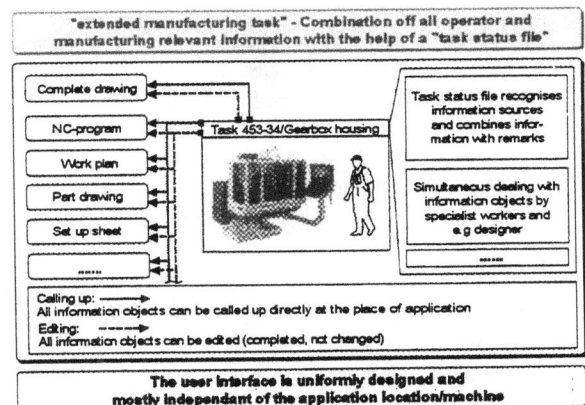

Fig. 4. Extended production task

Comments, suggestions for improvements, feedback or questions concerning the production documents are then forwarded to the appropriate professional by the shop floor information system on the basis of an analysis of the task status file.

Sequence example (Fig. 5):
The specialist worker would like to see the current work plan and maybe make suggestions for improvement. The shop floor information system finds the reference from where to load the current work plan and who created it in the task status file. After the specialist worker has entered a remark concerning the work plan, this is entered in the task status file. The entered remark is combined with the work plan in the task status file and the storage location of the remark is included. The status of the task status file now changes from "unchanged" to "changed". If the task has been completed, in this case the task status file is analysed and the additions or comments are sent to the appropriate professional, in this case the work planner. If the status of the file is "unchanged" after completion of the task, no analysis is necessary.

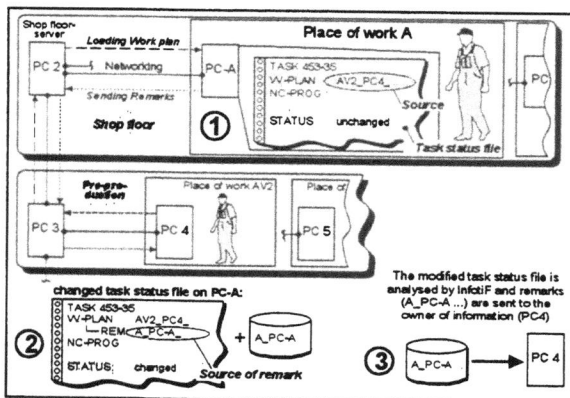

Fig. 5. Sequence example

This shop floor information system gives the user access to all relevant information for the task at his place of work. The system offers the possibility of controlled feedback in pre- and post-production areas. Remarks concerning production documents are not lost, but are rather made available to the appropriate professional. Organisational aspects have as yet not been treated. However, the problems are known and will be solved in further work stages.

3. SPECIALIST WORKER SUPPORT DURING FIVE-AXIS MILLING

In comparison to three-axis milling described above, other critical points must be taken into consideration during the machining of coon surfaces using five-axis machines. Whilst the creation of the NC programs in the pre-production leads to problems for the above mentioned milling operations, especially for single and small series production, amongst other

things the high degree of complexity of the coon areas is a hindrance for programming directly on the machine. However, the type of program that is currently transferred from pre-production to the machines degrade the specialist worker to a simple process observer and controller. Due to information regarding the workpiece geometry and the complex feed movements of the machine missing from the program he has no manual operation or interaction possibilities.

The description of the workpiece to be manufactured is generally delivered via a CAD drawing. The CAD data set contains the complete information regarding the contour of the finished piece. One result of the work of the pre-production is a machine-independent NC program describing the feed movements of the tools. The feed movements of the machine axes available after the post-processor run are dependant on the type of machine tool. Over and above this, they are always very small linear segments in axis coordinates.

Due to this methodology, only low-level information regarding the axis movements is available at the machine, from which the actual finished contour of the part can only be calculated with difficulty. The specialist worker can not check the correctness of the feed movements, as information regarding the connection between the axis movements and contour creation is missing. Any change of the programmed parameters - e.g. the replacement of an unavailable or broken tool, the application of a reworked tool with resulting changes in the tool data or the remedy of program errors that were noticed during running-in - necessitate the return of the program to pre-production. The specialist worker often has only the override switches for revolution speed and feed speeds as an engagement possibility, which leads to the degradation of his workplace and his qualifications.

In order to combat these inadequacies, a concept was developed within the framework of the MATRAS research project of the European Union and with the help of the WZL, some main elements of which will be described in the following. In doing so, the problem of the low-level information at the machine control is considered first. The information concerning the workpiece geometry is available in full in the CAD system. The CAD system uses complex spline functions to describe the workpiece surfaces. In order to be able to use this format at the control, the NC must be able to process spline functions. The NC functionality achieved in MATRAS permits the use of splines. In doing so, the model of a workpiece surface is available in the control. Fig. 6 shows the information chain from the CAD system via the program creation (CAM) to the machine.

The central question during the search for a suitable approach for the creation of the final workpiece ge-

172

ometry is which areas of the contour can be machined with identical technology data. The main focus in MATRAS is on final working. This leads to the subdivision of the total contour into surface elements that can be displayed by the workpiece model in the control using spline functions.

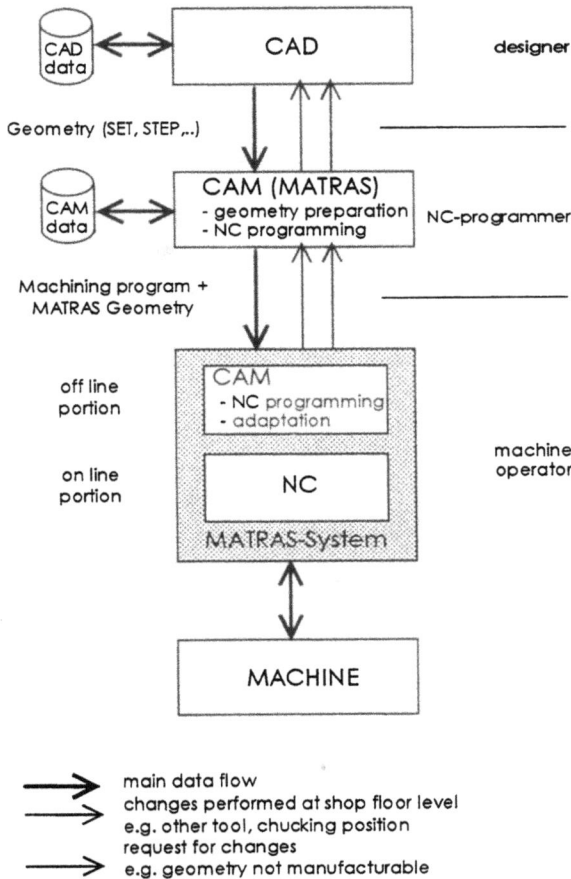

Fig. 6. Information flow and feedback loops

So-called WorkingSteps are defined for describing the partial surfaces. A WorkingStep is a region of the coon surface, which is machined with *one* tool and *one* set of technological parameters. As shown in Fig. 7, a workplan for the creation of a total surface is generated by sequencing the WorkingSteps, thus breaking down the highly complex task of programming coon surfaces into comprehensible units.

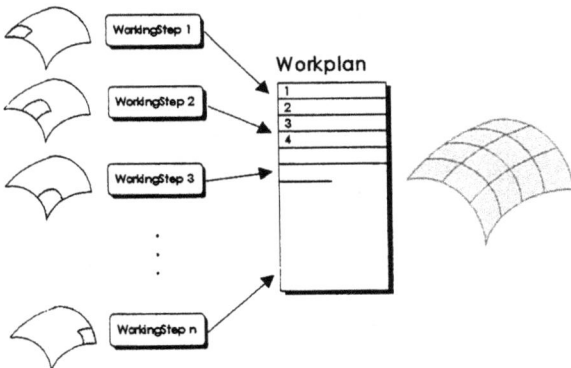

Fig. 7. Building a Workplan from WorkingSteps

This method has two great advantages for the specialist worker: On the one hand the availability of geometry data at the control opens new possibilities to understand the process sequence, to engage in and to modify it. For instance, a different tool than the programmed one can be utilised, as the control has the means to carry out the tool compensation and the transformation of the miller movement in the direction of the axis necessary in this case. On the other hand, the specialist worker is put in a position to be bale to choose or fade out individual milling tracks. This, for instance, is sensible in the case of model milling in sift materials if the programmed overlap angle is defined for a hard material. In addition, on this level of WorkingSteps the worker can make further changes to the program. He can change the sequence of the WorkingSteps, fade out individual steps or carry them out individually. Changes can also be made within a WorkingStep, e.g. if the worker, due to his experience, wishes to change the technological parameters or to react to unexpected situations.

The concept introduced here shows how, by improving the access to information regarding the workpiece contour and the machining strategy, a stronger integration of the specialist worker into the manufacturing process can be achieved even in complex machining situations such as coon surfaces.

4. CONCLUSION

This paper introduced examples of realised solutions for man-machine-interfaces and extended functions on numerically controlled machine tools carried out at the WZL. It was shown that the availability of information about the machining process and its sequencing is an important factor for the design of such systems.

As opposed to conventional concepts, advantages during three-axis milling are detectable for the specialist worker by the utilisation of operating elements with force feedback, as these present an additional help for controlling the process. The introduced concept of a shop floor information system offers the specialist worker extended possibilities for the utilisation of manufacturing information as well as the inclusion of his experience.

The control concept for five-axis milling introduced here shows the way in which the specialist worker is given direct access at the NC machine to valuable information regarding the workpiece geometry and its sequential programming even for coon surface machining. This puts the worker in the position to react flexibly according to his experience to unexpected situations in these complex applications.

Apart from an improvement of the workpiece quality and an increase in productivity, the result of such an improved man-machine-communication is the revaluation and increase in attractivity of the specialist work.

REFERENCES

Carus, U., Schulze, H. (1995). Leistungen und konstitutive Komponenten erfahrungsgeleiteter Arbeit. In: *CeA - Computergestützte erfahrungsgeleitete Arbeit* (Martin, H. (Ed.)), 48-82. Springer-Verlag, Berlin Heidelberg.

Fleig, J., Rundel, P., Schneider, R. (1993). Auftragsdisposition im FacharbeiterInnen-Informations-System CeAFIS. In: *CeA - Flexibilität durch Erfahrung* (Bolte, A., Martin, H. (Ed.)), 78-100. Verlag Institut für Arbeitswissenschaft, Kassel.

Mertens, R., Rose, H., Ligner, P. (1993). Fräsen mit Kopfhörer und Joystick. *Technische Rundschau*, Sonderheft CNC-Steuerungen, 10-14.

Mertens, R. (1994). Facharbeiterorientierte Möglichkeiten zur Prozeßführung und -beobachtung an NC-Werkzeugmaschinen. Verlag Kühner, Aachen.

N.N. (1994). CNCplus - Neue Wege in der CNC-Programmierung. Mercedes-Benz Technologie-Trendworkshop, 11.11.1994.

Rose, H. (1991). Bedeutung des Erfahrungswissens für die Bedienung von CNC-Maschinen. *ZwF* **86 (1991) 1**, 45-48.

Weck, M. *et al.* (1990). Wettbewerbsfaktor Produktionstechnik. VDI-Verlag, Düsseldorf.

USER INTERFACE DESIGN FOR MACHINE TOOL CONTROLLERS USING OBJECT-ORIENTED DESIGN TECHNIQUES

D. Zühlke, G. Schneider, M. Wahl

University Kaiserslautern, Institute for Production Automation

Abstract: The structural changes in manufacturing technology and the increasing complexity of modern production equipment will have a significant influence on the communication between the technical system and human. As a consequence of the lean production philosophy, the human operator has regained more importance. This paper describes the current status in the development of user interfaces of machine-tool controllers. Based on this analysis, a systematic approach to design task-oriented user interfaces is presented using object-oriented design methology.

Keywords: Design-systems, ergonomics, man-machine interaction, methodology, object-oriented modelling techniques, user interfaces.

1. DEVELOPMENT IN THE MANUFACTURING TECHNIC

Due to the advances in microelectronics, today's production facilities and machines are not only more powerful but also more complex. As more and more functions are automated, the tasks of the operator are changed from process control to process monitoring. This development as well as the use of the computer and visual display unit must be taken into account for the design of an operating system. The increasing presentation possibilities require the adaptation of the application to the operator, to his tasks and the machine. This is getting more important through the introduction of new work structures e.g. teamwork. An operator not only has to be able to handle several machines without specific training but is also responsible for the detection and elimination of errors (Zühlke and Schneider, 1994). The operator needs support especially for this task because he has to act quickly and systematically, which is of increased importance in systems with many informational or mechanical links and reduced buffers. It is therefore necessary to adapt technology to the human (human-centered design) with the goal to fully use ledge and skills. In the follo-

wing a guideline is described based on the analysis of user interfaces of modern machine-tool controllers. It contains object-oriented-design-methods which support the designer of user interfaces to consistently structure and present information.

2. USER INTERFACES OF MODERN MACHINE TOOLS

The trend in control technique towards "open systems" allows the producer of machine tools to configure control systems according to his own requirements by using quasi-standards, that already exist on the market. This also includes the user interface.

To hold costs down, more and more operation systems are developed using MS-DOS with WINDOWS and special realtime extensions. UNIX-based controllers are also available, allowing the installation of the corresponding hardware from the PC- and workstation-world. Color monitors or color-LC-displays with VGA resolution are often used as indicating element. This puts graphical presentation methods at the disposal of the engineer for the communication between human and machi-

175

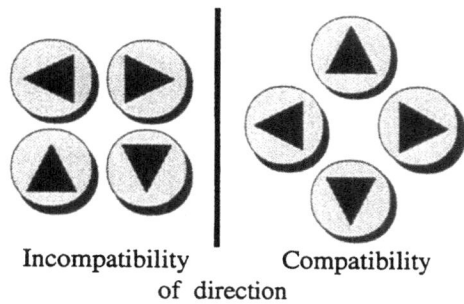

Incompatibility   Compatibility
of direction

Fig. 1. Group with cursor keys

ne.

Further, an operating hardware is often installed, such as a hand-held panel for setting the machine tool with additional operating elements for starting and halting, a NC operating unit with QWERTZ or ABC keyboard for programming and operating elements for axis motion and override functions. Usually a group of cursor keys is integrated in this operating unit. However one can often observe an unfavorable arrangement of the cursor keys (incompatibility of direction). Fig. 1 shows such an arrangement and in comparison a favourable arrangement using the design rule of compability(Norman, 1988).

Electronic handwheels are also used for setting the machine tool or for manual positioning. One can often notice, that the missing lettering complicates the allocation to axis and sense of rotation. Some controller producers also use the hand-wheel for setting the override values. In order to do so an additional switch toggles between axis and sense of rotation. This implies the danger of confusion, because the same operating element is used for different operating functions. In the majority of the cases, though, the override values are set by means of separate keys or potentiometers. However, when using these keys, the monitor must be observed because the exact position values are presented here.

The monitor is installed as an indicating element for the presentation of information, programming and parameterizing. The method for selecting the available information determines the user interface

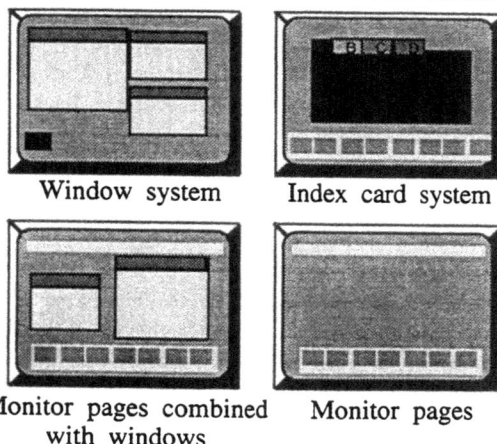

Window system   Index card system

Monitor pages combined   Monitor pages
with windows

Fig 2. Typical user interfaces

(s. fig. 2). Most controllers show a window-oriented interface, in which single windows are arranged next to one another and do not oberlap.

Individual windows cannot be extended, reduced or modified in their position because the pointing devices, common in office communication, are unsuitable for the rough environment. The invocation and the closing of a window is effectuated by push of a button. Windows for the input of data and for messages appear "in front of" the other windows.

Contrary to the applications used in office communication, where only the user opens the windows, a machine tool controller can present e.g. notice or error spontaneously and directed by events. It is therefore possible that an operator is hampered in the execution of his tasks by opening of such a window or that important information is suddenly hidden from view.

In order to avoid this, interface areas for important error and notice messages should be defined during the development of the user. These areas should not be hidden in any operating status.

One of the analysed NC controllers offers the operator the option to configure the user interface himself: he can determine the windows with the information which is important to him. The resulting problem is that by operating several machines as a workgroup, an additional operator does not instantly recognize the individually configured user interface.

In contrast to this system user interfaces with monitor pages replace the entire screen and its content. This is a result of a staticly defined menu structure which can not be configured by the operator. On one hand the disadvantage of such a system is that the operator is unable to invoke additional information. On the other hand the advantage is that the strict hierarchy, when well made, enables the user to get aquainted with the system quickly.

One of the controller producers has implemented an index card system where the information is attached to different areas. Every area is represented by a index card with a defined color. The index cards are visible through their labeling flap. A individual index card is selected by pressing a special key, bringing it to the front of the stack.

All of the analysed NC controllers have a similar screen structure. On the upper edge of the screen a line with general information is featured, such as main mode of operation, date and time. The part with the essential information is placed in the middle of the display. The description of the softkey assignment is located at the right or lower screen edge.

In window-oriented systems the input of data is carried out either in special input windows or directly in appropriate input areas in the window.To do so the window must first be activated. Activ windows are displayed in the foreground, having a title line or a frame with a different color. Analysis has shown that these methods of emphasising a selected window are not sufficiantly explicit, as it remains difficult to quickly determine which window has been chosen.

Softkeys are mainly used for selecting information, due to the fact that there are no pointing devices. They can be found either at the right side or below the screen of the monitor. Through the spatial arrangement of the assignment line of the softkeys, the direct allocation is given between the assignment and corresponding softkey.

Words, (not icons) are mainly employed in the assignment line of the softkeys. Because of the limited space, abbreviation and word separation may hinder correct interpretation. Difficulties might occur in understanding the icons, which every controller producer often generates himself. To achieve a greater menu width, scroll functions are integrated in the assignment line of the softkeys. By doing so, the time for searching a definite menu option is increased and the overview is made more difficult.

A selection is also possible, but without a pointing device or softkey. One controller producer has arranged a menu bar on the upper edge of the screen. The options and suboptions can be activated by means of numeral keyboard (s. fig. 3). For this purpose a numeric keyboard is schematically illustrated on the left side of the menu. In this illustration the numeral key is high-lighted using a red rectangle, (in fig. 3 as a black rectangle) which has to be pressed to select the menu option.

In many of the surveyed control interfaces the design of monitor pages and window lacks clarity because a structured information presentation is not employed. Criteria for this structuring can be e.g. the position of axis, technology or characteristic values.

Existing design attributes such as e.g. color, font size and font type are not used to demonstrate that the information can be divided into groups, depending on its content. Such a classification helps the operator recognize and sort information. This is especially necessary in situations where a large quantity of data is presented. Here one must verify

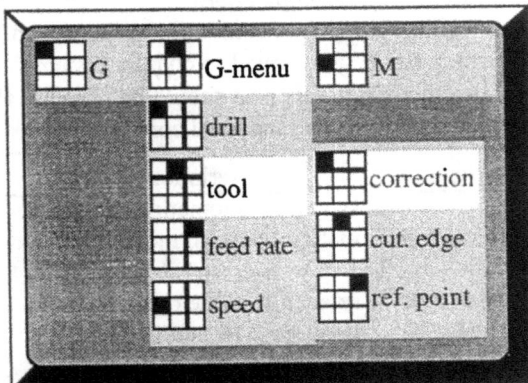

if all this information is really necessary. E.g. in the automatic operation mode the actual positions of axis and the lines of the NC program code are displayed. Because machine-tools and consequently the program run at high speed, the data shown changes so rapidly that it is no longer readable. An information presentation is required which is adapted to the actual situation.

An analysis of the menu structure has shown that it is not oriented to the tasks of the operator, but focuses on machine functions. The operator must therefore often carry out navigational movements in the menus from one step to another in the operating cycle.

The use of color as a attribute for information codification is frequently not adjusted to the general conventions and norms. E.g. the color red is in one case used as a background color for operation notices with white fonts and in another case or for identification of an activated input area. The use of many colors and color gradation in user interfaces cannot catch the attention of the operator. Additionally the bad contrast of the foreground and background color frequently reduces the perceptibility. As already remarked in reference to the use of fonts, a clear relation between the classes of information displayed and the color used can not be recognized.

The study of user interfaces of modern NC controllers presented above, clearly underlines the necessity of supporting the operator with a structured system. Additionally, fundamental design rules should be used with detailed statements to screen structure, tabular format, color use, and object selection as well as homogeneous designations and icons. This enables the operator to rotate easily from machine to machine without the demand for a "generic user interface".

In the following a systematic approach is presented which considers among other things the tasks of the operator and necessary information (in regard to the particular situation). Object-oriented-design-techniques are used for structuring the information and linking them to design attributes. This guarantees consistency and implemented as a software tool supports the development engineer of user interfaces.

3. STRUCTURING OF INFORMATION USING OBJECT-ORIENTED-TECHNIQUES

By establishing classes of information all objects of a class can be allocated the same design attributes, this is a consequence of an object-oriented method. Each class contains objects with common characteristics (Hickersberger, 1994). Every object of a class is described via attributes and methods. The attributes thereby represent inherent characteristics of the object, the methods describe the functions its

Fig. 3. Selection of menu options without a pointing device

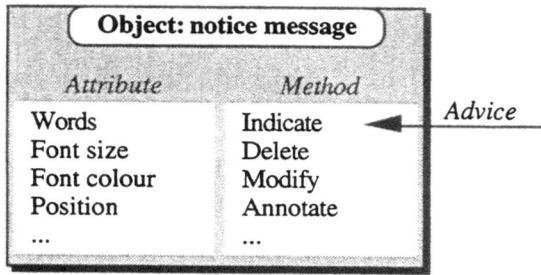

```
┌─────────────────────────────────┐
│    ╭──────────────────────╮      │
│    │  Object: notice message │    │
│    ╰──────────────────────╯      │
│   Attribute        Method        │
│   Words           Indicate  ◄──── Advice
│   Font size       Delete         │
│   Font colour     Modify         │
│   Position        Annotate       │
│   ...             ...            │
└─────────────────────────────────┘
```

Fig 4. Object and corresponding attributes and
methods

performance offers. The attributes of an object can
be read or modified only by the methods which belong to the object. Therewith an object is an encap-
sulated unit. The advantage is that further objects
can be integrated without changing of the defined
structures. Thus existing systems can be supplemented without expense. Functions of an object are
requested or triggered by message from other objects.

When speaking of the design of operating systems
an object represents a special information, such as a
special notice message. Fig. 4 shows the object
"notice message" with the corresponding attributes
and methods.

The relationship of objects to groups results in a
classification. All objects of one class are unique
representatives of this class. Common characteristics are used for the classification. Specialized
classes adapt or inherit the characteristics of general classes. By doing so all objects of a class have a
set of variables. Each individual object will then
have different values for that variable, e.g. different
words in the class "notice message". The inheritance principle allows all objects of a class take over

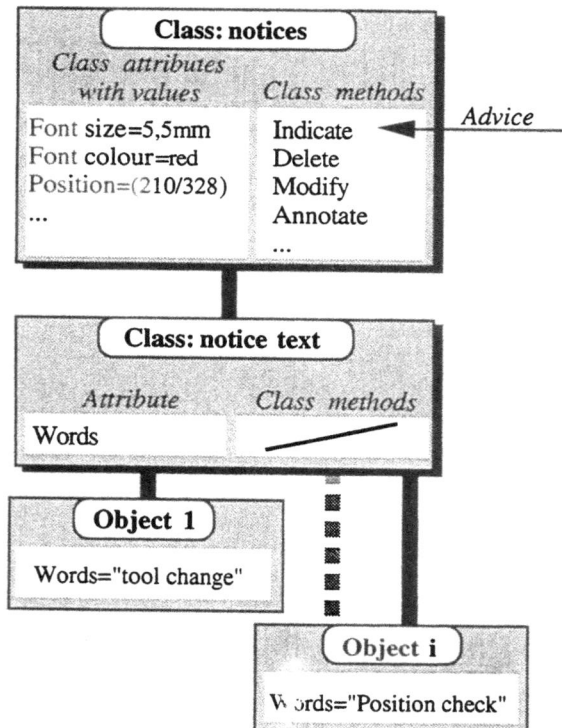

Fig. 5. Classification and inheritance

all characteristics from the higher class(es). These
are called class attributes or class methods. They
are linked to the class and as such to all objects of
the class. This assures that all objects of a class automatically dispose the same common attributes,
values, and methods. E.g. fig. 5 shows the class
"notices" with the subclass "notice text". A special
notice message would be an object of the class "notice text".

Consistent design of operating systems is guaranted
through the object-oriented methology. The implementation of this approach supports the development engineer and is presented in detail in the guideline below.

4. A DESIGN GUIDELINE

In the following a guideline should be given which
will lead designers to a human-oriented design of
man-machine-interfaces.

Definition of possible process / device states
First, the possible states of devices and processes
have to be determined. For these, the characterizing
variables and values of the process or device must
be defined. If available, additional information on
the status may be added. This is especially important in case of error handling.

status	variables and values	additional information
error 1	oil pressure < set value	tool position, operation mode (automatic or manual)
error 2 ...		...

By this action the development engineer is forced
to structure a problem into definite states which
may later be linked to display objects and actions.
If these states are defined as objects, they only have
to send a advice to a display object (e.g. the advice
"indicate" to the additional informations which are
necessary for the operator in this moment) or to other actions in case of appearance of the status.

Definition of operating functions
Next, the operating functions have to be determined which can be performed to change system states. After definition this data should be grouped already according to its functional correlation.

functions	description
select operating mode	switch to automatic, semi - auto- matic or manual mode
select feedrate override	select a feedrate override value between 20% and 120%
...	....

These first two steps may be supported by applying
formal description methods like Petri-nets or equivalent state-transition nets.

178

Task analysis and definition
As pointed out already earlier, the operator has to perform specific tasks connected with definite goals to achieve. Therefore, in the next step those tasks should be carefully analyzed and defined in detail based on the status and function definition.

Technical task: error recovery

operation sequence	necessary information	user action
1 display error msg.	error text	select error report and recovery page
2 display error report and PLC status signals	detailed error description and PLC status ...	check device

Normally, this requires the integration of the later user as he knows best which tasks must be performed and in which manner. Typical tasks in the field of production systems are e.g. tool monitoring and changing, reentering automatic mode after process interruption, cold-start of the production system. For each task the sequence of operations and the necessary or at least helpful information to be displayed as well as the required user action must be determined.

Definition of requirements to the operating functions and information display
As a result of the task analysis, operations and the technical functions have been described. Next, these have to be analyzed with respect to the importance, the time, the interdependancies etc. as shown examplary in the following list.

requirement	conclusion
frequent operation	arranging controls in the panel center
parallel operation	give information at the same time, arrange controls close to each other
sequential operation	give information sequentially, arrange controls logically

The analysis will then lead to design conclusions indicating menu structures, where to place e.g. function keys etc.

Grouping of information with time correlation
Here mainly for display reasons, the information is grouped into categories which have to be displayed

group	information	status of machine or process
group 1	value of the tool lip radius, position of the tool, operation mode, quality of work piece, status of the pump	tool lip radius < set value
group 2	position of the work piece, time rates	normal status

at the same time.
For this object-oriented-design-techniques may be used to classificate the defined groups. So a single advice is sufficient to display all the informations containing to the group.

Definition of information classes
Here information is grouped into classes according to the same type.
Classes could be e.g.:
- *error messages*
  The error message class itself should be divided into subclasses, like *fatal errors* which require immediate action, *warnings* which require attention and action. These classes should be connected with a priority to allow for proper sequencing during display. The priority can also be process state dependant such that messages with no actual relevance will be stored helping the operator to concentrate on the important things.
  Error messages must be visible in every process status and can only be eliminated by a distinct action. There should be a unique area reserved for error messages on a display such that a blank area is indicating that no errors are present.
- *general messages*
  General messages may be subdivided into e.g. status messages, recommendations on further actions. These messages should be given context sensitive and also in a reserved display area, which is visible all the time.
- *help messages*
  The help messages explain each function and may give the operator additional advice if he askes for it. They should be given also context sensitive.
- *process parameters*
  Process parameters like set values or actual values are of great importance, as they indicate the process status and possible deviations. Their display should be carefully chosen in respect to size, color and other display attributes. In highly dynamic processes the software should hold each data output to at least 5 to 10 s to allow the operator to read the data value.
- *input information*
  In this class all information is grouped which represents an input to the process. All input operation should follow the same rules, i.e. by opening an input data window or by selecting and then changing parameters.

class	presentation form	presentation location
error messages	text	display > error area
general messages	text	display > msge area
help	text and graphics	help window ...

179

By using object-oriented-design all members of a class have the same attributes like e.g. position, colour and font size. Also the operator may attach very quickly a information to a class and so to its meaning.

Definition of feedback
To prevent the operator from performing unintended actions a good feedback is necessary (Shneiderman, 1987). First, the best feedback form for each action should be determined. This will start with tactile feedback given e.g. by different button shapes. Aural feedback is often to be found in computer environments by using different beep sounds.

action	form	feedback
input of a value	visual	visibility of input data on the display
pushing a button	aural	short beep
error detection	aur+vis	sirene+red flashlight

In future, artificial voice will surely become more popular, as cheap digital voice message systems are now available. But, aural feedback is only applicable in environments with not to much noise. Therefore in most production applications, it is unacceptable. Visual feedback given by e.g. lamps or the display is also a good advice and above all used to give complex feedback information, like feedback on the process status when the transition between initial and desired state takes some time.

Definition of a failure strategy
In the case of operating system failure (i.e. display or keyboard failure) a strategy must exist which enables the operator to still run the process safely until the problem can be fixed.

Selection of display form and location
Here, for the already defined information classes and groups the display form and location must be determined. Possible solutions are indicator lamps, text displays up to high-resolution color graphic displays.

Definition of the communication form
Especially if computer displays are used as control panels, there are many possible forms of communication, e.g. direct manipulation using a mouse or another point-and-klick device like trackballs or touchscreens, menu - driven or command language driven.This will have influence on the input device hardware.

After having performed the presented steps, the development engineer has a well-structured concept for the man-machine-interaction. Next he should proceed by designing a coarse control panel.

5. REALIZING THE CONTROL PANEL

Following the guideline the development engineer will now make a detailed design of the control panel. This will include a detailed communication description, the hardware arrangement and the software design. The structuring into classes and display features will automatically lead to an object-oriented software design which will itself enforce consistency in information display.

After the design, the system must be tested and validated by future operators. Here the rapid-prototyping approach has shown good results. Using rapid-prototyping software an interactive user-oriented communication design will be possible at economical cost. In direct interaction between user/customer and developer the realized concepts can be tested and validated very easily.

6. SUMMARY

The advances in modern microelectronics had a tremendous impact on the performance and complexity of production equipment. This has also an impact on the man-machine-communication. Here, the designers often demand to much from a human. The designers must be aware of this situation and recognize that a good conceptual design can only be achieved in a close relationship to the later user. In order to support the mostly unexperienced designers, a systematic approach to the design of the man-machine-communication is required. The base of the presented guideline is a strict task orientation supported by an analysis of the process and operator requirements. By using object-oriented-design-techniques information is grouped into classes. So every object of the class has the same presentation. This approach will lead designers to a structured design which can easily be applied in hard- and software.

REFERENCES

Hickersberger, A. (1994). *Der Weg zur objektorientierten Software.* Hüthig Buch Verlag, Heidelberg.
Norman, D. (1988). *The Psycholoy of Everyday Things.* Basic Books, New York.
Reason, J. (1991). *Human Error.* Academic Press, Cambridge.
Shneiderman, B. (1987). *Designing the User Interface: Strategies for Effective Human - Computer Interaction.* Addison - Wesley Publishing Company, Reading (MA), Menlo Park.
Zühlke, D., Schneider, G. (1994). Bedienung moderner Produktionsanlagen. *pa - Produktionsautomatisierung* 3, p. 36 - 39.

QUALITY MANAGEMENT IN
CHEMICAL PROCESS CONTROL EXPERT SYSTEMS

Peter Szczurko*, Kai Finke**, Dorothee Hoberg**

* RWTH Aachen, Informatik V, Ahornstr. 55, D–52056 Aachen, Germany
szczurko@informatik.rwth-aachen.de
** Henkel KGaA, TIS–I Engineering Systems, 40191 Düsseldorf, Germany
{kai.finke, hoberg}@cadcae.henkel.de

Abstract: The reliability of Process Control Expert Systems (PCX) can be guarantieed only by comprehensive test methods. The knowledge acquisition process as well as the carefully applied test methods are important factors in increasing the acceptance of automatical systems even in controling complex chemical processes. The quality of product and process knowledge and the quality of the software system from the technical point of view has to be considered with the same consciousness. In this paper we describe how formal test methods are used during a PCX development process and how the development process itself is guided intuitively by knowledge engineers in five steps. An investigation of a concrete PCX development process based on several interviews at a major German chemicals company is the basis for a quality process model which is described in detail.

Keywords: process control, quality control, automation, ISO 9000, testability, PCX

1. HIGH QUALITY PROCESSES

Increasing customer demands on products and processes cause appropriate quality management activities. According to ISO 9000 part 3 systematic, repeatable, and documented test procedures are required especially for highly automated chemical processes. Henkel KGaA, a major German chemicals company pursues the goal to raise the quality consciousness of every employee. Even in chemical processes the company invests a huge effort in receiving a quality certificate (Witzke, R., 1994) of the developed software for process control. The requirements for complexity, ease of maintenance, and the ability to manipulate large volumes of data are reasons why AI techniques, e.g. Process Control Expert Systems (PCX) have been successfully applied. The quality of the PCX software, the knowledge acquisition process, the development and applicability of an appropriate test environment as well as the development process of the PCX itself are special interests of the company.

An investigation of test methods for knowledge based systems has shown that conventional test methods used in software quality management have to be adapted for special expert system development requirements (Coulibaly, A., 1993). As we can see in specific literature on knowledge based process control, testing only plays a marginal role, mostly in connection with the question on how to embed PCX in a general CIM architecture (Wallmüller, E., 1990 and Yeomans, R.W. et.al., 1991). A comprehensive computerized environment for PCX testing and metrication called FAITH (Fault Analysis and Integrative Testing in Heprox) (Finke, K., *et al.*, 1994) has been developed for a PCX shell called HEPROX (Soltysiak, R., 1989) which is the standard development PCX tool for Henkel. The expert system shell is based on a host target architecutre (see Fig. 1) which enables full testing on the host system. The knowledge base is given in PROLOG which will be transformed in FORTRAN77 to run on the production control computer.

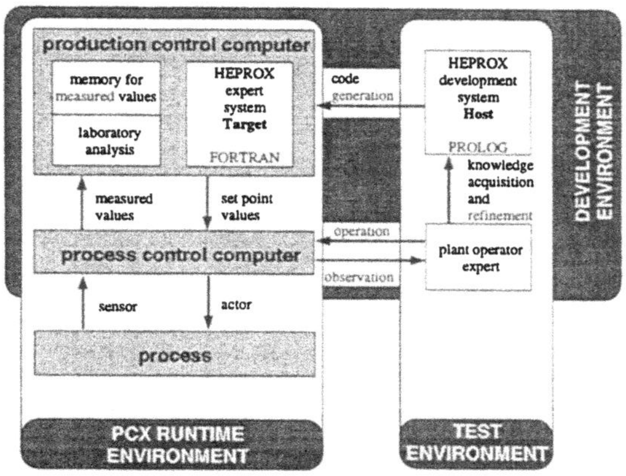

Fig. 1. HEPROX PCX Environment

Plant operators have to take decisions in any of essentially infinitely many possible process situations. To deal cognitively with this complexity, plant operators tend to categorize process situations into classes. This leads to discretization and is a prerequisite for the qualitative kind of process modeling used in expert systems (Kuipers, B., 1989). This is done in HEPROX using different kinds of language levels describing process situations (Soltysiak, R., 1989).

2. SYSTEMATIC TESTING

The faith in PCX is increased by embedding various verification steps in the development process. This stepwise testing should be executed with appropriate methods easy to use and as automatic as possible to facilitate the knowledge engineers tasks. Each verification step should be documented to show the correctness or to indicate errors and inadequacies. The verification process should be transparent and stepwise executable so that appropriate modifications would lead to the next consolidated development step. The general strategy is to detect any error as early as possible in order to minimize its impact, and to maintain the quality status achieved in the presence of change with as little effort as possible (regression testing). On the organizational side, the technical quality support achieved by FAITH has to be accompanied by a Total Quality Management environment as it is demanded by Deming, W.E. (1986).

In expert system for external specifications validation and verificationare often missing (Bologna, S., Vaelisuo, H., 1991). In contrast to the traditional evolutionary prototyping approach to expert systems development we emphasize the use of predefined external specifications as documents to test against (see also Lane, N.E., 1986). It should be avoided that the first prototype is only refined and expanded into a new prototype without specifying changes in a requirements document. This would be unacceptable from the viewpoint of dependability in a chemical production setting.

Although PCX are too complex for a comprehensive specification, at least essential parts can and must be specified efficiently. In FAITH this is done by exploiting the deep engineering knowledge encoded in measurement, control and regulation standards like (ISO 3511, 1977).

Five special adapted testing techniques, exploiting external specifications gained from the process control environment have been developed and were implemented as integral parts of the PCX quality management process. A detailed description of the test methods can be found in Finke, K. (1993).

- Consistency check: includes syntax check according to the specific knowledge representation form in HEPROX, formal verification of a set of rules to find out redundancies, subsumptions, contradictions and circularities.

- Knowledge inspection: means tracing the PCX strategies by a knowledge engineer and at least one expert guided by external specifications. With this, nearly complete automation of the inspection has been established.

- Whitebox test: covers inference paths considering the content of modules passed by the inference. All possible combinations of actions influencing the process will be simulated checking the inference following the strategy as intended.

- Blackbox test: is used in traditional manner. Some important limiting values and equivalence classes can be derived from the external specifications.

- Regression test: each test case is archived in a file system so that they can be read selectively or as complete test package. The overall test effort for a PCX has been reduced to 50%, first because it runs completely automatically, and second because the regression test ensures that no side effects have been worked in during the improvement or refinement phase of the PCX development process.

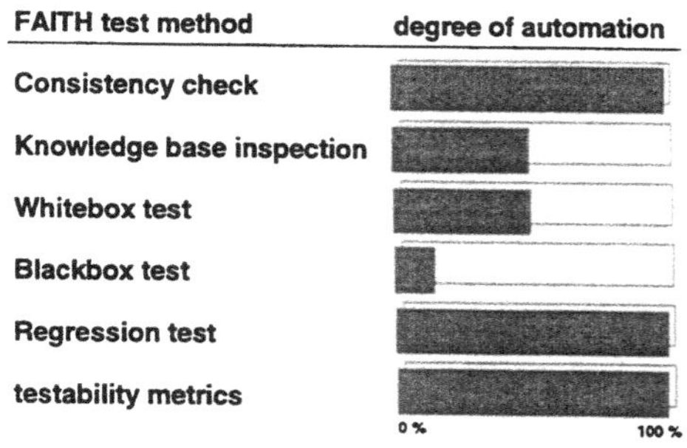

Fig. 2. Automatism of FAITH test methods

3. TESTABILITY METRICS

The test environment FAITH also comprises a set of easily computable metrics for analysing the knowledge from a structural viewpoint and enables estimating the test effort. The full paper will conclude a detailed description of test methods and metrics.

- knowledge base statistics: to give a general survey of size and growth

- largest substrategy: as a measure for the degree of modularity

- procedural share: gives hints whether exhaustive whitebox testing is possible

- maximum breadth, maximum depth: as measures for the complexity of PCX decisions. This are very important values for knowledge base transparency.

- longest possible inference: is defined as the inference path taking the most runtime in execution. The estimation is important to ensure that real time requirements are fulfilled.

Table 1 Testability metrics for real PCX

expert system / kind of entry	ester inter- change reactor	sewage preparation
nodes	17	231
substrategies	2	27
variables	29	178
modules	12	137
conditions	14	258
actions	17	139
explanations	30	515
maximum depth	7	4
maximum breadth	2	8
proc. share	13 %	50 %
largest substrategy	11	26
longest possible inference	19	348

4. MODELING QUALITY PROCESSES

Several applications have been developed successfully using the facilities of the development environment HEPROX. With the establishment of automatic test methods using FAITH the effort of PCX maintenance has been reduced to merely 50 % than before. The most important help is given by the regression test which relieves the developers of boring repetitions after knowledge base changes to avoid error propagations. This led to a reduction of the personnel which is involved in the PCX development process, while carefully ensuring the highest possible quality.

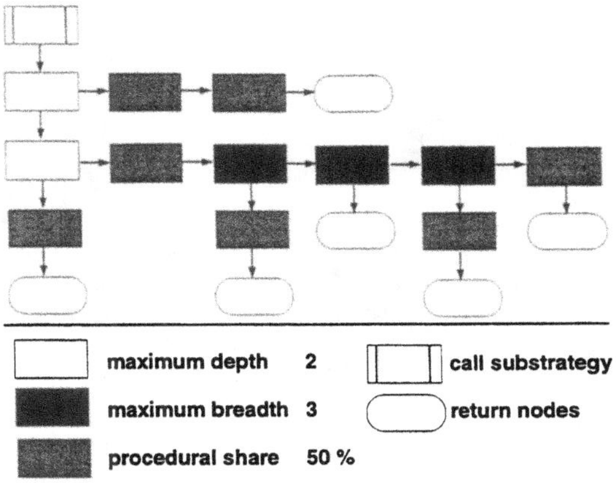

maximum depth 2   call substrategy
maximum breadth 3   return nodes
procedural share 50 %

Fig. 3. Metrics of an example strategy in HEPROX

The continuous call of the companies management on the improvement of product and process quality led to an intuitively applied PCX development process model of the consciencious knowledge engineers. An explicitly given quality process model is not yet given though it is demanded for software engineering processes in (Stucky and Oberweis, 1992). The following steps are described as a result of several interviews with the involved staff during the development of a concrete PCX.

The sewage control PCX has been developed by only one knowledge engineer. Within two years from the beginning to the application of the sewage control PCX we can distinguisch five development steps. In Fig. 4 an approach to a quality process model for PCX development is given. The knowledge acquisition process as well as several quality steps can be separated by certain characteristics of documents.

We noticed that social impacts are also important facts in achieving high quality expert systems as the consideration of technical details and knowledge formalisation. The development of high quality knowledge and inference processes for PCX is both a result of carefully composed specification documents and the capability of motivating the development team considering technical, organisational and personal aspects simultaneously.

Fig. 4 shows that the first phases of cooperation between experts (experienced engineers), operators, and knowledge engineers have the main effect of reducing both exaggerated expectations and fears on the user side while bringing the knowledge engineer up to the needed threshold of domain understanding. Together with increasing trust, the operators then acquire the understanding of PCX necessary to actually participate in the functionality, the expectations to the system grow again but the trust in the system oscillates as serious errors are detected, until the basically error-free operation of the system in parallel to human control finally establishes the ground for phasing in actual system usage.

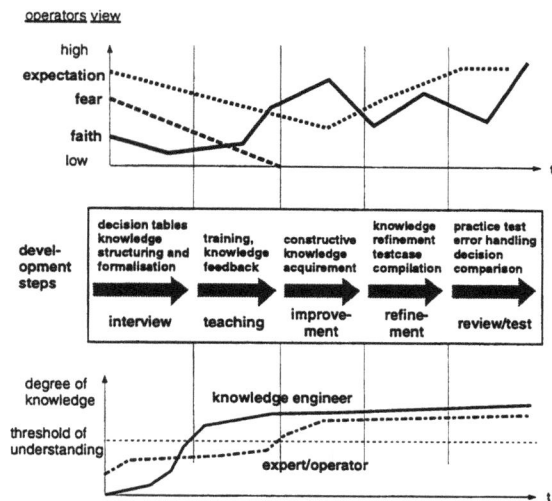

Fig. 4. Steps in the PCX development process

5. PCX DEVELOPMENT STEPS: A CASE STUDY

In the following we give a detailed description of the development steps derived from an example of a sewage control system.

- Interview phase

At the beginning the knowledge engineer did not have any knowledge about the PCX to be modeled. Process documents for training are not used because of their less detailed and illstructured form. On the basis of informal interviews between the operators and the knowledge engineer, documents like decision tables and rules in certain problem situations were drawn up. These tables consist of combinations of measured values and necessary actions to improve the process accdording to resources optimization, quality control and avoidance of pollution.

A listing of all possible combinations of measured values was drawn up as a first description of correct process states and rules to detect alarm situations. Within this phase the social competence of the knowledge engineer and the talent to motivate the operators to give expressive process knowledge is the key of increasing the mutual understanding of the operators process knowledge and the formal and abstract way of thinking of the knowledge engineer required to build up an expert system.

- Training phase

With respect to the formal representation of the PCX the operators knowledge will be transformed in modules and strategies as basic elements of the PCX knowledge base. In special training sessions the knowledge engineer explains the representation

which is used in HEPROX to hold the process knowledge. The formalized knowledge which has been derived from the decision tables and the informal interviews is the basis for discussion.

The replay of the operators knowledge is an important step in PCX software quality because of the following aspects:

First the knowledge engineer and the operators form a team during the PCX development process. Building up the team spirit in periodical meetings where the current state of the PCX development can be discussed is beneficial for new ideas and knowledge improvement from both sides especially if the discussion takes place not at the usual place of work. The motivation increases visibly giving the operator the certainty when the companies philosophy of total quality management is realized through have the operator participate in the PCX development process actively, involving him in training processes and make his work more interesting by varying the scope of work.

Second the given informal documents are processed by the knowledge engineer for the use of computers. A comparison and verification of this knowledge can only be performed by the experts themselves. The goal is to achieve a complete view on the necessary process knowledge and its correctness and completeness. The discussion of process situations and decision rules is already an improvement of the process quality without the PCX because of the complete gathering of experience of all involved experts and operators.

Third the operators will be aware of the PCX's area of authority explaining which kind of work the PCX will take over. It is not rarely the fact that operators are afraid of the expert system because they don't know exactly how their scope of work will be changed when the PCX is used instead of the human process control. It is a must to explain the operators how a PCX works and in which situation, namely the routine work, decisions and actions will be taken over. On the one hand side the operators should lose their doubts and fears and on the other hand side they should be aware of their interesting future work and their responsibility for a high quality PCX.

The need of abstraction mechanisms and formal representations as a basis for automatism confronts the operators with a new way of thinking. The consideration with knowledge representation and inferencing mechanisms reduces existing cognitively and intuitively used actions. Although the experience knowledge will be structured and formalized the often praised instinct for the optimal regulation of processes will be replaced by rational ways of thinking and behaviour.

- Improvement phase

On the basis of the explicit abstract representations the experts give constructive remarks to improve the knowledge base. The correctness will be proved in further team sessions and meetings together with the operators. Detailed questions like, which values are measured by an instrument or which states can be calculated from combined measurements, lead to a considerable improvement of the existing knowledge. Furthermore the consideration of so called external specifications which consist of explicitly given documents of PCX requirements as well as the behaviour of the PCX in certain alarm situations is a prerequisite for the knowledge base inspection as well as the blackbox test in the test environment FAITH.

- Refinement phase

The PCX development team generates a lot of test cases which should cover all decision paths as far as possible. The number of test cases depends directly on the number of decision nodes within a PCX decision tree. As an example for the sewage control PCX exist about 200 test cases documented and executed (see also table 1). The results of the PCX in every given test case are discussed together with the operators with the goal to refine the knowledge base using the abstraction mechanisms offered by HEPROX namely using strategies and substrategies. In case of obviously wrong PCX decisions the traceability of decisions plays a main role in the refinement phase. Only a comprehensive understanding of the whole decision process can reduce the operators breach of trust which arises because of the detected errors. During the development of the sewage control PCX about 50 errors have been detected and eliminated.

- Review / test phase

During this phase the already existing PCX is integrated in the process. Although this is done without any competence changing the process, all the measured process values are given to the PCX. The process situation as well as the decisions are documented and discussed later with the operators. For an evaluation of the PCX in problematic situations the behaviour of the PCX is observed by simulating known critical situations. Especially for this reason test cases are generated and executed in the process environment. Blackbox test cases derived from the external specifications as well as whitebox test cases which are generated from the internal structure of the PCX knowledge will be executed equally.

We know from experience that in this phase only few errors are detected because even the problematic situations are covered by a test case which is already investigated in the previous phase. Additionally to the validation of the PCX decisions the usual interface

and integration tests within the process environment are performed (see Fig. 1).

6. EXPERIENCE

In interviews with the knowledge engineers and users we established the fact that during the PCX development using the test environment FAITH the prototyping process was supported appropriately. It was easy for the developer to get a feeling for the knowledge base and its quality looking at the metric information generated during the various test phases. Feedback information like error messages and improvement hints rises step by step the trust in the system even because the boring and lengthy regression test was automated. In the development of a sewage control PCX, FAITH reduces the regression test effort to 30 %.

As a side effect, the implemented metrics give adequate help in estimating the knowledge base quality from a structural point of view. We found out that the procedural share is the most important metric. A value higher than 10 % disables exhaustive tests in most cases.

The very success of PCX is not only a consequence of the establishment and carefully application of software quality methods but also a service of the knowledge engineer. The goal of the PCX development team should not only be reduced to reach the companies goal but also to consider the interests of every involved team member. The consideration of organizational and personnel aspects was remarked as very important for a successful PCX development (see Fig. 5). The appropriate technical environment which is given by HEPROX, FAITH and several techniques for knowledge acquisition, visualization and refinement is only one layer for the quality concept which is followed in developing successful PCX at Henkel.

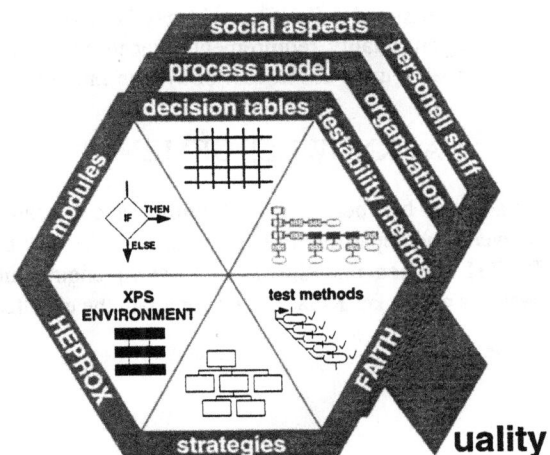

Fig. 5. Quality facets in HEPROX PCX development

The application of an expert system in process control causes changes of work spaces which requires additional qualifications of the involved operators. It has been shown that fears and doubts have to be eliminated to ensure that the operators understanding of a PCX means positive work changes with more responsibility and less routine work. This is a prerequisite for an efficient cooperation between knowledge engineers and plant operators.

The acceptance of PCX has been increased because both developer and user have the opportunity to participate in all steps of PCX development. In fact, we noticed that with the help of FAITH, the work itself was made more interesting. Generally, the use of FAITH substantially reduces the human test effort while improving the quality of PCX significantly.

Is has been shown that the development process has been guided in an excellent manner by the knowledge engineer. This includes the carefully preparation and execution of necessary development tasks as well as the ability to motivate the involved plant operators creating a high quality PCX. In fact the knowledge engineer at Henkel is female, which may be a reason for the successful collaboration in the development team. It can be shown by several empirical studies that women are thinking more realistically, that they are more pragmatic and that they lead a conversation with more feelings for important issues than their male colleagues (Schinzel, B., 1991). In this given case it can be confirmed that the social competence of the knowledge engineer was a main factor for efficient and successful work.

Until now several PCX have been designed and implemented with the use of the above mentioned development environment facilitating the knowledge acquisition process and prototyping. The given quality process model has not been described until now in detail, but has been applied intuitively by the involved knowledge engineers with great success. Although we hasten to acknowledge that this case study is in no way statistically significant, it may help to explain the stringent quality requirements placed on PCX in the company, even at the cost of system *intelligence*.

ACKNOWLEDGEMENTS

This work has been realized within a cooperation between the department of computer science V at the RWTH Aachen and the department of engineering systems at Henkel KGaA. We appreciate the excellent cooperation with this team and the willingness to give expressive information during the held interviews.

REFERENCES

Bologna, S., Vaelisuo, H. (1991), Deep knowledge and rigorous engineering practice, In *Dependability of Artificial Intelligence Systems (DAISY-91)*, (Schildt, Retti, Ed.), North-Holland, 73–90.

Coulibaly, A. (1993), Development of test methods for knoweledge based systems, (in German), Diploma Thesis, RWTH Aachen, Informatik V.

Deming, W.E. (1986), *Out of the Crisis*, MIT, Cambridge.

Finke K. (1993), Systematic Testing of Process Control Expert Systems, (in German), Diploma Thesis, RWTH Aachen, Informatik V.

Finke K., Jarke M., Szczurko P., Soltysiak R. (1994), FAITH in Process Control Expert Systems, *Proc. 11th European Conference on Artificial Intelligence (ECAI94)*, Amsterdam (NL), 8th-12th August 1994

ISO 3511 (1977), Process measurement control functions and instrumentation – symbolic representation, Part I-II.

Kuipers, B., Qualitative reasoning – modeling and simulation with incomplete knowledge, *Automatica 25, 4*.

Lane, N.E. (1986), Global issues in evalutation of expert systems, *Proc. IEEE Intl. Conf. on Systems, Man and Cybernetics*, Atlanta 1986, 121–125.

Schinzel, B. (1991), Frauen in Informatik, Mathematik und Technik, *Informatik Spektrum*, 2–1991.

Soltysiak, R. (1989), HEPROX, eine Expertensystemshell für Prozeßführungsaufgaben", *Automatisierungstechnische Praxis*, Nr. 31, 2/1989.

Stucky, W., Oberweis, A. (1992), Zur Beherrschbarkeit des Entwicklungsprozesses komplexer Software-Systeme, *Forschungsbericht 242*, Univ. Karlsruhe.

Wallmüller, E. (1990), *Software-Qualitätssicherung in der Praxis*, Hanser-Verlag.

Witzke, R. (1994), Zertifizierung von Qualitätsmanagement-Systemen bei Softwareherstellern, *Theorie und Praxis der Wirtschaftsinformatik*, Heft 175, Jan. 1994.

Yeomans, R.W., Choudry, A., Ten Hagen, P.J.W. (1991), *Design Rules for a CIM System*, North Holland.

INTERNATIONALIZATION OF PRODUCTION AND DEVELOPMENT OF INDUSTRIAL LABOR

Hartmut Hirsch-Kreinsen

Institute for Social Research, ISF Munich, Jakob-Klar-Str. 9, D-80796 Munich, Germany

Abstract: The current industrial situation in all western nations is characterized by a broad spectrum of developmental paths of labor. An important cause for this uncertainty lies in problems resulting from the internationalization of production and the different strategies of internationally active companies. The strategies move in the direction of either a "global" or a "transnational" strategy according to their market situation and industrial branch. Which forms of labor get implemented in companies is dependent upon a whole range of locally specific socio-economic factors to which the internationalization strategies of a company have to adjust. Such factors, e.g. the labor market conditions and the "technological infrastructure", mark central areas of politcal action to influence the strategies of international companies.

Keywords: Industrial production systems, human centered design, work organization, social sciences

1. THE UNCERTAIN FUTURE OF INDUSTRIAL LABOR

Since the middle of the eighties the develoment of industrial labor is characterized by a departure from the former - "Tayloristic" - rationalization principles in almost all western nations and industrial sectors. As the current debate on "lean production" or "business reengineering" shows, management concepts are being pursued that strive towards such goals as the flattering of hierarchies, a company wide decentralization process, a more systematic use of skills and team work structures, and a continuous improvement of production processes. In the framework of these concepts the term "skilled labor" indicates the main feature of a new consistent "post tayloristic" rationalization pattern.

Contrary to the claims and hopes of this debate, however, these new management concepts do not get implemented wholesale in company reality. As an entire series of national and international studies demonstrate, a variety of new rationalization patterns can be found in nearly all western industrialized nations at present, indicating an irregularity and nonconcurrence of the current situation (Altmann, et al., 1992).

The result is that the development of industrial labor is characterized by a broad spectrum of different developmental paths with varying consequences for work organization, tasks, and skills. As we pointed out formerly at least three such paths can be identified: first, a "neo-tayloristic" path, second a path of "integrative" industrial work and third the path of "polarized" industrial work (Hirsch-Kreinsen and Schultz-Wild, 1992).

The reasons for this situation lie in the numerous contradictory influences that companies are being confronted with at present. As a result diverging orientations and reactions present themselves to the companies which are more or less mutually exclusive. For example, opposing requirements result out of the contradiction between the turbulent demands of the sales markets which exert enduring pressure for changes in company structures on the one hand, and the persistence of deeply rooted hierarchical organizational structures in companies on the other. Another example is the unpredictable development of the labor market, especially the very uncertain supply of skilled production workers in the future. Finally, only a few companies have either the necessary personnel or financial resources, nor do they posses the know-how which would be required for a fundamental reorganization of their traditional structures and the realization of new management concepts.

2. THE CONTRADICTIONS OF INTERNATIONALIZED PRODUCTION

An important cause for the uncertainty of the developmental trends of industrial work lies in problems that result from a continuing internationalization of production. Normally the question for the future of industrial work is being discussed in a national perspective; or this question has been dealt with in international comparative studies. However, the question of the internationalization of production and the corresponding company strategies are playing only a minor role in the discussion on current rationalization trends.

Internationalization is an increasingly important aspect of industrial rationalization. Internationalization of production means the relocation of production sites to other countries, the partnership in foreign companies in the form of direct investment and joint ventures, as well as the integration of individual national companies into cross-border buyer-supplier relations.

The growing importance of internationalization of company strategies is amply documented by the sharp rise in worldwide direct investment in the past years. According to OECD figures international direct investments made by companies based in OECD countries climbed by an annual average of 31% between 1983 and 1989, while worldwide trade increased by a mere annual 11% during the same period of time. As these figures suggest, international direct investment, especially since 1985, is currently in a "take-off-phase" (OECD, 1992). Following a decline due to the worldwide recession at the beginning of the nineties, international direct investment has been on the rise again since 1993 (UNCTAD, 1994).

The contradictions connected with these trends can be described as follows: On the one side, saturated domestic markets, growing competition, and increasing costs, particularly in research and development, push companies in many industrial sectors to a global expansion and standardization of their sales and production strategies; this calls for a resolute and worldwide orientation towards a strategy based on the principles of an "economies of scale". On the other side, the globalization of sales and production creates pressure for intensified flexibility, closeness to specific market segments and customers; this calls for a strategy based on the principles of an "economies of scope", and the acceptance of increasingly specific market conditions. This pressure comes from the increasing segmentation and regionalization of the world market, which, among other reasons, is caused by the large trade blocks as the EU, Nafta and Mercosur and even more by a number of national policy measures designed to protect indigenous industries (Emmott, 1994).

3. CHANGES IN PRODUCTION STRATEGIES

The altered market conditions have resulted in considerable consequences for companies, their strategies and the development of industrial work. Strategy changes became imperative in order to secure or increase sales. A considerable increase in foreign direct investment, i.e. the internationalization of production was perceived as a way out of this situation.

While direct investment continued to rise the traditional corporate internationalization strategies, many of which can be followed back to the twenties, began to lose their former significance: For one, this referred to the "export strategies" of companies with a basic national orientation which exported their products without adapting them to foreign market conditions to any considerable extent. Typical examples are US companies from the office machine, computer or machine tool sectors, which formerly encountered relatively few problems exporting their products thanks to productivity advantages and their technical lead. Another strategy once frequently encountered can be referred to as "multinational" and is characterized by a loose conglomerate evolving in the course of decades and consisting of corporate headquarters and strong foreign subsidiaries. In Europe these strategies are primarily found in companies in the electro-technical and food industry, or in American automobile corporations such as General Motors and Ford.

A process of change in company strategies now stands out. This process is unfolding in a continuum limited by two types (e.g. Bartlett and Ghoshal, 1989):

(1) The first type is the "global strategy" that targets worldwide homogeneous market segments and integration on its course to a more or less worldwide standardization of production and products, while aiming for the most extensive centralization of decision making and functions. According to this approach, competitive advantages are to be achieved on the basis of the "economies of scale", a considerable reduction of vertical range of integration at the individual production sites and a specific utilization of regional and country specific cost advantages by building up global supplier relations. This type of internationalization strategy was primarily pursued by Japanese photo industry and consumer electronics companies since the end of the seventies. To a certain extent they are also evident in the automotive industry (Jürgens, et al., 1989).

(2) At the opposite a type of "transnational" strategy is discernible (Bartlett, 1986). Here, rising internationalization at best denotes a lower priority given to the utilization of standardization benefits, for example in the manufacture of certain product components, in connection with the centralization of individual key functions such as research and development, as well as procurement. This strategy places emphasis on pronounced regional sourcing and differentiated product and production strategies. This type of transnational strategy is primarily evident in branches of the investment goods industry that are characterized by distinct market differentiation and innovative products.

In this case, company wide integration is not achieved primarily through centralization, but more by way of network-like governance processes of regionalized and decentralized company units that cooperate and compete with each other at the same time. In terms of company organization, formally independent companies are linked together, and they are differentiated into "cost centers" and "profit centers" with a relatively high degree of autonomous decision making. This approach seeks to attain competitive advantages through an "economy of scope", a closeness to market, as well as by a world wide, yet synergetic utilization of external and internal local conditions and also by drawing on the flexibility and innovative strength of small company units. In a general context, this relatively open structure seeks to enable continuous rationalization processes and rapid adaptation to changing market conditions. On the long run, this structure creates an organizational dynamic of change, powered by the interplay between the strategic objectives specified by corporate headquarters and the scope for autonomous action wielded by the decentral units, and is also propelled by the cooperation and competition processes between the decentralized units.

The strong local orientation means that this strategy hinge particularly strongly on regional and national factors that can be summed up by the term "technological infrastructure". This refers for one to special external experience and competence in the form of service, maintenance and readily available technological know-how that play a key role for investment and smooth operation of particularly complex production systems. This also refers to the relationships and interaction between machine manufacturers, consultants, scientific institutes and machine users. Primarily, this network provides the opportunity for mutual utilization of specialized know-how and the exchange of knowledge. Also significant in this context are the contacts to the institutions of the system of vocational training and generally the access to special segements of the labor market such as skilled industrial workers, engineers or experienced managerial personnel.

4. CONSEQUENCES FOR THE DEVELOPMENT OF INDUSTRIAL WORK

The diverse trends in the development of industrial work are particularly linked to the change in internationalization strategies. The consequence is that paths of development do not only differ between companies, industrial branches, and nations, but also within an internationally active company.

(1) In the case of global strategies of international companies the corporate wide unified and centrally stipulated rationalization strategies aim for an alignment of work organization in the various companies under the Group umbrella. However, the course laid down by corporate headquarters, a course of action that may comprise the relocation of production functions, plant shutdowns and personnel cutbacks, also meets with a certain degree of resistance and barriers. The specific locational conditions of individual company sites may exert a strong influence on central rationalization concepts and guide them in a certain direction.

As a study on the rationalization strategies of a major international corporation in the consumer electronics branch showed (Düll and Bechtle, 1991) the established structures and local conditions of corporate units exerted considerable inertia against corporate wide planning. For example, distinctly different task scope and personnel utilization structures were evident at technically similar assembly facilities in different plants on a national and international level. These differences were especially dependent on the labor market conditions at the respective plants. In urban areas where skilled workers are available, comparatively polyvalent forms of personnel utilization with a low division of labor were possible. In rural areas with predominantly semiskilled workers the work structures realized were based on a decided division of labor. Similar differences were found in terms of wages and company performance policies, which, in the final instance, were also dependent on the given company system of industrial relations. As a result of this situation similar or recurring forms of work organization were encountered within the corporation, however, also differed with regard to such important dimensions as division of labor, the utilization of human resources and performance policies.

(2) In the case of transnational strategies the corporate headquarter only plays a limited role concerning the development of the organizational structures and the development of industrial work. Here, the processes of competition and cooperation between the units within the corporate network play a decisive role for the development of industrial work. It is up to the individual units and companies to secure their efficiency with the help of appropriate work organization structures based on their specific local conditions while realizing corporate objectives for

example in terms of reducing cycle time and managing quality. In these cases, it is apparent that not only the labor policy constellation and the labor market situation of the individual company plays an important role, but also the conditions on the national or local sales markets which the transnational type of strategy is targeting to a considerable extent. It is the national and local sales markets that decide on the actual product range and the associated work requirements. Therefore diverging development trends of industrial work are encountered within the corporate network not only on international, but also on national levels.

This diversity of industrial work will even increase because - following the current discussion of management experts (Bartlett and Goshal, 1989; Emmott, 1994; Wooldridge, 1995) - the transnational strategy can be recognized as a strategy of growing significance. The reasons lie in the intensified competition and the ongoing process of segmentation and regionalization of the world market.

5. A CASE STUDY

This type of transnational strategy can be illustrated by preliminary findings of an ongoing research project at the ISF Munich dealing with the strategy pursued by an international corporation of the European capital goods industry (Hirsch-Kreinsen, et al., 1994). The first findings refer to the special features of this type of corporate strategy and the contradictory consequences for the develeopment of industrial work in the corporate companies.

Since the end of the eighties substantial changes in the organization of this corporation took place: The introduction of matrix organization, a major tigthening up of the central administration, the creation of small, clearly identifiable profit centres, relocation of production elements and the repeated acquisition of individual companies are the main examples of the reorganization process. Structurally, the corporation is now federal in its principles, regionally oriented and largely decentralized. In 1994 the corporation constisted of about 1.000 individuel companies and had about 5.000 profit centres.

The organizational development of the corporation is process-oriented to a very high degree; there is no evidence of a final objective of the reorganization measures in either technical or organizational terms. Rather, coporate-wide - i.e. under a wide range of location-specific structural conditions - the organizational development is supposed to permit a continuous, ever-improvable and above all flexible optimisation between the contradictory objectives of cost cutting, flexibilisation and reducing cycle times.

This implies that the rationalization process is in many respects contradictory. Firstly, rationalization objectives which usually stand in contradiction to one another must be reconciled. Examples are the twin aims of a reduction in cycle times and a substantial improvement in quality; the reorganization measures aim to achieve both a standardisation and a flexibilisation of processes, while the extent of central process control must be balanced against the degree of decentral autonomy.

Secondly, the realisation of these aims is tied to the concrete conditions prevailing at each individual plant. Of major importance are thus the conditions within the plant, such as its technological and organizational structure, personnel, managerial capacaties, and specific labor relations. These differences point directly to regional, and in particular national-specific socio-economic conditions.

Furthermore, the scope for possible reorganizational measures is being determined by the market conditions. Particularly, in the case of the capital-goods sector, the conditions on the sales market and the established relations with customers are an very important factor for the succes and the profitability of the corporation. This conditions management has to take into consideration deciding on production volumes, the innovation of manufacturing processes and the relocation of plants.

It ist within this multi-dimensional tension between contradictory rationalization objectives and local production conditions that actual organizational structures develop. So far, the following organizational trends have become apparent:

(1) Complementarity of production locations: It seems that it is all but impossible to optimise rationalization objectives by ensuring that operating conditions at the various plants are as equal as possible. Attempts are being made to reduce local differences in production sites in the various European countries by means of technological, personnel and know-how transfer. However, there are clear limits to the extent to which this can be achieved. In terms of corporate-level organization, the only objective can be to utilise in a complementary way the specific conditions of each production site within the framework of an overall production network. In many cases this means that complementarity must first be consciously created by adjusting and modifying specific local conditions.

(2) Process differentiation: The implementation of a production network is associated with a strategy of process differentiation. On the one hand high automated and know-how intensiv processes can be found in the so called "Lead center" plants. This plants regularly are located in the old industrialized countries like Sweden, Germany or the U.S. Measures of standardisation of the process and the manufactured parts and componenets aim to achieve a maximum capacity utilisation. The concept underlying the Lead center plants is that of concentrating on "Core

components" and an company organization called "Focused factory". On the other hand low automated and labor intensive processes can be found in the so called "Satellite" plants where simple parts are manufactured. Such plants can be found mainly in low-cost countries, particulraly in Eastern Europe.

(3) Opennes of work organization: At the individual plant within the production network the organizational changes are not uniform and unidirectional. The most that can be obseved is a trend towards decentralisation, the implementation of cost- and profit-centres, and segmentation, in which planning functions are split up.

Work organization on the shop-floor varied, in some cases considerably. The different forms of work organization range from the sophisticated tasks focusing on securing, optimizing and maintaining automated manufacturing processes to simple assembly processes with restrictive forms of manual labor. The latter are characterized by distinct division of labor, short cycle work sequences, low skill requirements and high exchangeability of workers

Also in an international perspective the first findings suggest significant differences, even between plants which are rather similar in technological terms; at present the following findings can be stated: The corporate plants in Scandinavia have gone furthest in realising new principles of production management at all organizational levels. This is partcularly apparent with respect to work organization, where the principle of group work has been conciously realised in many areas, not only on the shop-floor. In both countries Germany and Switzerland on the level of the plant organization considerable change has been made in realising new rationalization principles; so far the changes on the shop-floor have tended to be contradictory. For example, organizational forms as group work are very seldom and have been realised only in specific areas.

At the moment little can be said regarding the reasons behind these differences. They do, however, tend to confirm general hypotheses concerning the different nationally specific conditions for developmental paths in industrial work. It seems that in Sweden the traditional co-operative relationship between management and the trade unions, the predominance of long-term rationalization strategies within industry and the particular ability of the system of vocational training to provide skilled, flexible and adaptable workers are all conducive to innovativness at the company level.

In the German-speaking countries the organizational measusres are based on skilled-worker structures, which in the past have proved to be relatively flexible and conducive to organizational and technological change. Their quasi "automatic" utilisation for the rationalization process would therefore seem to

make sense, although it implies only gradual changes with at most only minor break with existing work structures.

6. AREAS OF POLITICAL ACTION

Summarizing the findings: Which forms of work get implemented in companies is dependent upon a whole range of locally specific socio-economic factors to which the internationalization strategies of a corporation have to adjust. Particularly important in this regard are, firstly, the regional labor market conditions and the availibility of specific qualifications, which, among others, are effected by the regionally or nationally specific system of vocational training. Related to this, secondly, the regionally specific "technological infrastructure" is of major significance which comprises company external resources such as know-how, service, and relations with consultants and suppliers.

Beyond these direct labor and skill related conditions, existing labor policy constellations, wage agreements, and the position and power of unions play an important role for the strategies of transnational companies. They, in turn, are based on the particular national system of industrial relations and its long running practices and traditions.

This factors mark central areas of political action to influence the corporate strategies in future (Porter, 1989; Reich, 1991); particularly to influence corporate investment decisions, to secure industrial jobs in a country and to improve the quality of the jobs. It makes only a limited sense that political measures aim at the microlevel of a single transnational company. The main area for political action is no longer the level of the individual company because a transnational companay has in the long run no "home country"; it is able to shift its headquarter as well as its units from one region or country to another in a very short time. That means, political institutions have still a limited chance to influence company strategies directly. On the long run, the more effective way to influence company strategies is to develop and to improve the above mentioned socio-economic conditions the companies are interested in. This refers to technology, industry or education policy measures enhancing the "technological infrastructure" of a country or a region. These factors are undoubtedly gaining increasing significance for the investment decisions of such companies.

Political action, however, can hardly remain restricted to the national state level. On the one hand the growing influence of supranational organizations such as the EU and NAFTA has to be seen that will exert more and more influence on industry and technology policy measures. In view of the development and utilization of infrastructure conditions, on the other hand, sub-national units in individual industrial districts will also assume increasing importance. This is primarily borne out by the results of various studies on "industrial districts" (e.g. Piore and Sabel, 1984) that underline the significance of regional networks and their historically evolved special aspects for industrial rationalization processes.

REFERENCES

Altmann, N.; Köhler, Ch.; Meil, P. (1992). No End in Sight - Current Debates on the Future of Industrial Production Work. In: *Technology and Work in German Industry* (Altmann, N.; Köhler, C.; Meil, P. (Eds.)), pp. 1 -11. Routledge, London/New York.

Bartlett, C.A. (1986). Building and Managing the Transnational - The New Organizational Challenge.In: *Competition in Global Industries* (Porter, M.E. (Ed.)), pp. 367 - 401. Harvard Business School Press, Boston.

Bartlett, C. A. and S. Ghoshal (1989). *Managing Across Boarders*. Harvard Business School Press, Boston.

Düll, K. and G. Bechtle (1991). *Massenarbeiter und Personalpolitik in Deutschland und Frankreich*. Campus, Frankfurt/New York.

Emmott, B. (1993). Multinationals. Back in Fashion. *The Economist* Vol. 326, No. 7804, March 27th.

Hirsch-Kreinsen, H. and R. Schultz-Wild (1992). Chances for Skilled Production Work in Computerized Manufacturing Systems. In: *Ergonomics of Hybrid Automated Systems III* (Brödner, B. and W. Karwowski (Eds.)), pp. 147 - 154. Elesvier, Amsterdam

Hirsch-Kreinsen, H., Altmann, N. and R. Schultz-Wild (1994). *Nationally Specific Development Trends of Industrial Work. Concluding Report on the Preliminary Study to an International Comparative Research Project*. mimeo, ISF Munich.

Jürgens, U., Malsch, T. and K. Dohse (1989). *Moderne Zeiten in der Automobilindustrie. Strategien der Produktionsmodernisierung im Länder- und Konzernvergleich*. Springer, Berlin.

OECD (1992). *Technology and the Economy. The Key Relationships*. Paris.

Piore, M. and C. Sabel (1994). *The Second Industrial Divide*. Basic Books, New York.

Porter, M. E. (1989). *The Competitive Avantage of Nations*. The Free Press, Boston

Reich, R. B. (1991). *The Work of Nations*. Alfred A. Knopf, New York.

UNCTAD (United Nations Conference on Trade and Development) (1994). *World Investment Report 1994. Transnational Corporations, Employment and the Workplace*. United Nations, New York/ Geneva.

Wooldridge, A. (1995). Multinationals: Big is back. *The Economist* Vol. 335, No. 7920, June 24th.

INTEGRATING DESIGN ENGINEERING AND THE WORK-FORCE

Frank Emspak

School for Workers, Department of Labor Education
UNIVERSITY OF WISCONSIN-EXTENSION, CONTINUING EDUCATION EXTENSION
610 LANGDON STREET, 422 LOWELL HALL
Madison, Wisconsin 53703

Abstract: This paper will discuss efforts to integrate workers into the process of design (collaborative). It will briefly survey Federal Government programs which support this process, including the development of software systems. It will describe examples of best practice firms, suggest some exciting directions for future developments, and initial steps in a multi-site projects whose object is to develop collaborative decision making tools. The development itself is being conducted with teams of skilled workers and engineers.

Keywords: Design systems, human centered design, flexible manufacturing systems, skill-based systems.

1. INTRODUCTION

In recent years more and more firms have been experimenting with design concepts that value customer needs as well as production concerns. A minority who have defined their efforts as customer driven have further refined their thinking to recognize workers as customers. In these firms the vocabulary and practice of "socio-technical" design has begun to emerge. Some firms, while adopting the principles of socio-technical design, have gone one step further and have taken measures to introduce non traditional personnel into the design process itself. Still the overwhelming design paradigm in the United States remains a combination of Taylorism overlaid with efforts to broaden job classifications, outsource aspects of production, and now, outsource and downsize design and engineering functions.

This paper will discuss efforts to integrate workers into the process of design. We will first discuss Federal government support for initiatives which encourage collaborative design. Then we will describe examples of design initiatives taking place in several best practice firms. Following a discussion of integrative design experiments, we will describe the development of tools that encourage collaboration. In conclusion, exciting directions for future developments will be suggested.

2. FEDERAL GOVERNMENT SUPPORT FOR A NEW DESIGN PARADIGM

In contrast to the situation in many other industrialized countries, governmental support for technology is primarily a function of the Department of Defense. Civilian agencies, such as the National Institute for Standards and Technology (NIST) and its Advanced Technology Program (ATP) constitute only about 1-3% of publicly funded research and development. This small contribution is currently controversial and will certainly not grow in the near future. ATP criteria for funding individual projects within a specific focused technology development program do not include requirements to include assessment of skill or ergonomics, nor do they include any suggestion that users should be involved in the design process. (NIST Proposal Preparation Kit, November 1994).

Military supported technology research is also resistant to overtly including the concerns of the workforce in its design projects. However, as the armed forces mission is dependent on an efficient procurement system, several initiatives are underway to encourage manufacturing innovation. For example, there is support by the Air Force for the National Center for Manufacturing Sciences, a program designed to assist small and medium sized manufacturers develop and deploy new manufacturing technologies. The Advanced Research Projects Agency (ARPA) is also a sponsor of research in manufacturing techniques through such programs as the "Lean Aircraft Initiative" and the Agile Manufacturing Program (Department of Defense). The Agile Manufacturing Program, among other initiatives, will collect and disseminate studies of firms who are involved in significant experiments with employee participation in manufacturing design efforts. In addition, the Department of the Navy in conjunction with ARPA, is sponsoring the Maritech Program. The Maritech Program is devoted to encouraging new methods of ship building so as to enable the United States to re-enter commercial ship building as well as to allow the United States to maintain an up to date military ship building capacity.

Within each of the manufacturing innovation programs there is support for one or two individual projects which encourage new reviews of man-machine interface issues. In some cases, the size of the projects and the extent of the changes are very significant. For example, the Maritech sponsored program at Bath Iron Works supports a high level of employee participation in all decisions affecting the shipyard, including bidding on new work, investment, technology and the organization of production. Few other examples exist of workers in the private sector having as much influence in design and organization of their work place (see BIW below).

3. THE FRAMEWORK FOR COLLABORATIVE DESIGN

Other than financing for specific projects there is no legal or educational infrastructure to encourage or support worker participation in the design of manufacturing systems. No private sector firm in the United States is obligated to meet with their unionized work force to discuss issues of design or any technological issue except in so far as it effects wages, hours and working conditions. Although this caveat seems like a broad invitation to discuss technology and involve workers in decision making, in practice this has not been the case.

Of equal importance to the legal framework is the intellectual framework workers, managers and engineers bring to the discussion. With very few exceptions existing apprentice or trade school education omits courses which teach team work development and collective problem solving skills. Although more attention is being given to interpersonal skills in the context of working within a team of people more or less at the same level of hierarchy, there is little attention given to assisting skilled blue collar workers with skill development to

allow participation in the design of their tools. No leading institution of engineering education has course work devoted to collaborative design methodology or skills building to encourage collaborative design. Some engineering schools do have some formal course work that encourages team building among engineers for the purposes of designing a system or product, but as of yet no leading institution has begun to focus on the skills and methodology necessary to incorporate non engineers into an engineering design team (Louis Bucciarelli, personal communication 1995). Education at all levels continues to maintain the separation between conception and execution. Even while there is considerable evidence that the context in which design will be used is of crucial importance in achieving the best design alternative, there are no legal or educational institutions which support such an approach. (Bucciarelli)

4. INTEGRATING THE WORK FORCE INTO DESIGN FUNCTIONS: INNOVATIVE APPROACHES

There are projects initiated by corporations often with considerable prodding from the unionized work force which have integrated blue collar workers into design functions.

4.1. *Bath Iron Works - I.A.M.*

One example of the integration process is underway at the Bath Iron Works in Bath, Maine. (International Association of Machinists (IAM) and Bath Iron Works (BIW). The Bath Iron Works is one of the few remaining shipyards in the United States. It has been in existence for over 150 years and was a successful producer of both commercial and military ships. It is now producing destroyers for the U.S. Navy. In 1993, the International Association of Machinists (IAM) and the company agreed to joint committees to design new methods of work, a training system and a technology review committee.

Unique in American heavy industry is the formation of a joint bid team. The joint bid team has a dual function. On the one hand, the team designs bids on new work. At Bath Iron Works bidding on new work entails detailed discussions regarding the design of potential new non ship projects. In addition, the team must possess a comprehensive knowledge of the production methods in the facility, and also be aware of processes that cannot be done at Bath.

The bid team, and all others operate in communication with the Technology Orientation Center. The Technology Orientation center's job is to provide a place to test, demonstrate and refine various technology options available to support a High Performance Work Organization (HPWO). As the firm and union state, "No technological solution will be considered valid until a majority of work force level reviewers agree that it is both needed and appropriate..." (Bath, p.14).

The parties go further with their innovative views regarding how new technologies can assist the formation of consensus. They have committed to the concept that "ring side seating" is required of everybody and therefore the communication system must enable participation. Thus the firm has made an effort to get support for the development of these technologies based on the following criteria: "Technologies that facilitate robust participation of individuals and groups are needed. All company employees must have access to a "ring side" seat that provides simultaneous views of the "Big Picture" as well as the "Critical Details". (Bath p.16).

The Bath Iron Works-IAM project is two years old. It continues to have growing support within both the management and work force at Bath. The shipyard is a modern facility, but Bath is also a large facility with several thousand workers. Its value for this discussion lies in the fact that a sizable organization has made a commitment to find ways to involve the work force in all levels of decision making at the BIW, but equally important, the IAM and BIW have recognized the existing technical systems that would really enable collaboration and participation on a scale (number of people) and a depth (strategic decisions) need to be developed.

4.2. *The National Labor Management Committee Of The Custom Woodworking Industry*

At the other end of the manufacturing scale we find the Mill Cabinet Industry. This group of firms includes those who make fine cabinets, furniture, one of a kind displays and the interiors of casinos and offices. Most firms have less than 50 people, are family owned and under capitalized. The United Brotherhood of Carpenters and Joiners, which represents the skilled apprenticed cabinet makers is one of the largest cohesive forces in the industry.

In May of 1993 the union and firms joined together a joint company and union technology committee. Expenses for the committee were defrayed by a grant from the Federal Mediation and Conciliation Service. The function of the committee, officially entitled the Technology Sub Committee of the National Labor Management Committee of the Custom Woodworking Industry, is to assess the technological level of the industry and recommend appropriate courses of action.

As part of its task, the committee embarked on an extremely innovative project. Owners and skilled workers determined the design criteria for the technologies they wanted and then set about attempting to finance the design and production of these technologies. The committee adopted nine criteria:
• machinery to be flexible - defined as capable of being used in production of one of a kind items;
• a technology system that would encourage self-learning, which is defined as cabinetmakers drawing on their craft skills and ways of conceiving work, utilize to use machinery, preferably without learning a whole programming language;
• systems that would improve competitiveness of the firms;
• equipment or systems that could decrease product throughput time;
• ergonomic considerations such as noise, fumes and repetitive motion syndrome to be dealt with in the design of the equipment;
• enhancing the skill of the craftsman "...the committee does not...want machinery that reduces the skill content of the craftsman's job..the jobs of workers using machinery should not be made dull and repetitive" (Carpenters);
• "simple, elegant" machines that skilled workers can use exercising their craft judgement;
• affordability (given the small enterprise nature of the industry) - machines should be priced at $100,000 maximum with payment over three or four years;
• machines that were durable and had low maintenance cost. (Carpenters).

The design criteria are to be given equal weight should a software system or machines be produced. In many respects the criteria developed by the committee mirrored the theoretical discussions that had taken place abroad in the late 1980s (Hirsch-Kreinsen et al., 1989; Corbett, 1987).

The criteria meant that equipment had to be quieter but also be able to cut and shape rapidly. In terms of CNC/NC equipment the software had to be designed in such a fashion as to encourage shop floor programming and to build on the tacit knowledge of the carpenter- especially as it related to the changing and variable nature of wood. This concept of building on the knowledge of the carpenter and designing controls and software has two aspects. One suggests using the knowledge of the carpenter to assess the conditions of the material rather than building expensive sensing and complicated software. This idea follows from the design criteria of low cost and agreement on the value of skilled work. The second aspect addresses the view that programming using existing software asks the skilled worker to learn a different language and conception of his/her work. The committee feels that it would be most helpful to develop software that enhances rather than replaces the thought process of the skilled carpenter. (Grant).

Overall there was a desire to enhance a work organization that is collective in content, rather than breaking up the work into discrete packages- such as programming, design, etc. The committee decided that the concept of "craft", that is the unification of conception and execution, needed to be enhanced because it is a positive good for the industry. As this is a custom business, often the carpenter as well as the owner or salesperson is interacting with the customer and as a result "designing" the works he will do. In other words the craft concept is the dominant view that most workers have of themselves. In turn that notion of craft is equated with quality and is valued by the owners. (Emspak, 1995).

Other companies such as New Balance, Allen-Bradley, Steelcase, Digital Equipment Corporation and Xerox also have initiated programs that support some form of integrating workers in the design of their tools and work place. (Kukla et al.) In fact, part of Steelcase's long term corporate strategy includes explicit recognition of the necessity to focus on collaborative design as a means to grow their business. "Technologies for collaborative learning and for collaborative design, jointly specified to support the above business process, would enable sellers to produce a discontinuity in key, existing markets that are now dominated by service suppliers. Sellers should penetrate these existing markets by engaging in partnerships among: (1) the suppliers of the jointly specified technologies and (2) the technologically astute service providers who already compete in these markets." (Bloomquist p.2).

Therefore, we can conclude that some tentative steps have been made to move beyond a socio-technical model; Engineers specify design criteria based on both "technical" and "social" needs to systems where the participants themselves specify their needs and work in collaboration with engineers and other design professionals.

5. TOOLS TO ENCOURAGE COLLABORATIVE DESIGN

While it is important to have training in collaborative design philosophy and techniques, meaningful government funding and a legal structure that could encourage collaborative design, it is also necessary to have tools that can aid the process. Most organizations who wish to pursue a program of integrating workers in the design process have only group process tools at their disposal. These tools are a necessity to inform the process of group decision making but inadequate to the task of allowing a joint team to assess a large complex system such as a factory or department in a factory.

An example of a possible tool which can aid a joint process is the ACTION system developed by a team from USC. (Blom p. 2).

The ACTION tool has a different focus and is more amenable at this stage to development by a collaborative skilled worker-engineering collective. It is almost ready for the market and is at the point of implementation by a series of labor management development teams. These joint teams will use the ACTION program to design production systems in their respective locations. They will use this assessment tool in conjunction with "traditional" collaborative design techniques.

The objective of the ACTION tool is to successfully integrate technology, organization and people systems. The developers of the systems have designed a tool which involves the conception of risk management - which they define "an effective design process is one in which all risks about possible design options, including those affecting workers are effectively understood and managed". (Gasser p. 3).

The developers of the tool assumed there is broadly diffused expertise among workers, engineers, technicians and managers and that different disciplines and stakeholders can work together to create an understanding of what "is" and then "what is to be". (Gasser p. 4).

The system encourages designs to evolve over time-based on continuing inputs of data. Thus, as one of the many variables which make up the system change, it is possible to see how other variables are aligned so as to maximize the objectives of the system.

6. CURRENT STATE OF AFFAIRS

We have already alluded to the fact there exists little public support for research and development of either processes or technologies that directly include skilled workers in their design and elaboration. Thus, although, there is considerable interest in meeting needs of workers, in most cases workers are not in a position to speak for themselves. None-the-less much progress was made in the last year in putting the issue of collaborative design before the NIST.

In May of 1994, several hundred participants discussed best practice design and implementation programs at the NIST facility in Gaithersburg, Maryland. Leaders of NIST participated, as did a large number of labor leaders, science policy experts and corporate officials. The conference was an opportunity to acquaint the technology and science bureaucracy with the reality that, indeed, workers are a meaningful part of the technology development and deployment process in the United States. Perhaps as important, major American corporations support this effort.

In a second initiative, NIST sponsored six regional meetings around the United States to assess support for a focused program integrating design engineering and the work force. About fifty of the Fortune 500 firms participated as well as numerous scientists, engineers and labor leaders. The papers presented, many of which are cited above, suggested that significant steps are underway wherein firms and their work forces are experimenting with ways to integrate workers in design efforts. However, after lengthy discussion, NIST ATP decided against a focused program in this area. The concerns cited were that a program of integrating workers into the design functions was too sociological and not technical enough, and that the mission of the ATP was more hard technology, not process oriented. In reality there was not a willingness to recognize that skilled workers acting in their own right and in collaboration with scientists and engineers could produce new designs and technologies on the cutting edge. There is also a continuing cultural bias in the United States which continues to see "technology" as a silver bullet able to solve problems, while people and the inaccuracies and imprecision associated with people are seen as part of the problem. Finally, NIST's emphasis on the marketability of a technology as a leading criteria does not lend itself to consideration of work force concerns such as skill, ergonomics and reduction of the toxic load on the environment. (Emspak-NIST).

7. CONCLUSION

Although the possibility of Federal funding for programs designed to support the integration of skilled workers in the design process seems remote, there are significant economic forces that are leading firms to consider the issue. At the same time organized labor has increasingly understood the importance of intervening at the design stage of investment and technology development. Several examples suffice to indicate the trend. As cited above, Steelcase sees the integration of non traditional personnel in design as a means to expand its market.

Integration of workers in design also reduces costs significantly. Estimates are that costs of error rise geometrically as a product nears the prototype stage and then exponentially after proto-typing. For example, the Boeing Company estimated that each error in a major new aircraft was equal to $10,000 and there were about 10,000 errors-prior to prototyping. (NIST Regional Meeting Transcripts and Notes).

Worker participation in the design of production equipment has also been encouraged by some firms. The John Deere Company of Horicon, WI is an example. Workers at its Horicon facility redesigned the control system for a metal press thus allowing it to be more flexible and produce stampings of considerably greater accuracy and quality. (McCubbin and Pappenfuss).

Last, but not least, the Steel industry, long a bastion of conservative governance and manufacturing has adopted a startling new approach to worker participation in design. Almost all integrated steel companies have agreed to have a form of employee decision making in the determination of their technologies. For example, the Inland Steel Corporation and the United Steel Workers of America have agreed to respond to technological change through joint mechanisms which will "cause technology to serve the interests of both the enterprise and the workers affected by change". (United Steel Workers p. 1).

The terms of the discussion of integrating workers into the design functions has moved from a theoretical level to the level of practice. In addition there is a development of computer based tools that will enable the expansion and deepening of the efforts to recombine tacit knowledge with conception.

REFERENCES

Bath Iron Works and the International Association of Machinists (1994). *The Implementation of a High Performance Work Organization*, A white paper prepared for the NIST.

Blom, Kristine A; and MacCriskin Jack (1994). *Design Engineering and the Workforce: A National Instruments Collaboration*, A White Paper for the NIST Advanced Technology Program.

Bloomquist, Lee G. (1994). *Company Specific, Degreed and Non-Degreed, Engineering Practices: Technologies to Optimize Tacit Know-How.* A White paper prepared for the NIST on behalf of the Steelcase Company, Boston, MA.

Bucciarelli, Louis L., Kuhn, Sarah (1995). *Engineering Education and Engineering Practice: Improving the Fit, in Technical Work and the Technical Workforce*, Stephen R. Barley and Julian Orr eds. ILR Press, Cornell University.

Department of Defense (1994), RDT & E Programs R1 Department of Defense Budget for FY 1996-1997, Washington D.C.

Emspak, F. (1994). *Focused Program Recommendation:Integrating Workers in Design Engineering*, (unpublished) Presentation to NIST-ATP.

Emspak, F. (1995). *Integrating the Work Force Into the Design of Production Systems*, IFAC Man Machine Systems Symposium, MIT.

Gasser,Les & Majchrzak, Ann, (1994). *Tools for Integrating the Workforce Into Design Engineering*, White paper prepared by Industry for the NIST, Los Angeles, California.

Grant, M. (1994). *Computerized Machinery and Work Reorganization in the Custom Woodworking Industry)*, Presentation to the NIST-ATP regional meeting, Seattle, Washington.

Hirsch-Kreinsen, H. and Schultz-Wild, R. (1989). *Implementation Processes of New Technologies-Managerial Objectives and Interests*, *Automatica* paper Number 87-29.

Kukla, Chuck et al. (1994) *Advanced Technology for Building a High Performance Work Place to Improve U.S. Industry Competitiveness*, presented to NIST on behalf of the Allen Bradley Company, Bath Iron Works, Digital Equipment Corporation, New Balance Atheletic Shoe Corporation, Xerox Corporation.

McCubbin,J.G.,and Pappenfuss, D (1994). *Involving the Workforce in the Design of New Process Equipment*, Presentation to the NIST-ATP Regional Meeting, Milwaukee, Wisconsin.

National Institute of Standards and Technology (NIST) (1994). *Advanced Technology Program Proposal Kit*. U.S. Department of Commerce, Technology Administration.

National Institute of Standards and Technology (NIST) (1994). *Guide to NIST*, U.S. Department of Commerce, Technology Administration.

National Institute of Standards and Technology, Advanced Technology Program, Transcripts, Regional Meetings to discuss a possible focused program (1994). *Integrating Workers and Design Engineering*.

Office of Technology Assessment, Congress of the United States (1994). *Electronic Enterprises: Looking to the Future*. U. S. Government Printing Office, Washington, D.C.

United Brotherhood of Carpenters and Joiners, (1994). *Minutes Mill-Cabinet Technology Committee*, Washington, D.C.

United Steel Workers of America/Inland Steel (1993). *Memorandum of Understanding on Inland/USWA Partnership*, Appendix G-1.

ORGANIZATION AS AN IMPORTANT SUCCESS FACTOR
IN APPLYING CONTROL TECHNOLOGY

Stanko Strmčnik, Janko Černetič, Matjaž Mulej*
and Dietrich Brandt**

J.Stefan Institute, Jamova 39, 61111 Ljubljana, Slovenia
**University of Maribor, School of Business and Economics and Institute*
for Systems Research, Razlagova 14, 62000 Maribor Slovenia
***RWTH-HDZ/IMA, Dennewartstrasse 27, 52068 Aachen Germany*

Abstract: In a research carried out recently in Slovenia, six important factors were
found to be relevant for success of a computer control application. In the paper the role
of the organization in comparison with other success factors is discussed. The analysis
is based on an inquiry about the relative importance of the success factors in design,
implementation and use of computer control systems. In the inquiry 86 experts from
end user companies, engineering companies and academic institutions have taken part.
An original approach for proper integration of technology and organization is also
suggested.

Keywords: control technology, organizational factor, socio-technical system design,
cultural aspects of automation, systems methodology

1. INTRODUCTION

The introduction of control technology into
industrial practice is a very demanding process.
One reason lies partly in the fact that introduction
of every new technology presents certain problems.
However, control technology as a part of
information technologies is in this respect even
more problematic due to its deep impact on people
and the organization of work. In accordance with
this, we are often faced with severe difficulties and
problems (Martin *et al.*, 1990). Many of them were
personally experienced by the authors during their
20-years long work in the area of control systems
applications (Černetič and Strmčnik, 1991).

Usually the reason problems arise is related to the
nature of the approach used. Very often, the
approach is partial, only taking into account the
narrow technical aspects of control system

implementation (Livingston, 1988c). People in
general are simply not aware of how crucial the
economic, organizational, cultural and human
aspects are to success (Livingston, 1988a, 1988b;
Martin, 1993).

The purpose of the paper is to show which are
according to opinions of a group of interviewed
professionals in Slovenia, the most important
factors for the successful introduction of control
technology in enterprises and which of them are
more and which less important.

One of the success factors is "organization".

The emphasis is given to its relative importance in
comparison to other factors and to the introduction
of an original approach which integrates
organizational and technical issues of large scale
control systems development.

2. DEFINITION OF SUCCESS FACTORS

The concept of success factors usually used in the expression "critical success factors" or "key success factors" is often used in the field of management as a top down planning methodology. Critical success factors are defined as "the key things which must go right for an organization to succeed" (Byers and Blume, 1994).

Identifying these factors and giving them proper attention has become an often used approach in introduction of information systems (e.g. Samek, 1986), microcomputer systems (e.g. McEniry, 1990; Schleich, 1990) and also control and automation systems (e.g. Tanaka, 1991; Tatlock et al., 1993). This discussion will be limited to only those success factors related to the practical use of control technology.

Though the complete spectrum of the considered success factors is quite wide, the majority of factors depends largely on the socio-cultural environment. In a sense, they reflect the main aspects and particularities of the state of development of a certain environment.

It is possible to define the success factors by means of experience from real-life projects, therefore such a definition is partially subjected to judgement. To avoid such bias in definition, a greater number of opinions and projects was taken into account in the research presented here.

The definitions given below are based on the practical experience of authors which has accumulated over a period of more than twenty years. The findings obtained from this experience were summarized, discussed and verified by a preliminary inquiry which involved a dozen control engineers in Slovenia, all having at least five years of practice in using control technology in this country. In the inquiry, the participants answered the following question:

"Which are, in your opinion and, according to your experience, the most important factors affecting the success of computer-based automation?"

About 60 answers were obtained which were arranged into 6 groups later on, each representing one success factor (Černetič and Strmčnik, 1991). The success factors were described by the following keywords: vision, knowledge, justification, organization, technology, human, which characterized the main points of the answers.

3. RELATIVE IMPORTANCE OF SUCCESS FACTORS

If effective measures for grater success in using control technology are to be developed, the success factors defined above must be prioritized according to their relative importance. Only by this, does one know where to concentrate the majority of efforts.

Based on the results of the preliminary inquiry and some additional insights, a special questionnaire was prepared. Here, each of the previously identified success factors was described by a few short declarative sentences. The respondents were then asked to assign an assumed relative weight (percentage) to each factor, all weights summing up to 100%. In the questionnaire the following success factors were listed:

- Clear vision of enterprise development and support of the management (MANAG)
- Proper knowledge and education (EDUC)
- Economic justification and monitoring of effects (JUST)
- Proper organization (ORGAN)
- Proper technology (TECH)
- Consideration of human, social and cultural aspects (HUSOCU)

In late 1993, the questionnaire was sent to approximately 110 professionals involved in control technology in some way: academics, researchers, engineers active in development and engineering, as well as to the engineers employed by user organizations. The distribution of the 86 received responses according to where the respondents worked is given in Fig. 1.

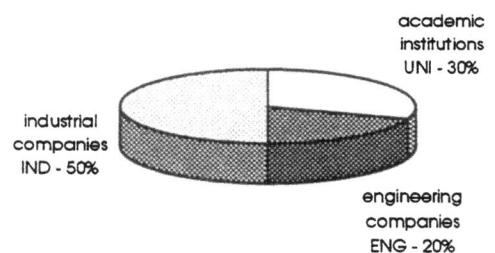

Fig. 1:   Distribution of respondents according to their working environment

It is considered that the number of the responses is representative enough for a country having a total population of about two million people and a relative low number of professionals active in control technology.

Results of the inquiry are given in Fig. 2.

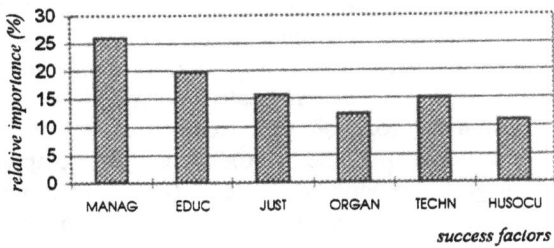

Fig. 2. Relative weight of importance assigned to six success factors.

First, it can be seen that the greatest weight of importance was given to the role of management and his vision associated with the introduction of computer-based control systems into production facilities (the keyword MANAG). This viewpoint is probably a consequence of a phenomenon very often observed in doing computer control projects in Slovenia: that managers in this country very heavily underestimate the possible contribution of control technology to the (economic) success of their enterprises. Therefore they do not pay enough attention to these projects.

Secondly, it can be observed that a relatively high weight of importance was assigned to proper knowledge and experience (the keyword EDUC). This observation can be explained by a relatively low level of education received by people employed in Slovenia. One of the most important reasons for this is a serious lack of continuing education in all fields of technology, which prevailed in all the former countries of Yugoslavia during the past decades. The consequences of this present a serious obstacle for the successful introduction of new technologies in these countries. It is only recently that some efforts in improving the situation surrounding control technologies can be observed (Černetic, et al., 1994).

The other success factors were not thought to be particularly important. For instance, the respondents assigned medium weights to the need for economic justification of computer control projects (JUST) and to the respective technical aspects (TECH). According to the experience from the previous decade, this is slightly surprising. Nevertheless, it is a very bad sign that organizational (ORG) and human, social and cultural (HUSOCU) factors are not seen as very important by the participants of this inquiry.

Now let compare the results of the inquiry from the aspect of where the respondents are working.
In Fig. 3, the viewpoint of engineers from the user enterprises can be seen.

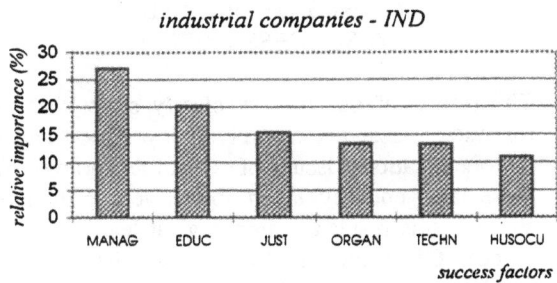

Fig. 3. Relative importance of success factors as assigned by the engineers from the user enterprises.

In general, the results are not very different from the average (as the engineers form the majority of the interviewed population), but it is interesting to note how low the engineers assign the importance of technology.

In Fig. 4, the viewpoint of academics is represented, i.e. university professors and research staff, mainly from R&D institutes.

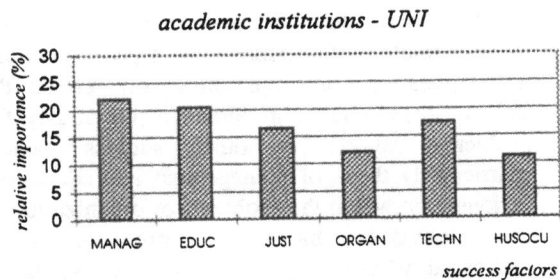

Fig. 4. Relative importance of success factors as assigned by the academicians and R&D people.

It is somehow logical that they assigned a great importance to technical aspects.

The answers of the professionals from engineering organizations, (Fig. 5), are different from the above in that they assign the lowest priority to the organizational aspects and that the differences among particular success factors are accentuated.

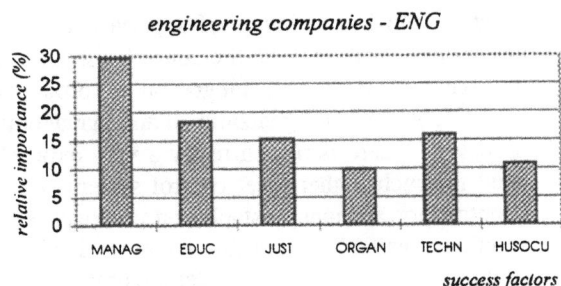

Fig. 5. Relative importance of success factors as assigned by the professionals from engineering organizations.

4. IMPORTANCE OF ORGANIZATION AS A SUCCESS FACTOR

The results of our inquiry clearly show that the respondents assigned a relatively low importance to the organizational issues of control technology. As these are generally taken to be very critical to success, this result is surprising, at least at a first glance. Of course, a direct comparison of our results to those from other countries is not possible because no similar inquiry is known to the authors. The general importance of organizational issues was surmised from the various references, for example Strina *et al.* from Germany (1993), by Tanaka from Japan (1991), and Samek from USA (1986).

In particular, organizational aspects are stressed to be important for control and automation systems by Livingston from USA (1988c). He argues that a proper interaction between the technical and organizational system should be the right solution for the majority of (complex) problems.

If one sets out to find the reasons for underestimation of organizational aspects, then one might guess that this opinion simply reflects the state of development in Slovenia. Obviously the problems related to other success factors (particularly those of management and education) seriously outweigh the problems of organization. It can be envisaged that the latter problems will be addressed when the present problems are better mastered.

There is still another reason why in Slovenia as a small country in the state of transition the awareness of organizational aspects is low. This is related to the natural progression of control technology use: the lowest level of control and automation is first implemented in factories, to assure basic (technical) functioning of production processes. At this lowest level, the control functions do not depend very much on production organization.

Later, at the higher levels of supervisory control and process management, the control systems must be able to support the integration of production with (higher-level) organizational and management processes. There is not currently a very expressed need for such higher-level control systems in this country, nor are many established approaches and solutions available. From this point of view, there is apparently no need to consider the organizational aspects, which explains the results of the questionnaire.

By no means this imply that no such need will occur in the future. This time must come naturally, along with the later phases of economic development. To these ends a new research direction was started in 1994, in order to get prepared for the indicated future. The main long-term goal of this research is to prepare guidelines for the introduction of computer-integrated production systems (CIPS) in Slovene enterprises. It is planned that these guidelines will largely deal with the organizational aspects of controlling and managing production processes. A new (integrated and interdisciplinary) methodological approach is being developed to cope with these aspects.

5. A NEW APPROACH TO INTEGRATING TECHNOLOGY AND ORGANIZATION

The search of a new methodological approach was based on the idea to take the existing approach to the development of control systems and completing it with a proper methodology for dealing with organizational, human and business issues. With this approach only those methodological elements are included which have proven to be successful in long-term practice.

Essentially, the proposed new approach is a combination of three parts: a hard (technical) part, a soft (organization/human) part and an integrating strategic part (Černetic, *et al.*, 1995).

The so-called "hard part" of the new approach is based on the analysis and design of process control systems by a structured systems engineering methodology, very similar to that used in software engineering (e.g. Yourdon, 1989; Shlaer and Mellor, 1992), supported by the commercially available CASE tools, such as System Architect or Excelerator. The initial results of this part of approach are, first, a semi-formal model of material and energy flows in the concerned production process, and second, a semi-formal model of information and control flows in the same production process. These are then the basis for preparing functional specifications of the control system, being the second result of the technical part of approach.

The so-called "soft part" of the new approach deals mainly with the analysis of organizational (working) processes running in the particular organizational units of the enterprise. This part of the new approach is rooted in a soft systems methodology, called Dialectical Systems Theory - DST, (Mulej, 1976). The DST is called "dialectical" to denote the way of looking at systems: a dialectical system is a system that changes through the interactions of its complementary parts. Their interdependence creates relationships between them and the environment, thereby resulting in synergy for achieving holism and improved chances for survival and overall success.

The implementation portion of this methodology is called USOMID; it can be seen as a problem-solving technique because it is based on individual and group interviews of the people concerned (Mulej, *et al.*, 1995). Its objective starting points are needs and possibilities, whereas its subjective starting points are values and emotions, knowledge on contents and knowledge of methods. Thus, the objective and subjective elements determine the dialectical structure of the methodology, corresponding to the "hard" and "soft" part of the new approach.

This way of looking at technological systems today closely corresponds to the Dual Design Approach suggested by Henning and Ochterbeck (Brandt, 1995). The Dual Design Approach is a set of principles to ensure appropriate development of both technical and human aspects of human-machine systems. The technological aspects of system design are represented by one triangle; the design strategy based on the human working process, is represented by another triangle. Both triangles are complementary to each other. The dialectical structure of both these triangles represents the complexity of the methodology of changing systems.

Information about the flow of working processes, as well as how corresponding information flows, obtained from the interviewed people, are put into a semi-formal paper-based model, called "programoteque". It describes the flow of input/ output information, physical input/output information carriers (e.g. documents) and the structure of work-flow processes running in the observed organizational (production) unit of the enterprise. This model can be used as a starting point for a wider optimization (rationalization) of organizational/business processes, which must be accomplished before the introduction of the computer support. The further steps of this methodology are done mainly iteratively and in the form of special working groups, similar to the well known "quality circles".

The "strategic part" of the new approach has two functions: first, it helps to deal with the manager-oriented business issues of introducing control technology into the enterprise, and secondly, it helps to integrate the previously mentioned hard and soft parts of the new approach. The level of integration depends on the degree of interdependence between the technological and

organizational processes concerned. This, in turn, depends on the level of the management/control pyramid where the new control system is to be introduced.

Some elements of the new integrated approach namely the "hard" part and the first (modelling) step of the "soft" part of the approach were recently used in two enterprises in Slovenia: one from the chemical and the other from the wood-semiproducts branch.

The main advantages of the presented integrated approach are wide-ranging. From the methodological and human (psychological) viewpoint, it is certainly beneficial that a single approach combines different methodologies and working styles. In this way, people with different personal interests and work-style preferences can more easily find something in which to enjoy and cooperate creatively. Another favorable consequence of the previously mentioned methodological diversity is that people from different departments of the same enterprise learn a common language for discussing problems of interfacing the three important systems in their organization: the business/management system, the production system and the supervisory/control system.

It was also observed that the involved people generally feel better with this approach, probably because it is not purely technical and impersonal. They become easier motivated, more cooperative and innovative - in short: synergetic. As they are asked for personal opinion in a positive and friendly atmosphere, they are willing to discuss creatively the most stubborn problems, thereby beginning to optimize business/organizational processes they are involved in.

Unfortunately, there are also some difficulties with this approach. As is generally understandable, it was observed in our project that interdisciplinary cooperation is difficult: it involves more people, more aspects have to be considered and it takes more time to accomplish something really useful. This adds to the complexity of systems analysis which has both positive and negative consequences. Furthermore, it was observed, that the soft approach easily seems strange to technical people involved, even to the point where they become distrustful ("Don't sell me empty nuts!" is a typical reaction of such people.) On the other hand, some would like clear, quick and simple results. Due to the complicating factors, managers in particular can rarely take enough time to be more seriously involved with the "soft" part of approach because they are preoccupied with company survival.

Fortunately, this difficulty is mainly a consequence of local circumstances and may not be taken as a disadvantage of the approach.

6. CONCLUSION

The critical success factors are important tools of strategic planning during the introduction of control systems into production. An analysis of these factors in Slovenia has shown that great importance is assigned mainly to two critical factors: first to management and to its role during computer control projects, and second, to education. Unfortunately, the importance of the organization is underestimated. Expecting that this situation must and will change in the comming years, the authors have begun to develop an integrated approach to the development of higher-level control systems. The new approach strongly considers the integration between the technical and organizational and business processes within an enterprise. Preliminary results of the reported approach, applied in two industrial case studies, are very encouraging. However, there were also abundant problems encountered that need to be solved in the future. This seriously motivates and challenges further international research and communication.

7. ACKNOWLEDGEMENT

The reported research was supported by the Slovene Ministry of Science and Technology.

REFERENCES

Brandt, D. (1995). Information, Automation and Chaos in Manufacturing. *Int. Colloquium on Organisational Innovation*. Melbourne, May.

Byers, C.R. and D. Blume (1994). Tying Critical Success Factors to Systems Development. *Information & Management*, **26**, No. 1, 51-61.

Černetič, J. and S. Strmčnik (1991). Automation Success Factors as Seen from a Developing Country. In: *Proc. of the 1st IFAC Workshop "Cultural Aspects of Automation"*, (P. Kopacek, Ed.), Krems, Austria, 34-41

Černetič, J., M. Mulej, F. Drozg (1995). Designing Large-Scale Systems by Integrating a Hard and Soft Systems Approach. In: *Proc. of the XII Int. Conference on Systems Science*, Wroclaw, Poland, (in print).

Černetič, J., S. Strmčnik, R. Karba (1994). A Continuing Education Project Supporting Technology Transfer. In: *Proc. of the 3rd European Forum for Continuing Education*,

Vienna, Austria, 265-270.

Livingston, W.L. (1988a). A Process of Elimination. *Control Engineering*, **35**, No. 4, 154-160.

Livingston, W.L. (1988b). What Organizations do in Complex Projects and Why They Do It. *Control Engineering*, **35**, No. 5, 220-230.

Livingston, W.L. (1988c). Design for Complexity, *Control Engineering*, **35**, No. 6, 144-156.

Martin, T. (1993). Considering Social Effects in Control System Design - A Summary. In: *Preprints of the 12th IFAC World Congress*, Sydney, Australia, **7**, 325-330.

Martin, T., J. Kivinen, J.E. Rijnsdorp, M.G. Rodd, W.B. Rouse (1990). Appropriate Automation - Integrating Technical, Organizational, Economic, and Cultural Factors. In: *Preprints of the 11th IFAC World Congress*, (V. Utkin, U. Jaaksoo, Eds.), Tallin, Estonia, **1**, 47-65.

McEniry, M.C. (1990). Potentially Important Factors in the Success of Microcomputers in Small and Medium-Sized Colombian Firms: An Explanatory Study. *International Journal of Human-Computer Interaction*, **2**, No.2, 153-172.

Mulej, M. (1976) Toward the Dialectical Systems Theory. In: *Proceedings of EMCSR, Progress in Cybernetics and Systems Research*, (R. Trappl, Ed.), Vienna, Austria, **6**

Mulej, M., J. Černetič, F. Drozg (1995). Dialectical Systems Theory Helps the Control Technology Consider the Organizational Aspects in Practice. In: *Proc. of the XII Int. Conference on Systes Science*, Wroclaw, Poland, (in print).

Samek, M.J. (1986). Integrating Systems into the Organization. *Information & Management*, **11**, No. 1, 9-12.

Schleich, J.F., W.J. Corney, W.J. Boe (1990). Microcomputer Implementation in Small Business: Current Status and Success Factors. *Journal of Microcomputer System Management*, **2**, No. 4, 2-10.

Shlaer, S. and S.J. Mellor (1992). *Object-Oriented Systems Analysis - Modelling the World in States*. Prentice-Hall, Englewood Cliffs.

Strina, G., M. Suethoff, S. Grinda, D. Brandt (1993). Automation without Organizational Development won't Do. In: *Preprints of the 12th IFAC World Congress*, Sydney, Australia, **7**, 339-342.

Tanaka, N. (1991). Critical Success Factors in Factory utomation. *Long Range Planning*, **24**, No. 4, 29-35.

Tatlock, R., R. Taylor, L. Noble, R. Manternach, M. Mueller. J. Salazar (1993). Key Success Factors in Automating a Pharmaceutical Process. *Pharm. Tech.*, **17**, No. 6, 54-60.

Yourdon, E. (1989). *Modern Structured Analysis*. Prentice-Hall, Englewood Cliffs.

OPERATOR REQUIREMENTS FOR MAN-MACHINE INTERFACES OF CNC-MACHINE TOOLS

C. Schlick, J. Springer, H. Luczak

Institute of Industrial Engineering and Ergonomics
Aachen University of Technology
Bergdriesch 27, 52062 Aachen, Germany
Email: cschlick@iaw-1.iaw.rwth-aachen.de

Abstract: An investigation of operator requirements for man-machine interfaces of CNC machine tools was performed in 13 small and medium enterprises of metalworking industry. For the purpose of a detailed analysis of interview contents a framework of description based on Rasmussen's model of human performance is introduced, considering task-oriented user behavior and software ergonomic aspects of NC-design simultaneously. This framework is illustrated on the basis of the task "Planning of Machining Strategy".

Keywords: CNC, Human-centered design, Man/machine systems, Manufacturing systems, Multimedia, Tasks

1. INTRODUCTION

In modern computer-integrated manufacturing CNC-machine tools play a key role as the origin of the line of production. In spite of all automation efforts CNC-controlled machines still need a skilled operator to optimize performance and ensure flexibility. It is the irony of automation that the more advanced the computer technology on the shop floor is, the more crucial is the contribution of the human operator (Bainbridge, 1988). Thus the design of user- and task-oriented man-machine interfaces of CNC-machine tools is an important goal for the development of efficient and reliable manufacturing systems.

2. INVESTIGATIONS AND METHODS

Funded by the German Research Foundation (Deutsche Forschungsgemeinschaft) several institutes of Aachen University of Technology investigate next generation manufacturing cells. Such systems are called "Autonomous Production Cells", which stand out due to an increased flexibility in view of changing boundary conditions at the shop floor. In an ideal state of affairs the entire information neccessary to perform the manufacturing task is available in place. The term "Cell" means the well-suited unit, comprising both skilled operator(s) and machine tool(s) embedded in the "organism" of flexible production plants. As part of the whole project the Institute of Industrial Engineering and Ergonomics cooperates with the Machine Tool Laboratory for the purpose of design and development of a user- and task-oriented man-machine interface. Focusing on human information processing, the target of research and development is the numerical control (NC). In order to integrate potential users of future technology in an early stage of the project, an investigation of skilled operator requirements for man-machine interfaces of CNC-machine tools was performed in 13 small and medium enterprises of metalworking industries. With reference to the manufacturing conditions the focus was on small batch production with milling machine tools equipped with a modern NC. From the user's point of view a holistic task spectrum was

205

considered: (1) Setup, (2) NC-Programming, (3) Running-in of the NC-Program, (4) Process Control, (5) Fault Management, (6) Quality Inspection and (7) Maintenance. Methodically task-structured, single and group interviews of 2 hours duration were performed. In total 22 machinists with at least 5 years of CNC experience and a holistic task spectrum participated in the investigation. The interviews were stored on tape and typewritten afterwards. Five interview phases were distinguished:

1. In a "warm-up" phase the research project including design proposals for machinists' support based on literature analysis (e.g. Bullinger *et al.*, 1994; Martin, 1992; Zimolong and Konradt, 1994) were presented to the experts.
2. The boundary conditions (personal, technological, organizational) were recorded.
3. To begin the technical discussion the operators were asked to name the most problematic items of their task spectrum and possible reasons.
4. The main part of the interview consisted of the technical discussion. It was structured along the task spectrum and was enriched by questions regarding user behavior or actions in critical, error prone situations.
5. Finally the experts were asked to comment on the interview.

For the purpose of a detailed analysis of interview contents a framework of description was developed, considering task-oriented user behavior and software ergonomic aspects of NC-design simultaneously:

- The user-related part of the framework is based on Rasmussen's model of human performance (Rasmussen, 1983). This model distinguishes three levels of control of human actions: skill-based, rule-based and knowledge-based behavior. Referring to knowledge (or model) -based behavior, Rasmussen furthermore introduced a means-ends abstraction hierarchy used for the representation of functional properties of a technical system. Five levels of abstraction are distinguished: functional purpose, abstract function, generalized functions, physical functions and physical form (for details see Rasmussen, 1986; Wirstad, 1988). A simplified diagram is shown in figure 1, right side.

- The software ergonomic (NC-related) part of the framework is based on the semiotic model of interaction (Cherry, 1967). In semiotic terms four levels of interactions are distinguished: physical, syntactical, semantic and pragmatic. From the point of view of human-computer interaction (Bullinger, 1985; Dzida, 1983, Foley *et al.*,1990) the physical level refers to display- and I/O-devices, the syntactic level to dialogue structures, the semantic level to software functions (and objects) and the prag-matic level to application models (and organi-zational embedding). This structure is shown in figure 1, left side.

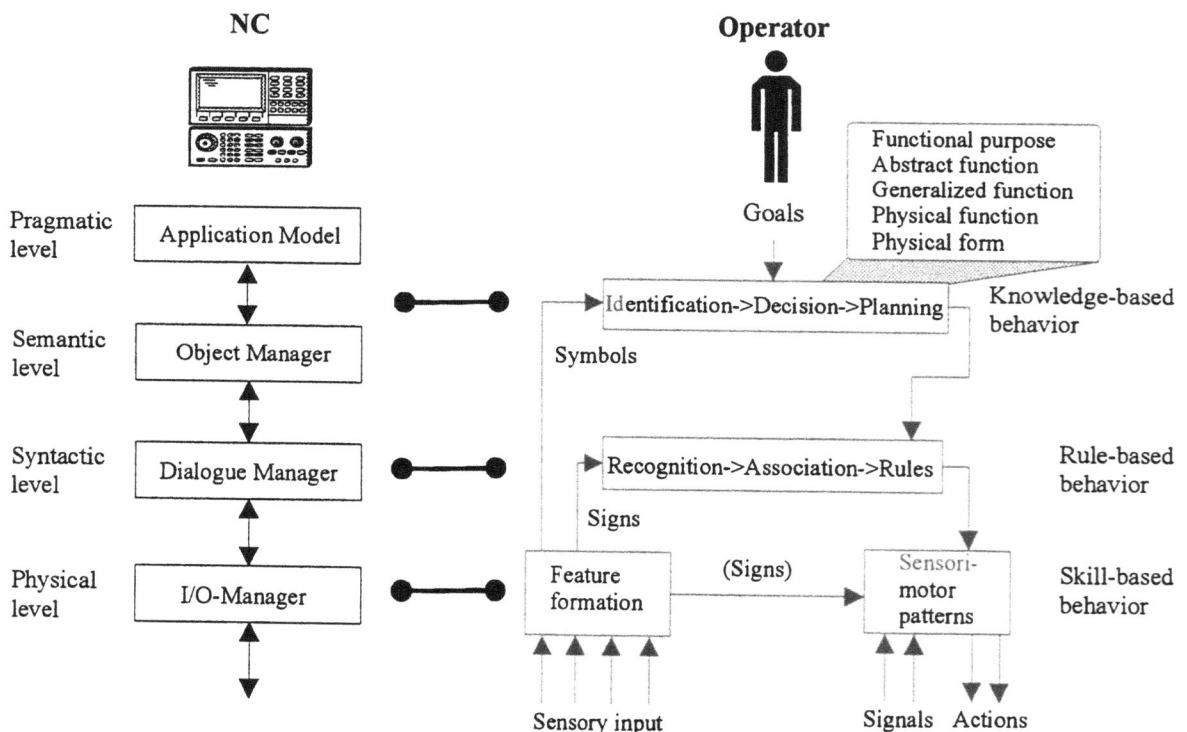

Fig. 1. The framework of description used for the analysis of the verbal protocols (see text).

3. RESULTS

With regard to the entry into the technical discussion (phase 3) quantitative results are shown in Table 1.

Table 1 The most problematic items of the operators' task spectrum as named in the beginning of the technical discussion (22 Experts, 3 double namings).

Task	Namings
SETUP	6
NC-PROGRAMMING	7
RUNNING-IN	6
PROCESS CONTROL	1
FAULT MANAGEMENT	5
QUALITY INSPECTION	0
MAINTENANCE	0

The table indicates that from the operators' point of view the planning (pre-process) tasks are regarded more problematic than others. The task of process control itself is rarely considered critical. According to expert statements, this is due to advanced control and machine design leading to a high degree of process reliability even in small batch production. On the basis of the introduced framework of description the item "NC-Programming" is selected from the task spectrum and the sub-task "Planning of Machining Strategy" is analyzed in detail (figure 2). This sub-task consists of three elementary tasks: "Planning of Clamping Strategy", "Planning of Tool Employment" and "Planning of Machining Sequence", which are very strongly linked and performed iteratively. From the user's point of view the task objectives are, e.g. the accuracy of dimensions and surface quality, a minimal amount of fixture rearrangings, optimal cutting conditions etc. Problem tendencies are, e.g. a complex geometry, a high degree of novelty, free clamping, varying contours (e.g. cast iron), thin or hollow parts etc. Based on an analysis of the verbal protocols the hierarchy which was applied for the representation of functional properties of the product in relation to knowledge-based behavior is shown in figure 2, right side. For user-oriented design of NC-software the different levels should be represented adequately by application models and software functions, which refers to the pragmatic and semantic level in terms of the framework of description. Thus the operator should be able to change levels of abstraction and aggregation simultanously or seperately. At each level alternative solutions should be explored easily. Analogies can help to transfer the representation to a category of model for which a solution is already known or rules are available to generate the solution. According to the experts, some of the useful software tools may be (figure 2, left side):

- Workshop-oriented CAD: Workshop-oriented CAD has a tight relation to the functional purpose and abstract function of mental representation. For an optimal planning of the machining strategy the operator should be able to recognize the intended functional effect of the workpiece as part of an assembly or subassembly. Therefore a workshop-oriented CAD system should clarify functional links and interfaces of the workpiece to other components of the whole product. If design and manufacturing are locally distributed this software tool may also serve the purpose of computer supported cooperative work. In this case operators and design engineers were enabled to discuss critical areas of the workpiece in terms of designing for production on the basis of a shared view of the CAD-model (application sharing, Dix *et al.*, 1993) .

- Workshop-oriented Programming: Workshop-oriented programming mainly relates to the general function. Therefore basic machining steps, e.g. centering, drilling, threading etc, should be selected, combined and aggregated easily. According to the experts switching between different modes of presentation should be supported (graphical, high or low level textual programming).

- Realistic Simulation of the NC-program: A realistic simulation of the NC-program has a tight relation to the physical form, but also refers to the physical function. Supported by a realistic simulation the machinist should be able to estimate machine and process behavior accurately. According to the experts it is a weakness of conventional simulations that in spite of a proper representation of the physical form, the dynamic of playback is only determined by the geometric complexity of the focal plane and the actual processing performance of the NC. Neither technological rates nor the machining states (e.g. resonance phenomena) are represented. This kind of implementation often leads to confusion.

- Multimedia Documentation: A multimedia documentation basically relates to the physical form. From the machinists' point of view it is useful to store certain manufacturing conditions like complex clamping situations in a pictorial way. In many cases this kind of storage is already done with conventional techniques (e.g. photographic images fixed to print outs of NC-programs). An important requirement is the direct link to lines or segments of the NC-program.

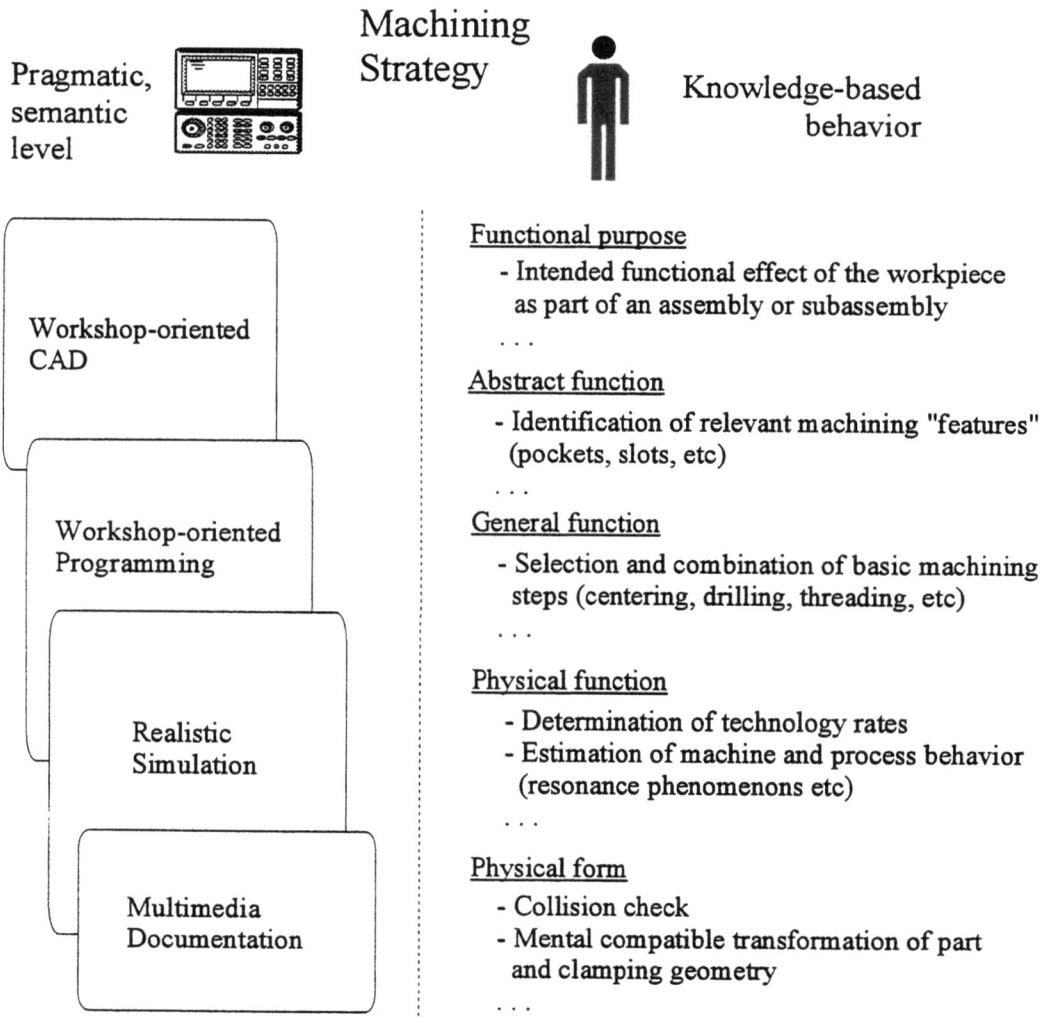

Fig. 2. Application example of the introduced framework of description. The analyzed sub-task is „Planning of Machining Strategy" (see text).

4. CONCLUSION

Due to the holistic task spectrum considered in this study various usability and support issues were identified that pertain to CNC-milling in small batch production. To analyze the results with respect to task conditions and user behavior a framework of description was introduced and applied to a certain sub-task. Regarding future activities in the research project the main question related to knowledge-based behavior is how to link the various software tools in an ergonomic infrastructure of information objects. For this kind of problem the concept of hypermedia seems promising with its claim for a design based on the ᵢₑₐ that multiple associative structures are possi⸱ ᵒr the same information content (Nyce and

Kahn, 1992). On the basis of the analyzed sub-task (figure 2) a possible infrastructure is illustrated in figure 3. This infrastructure links database tools, e.g. Tool Catalogues, Pictorial Sequences of Clamping Situations, with decision support systems, e.g. Technology Processors and applications for computer supported cooperative work, e.g. workshop-oriented CAD.

5. ACKNOWLEDGEMENT

The research is funded by the German Research Foundation within the framework of SFB 368: "Autonomous Production Cells". The authors would like to thank the enterprises and especially all participating operators for their engagement.

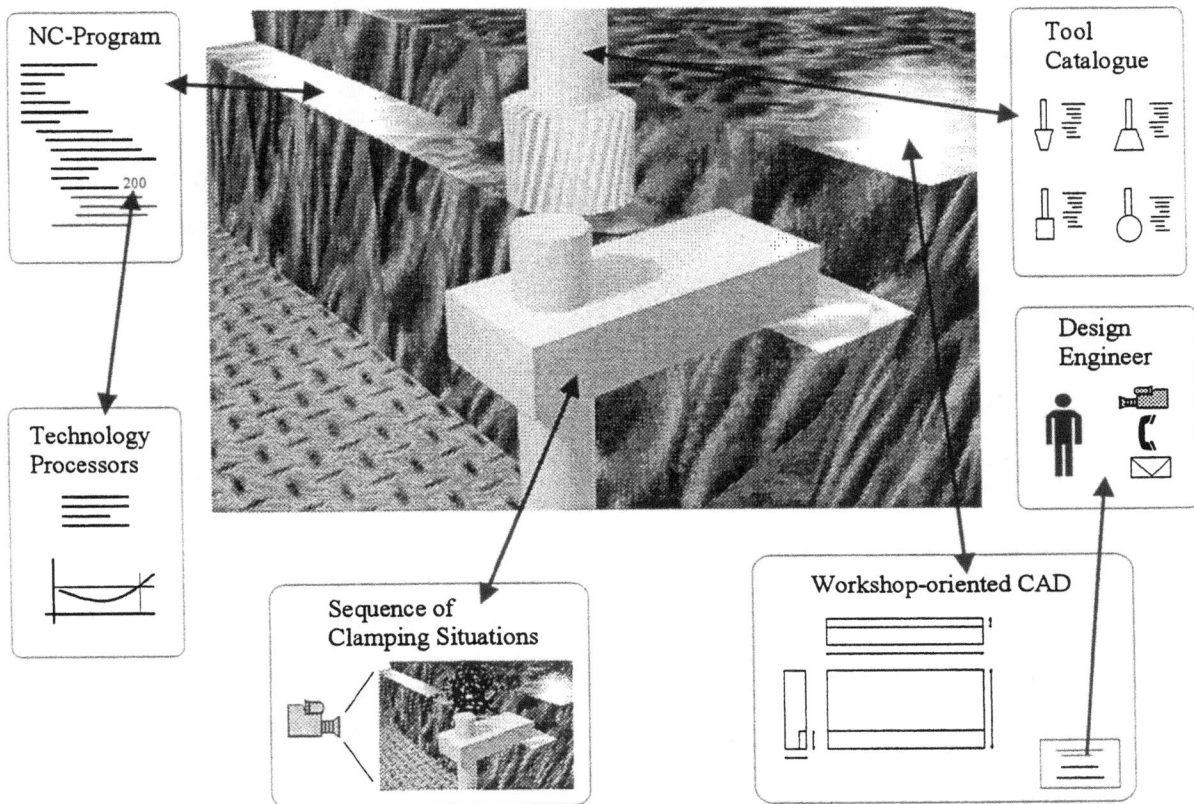

Fig. 3. Illustration of a possible infrastructure of information objects with reference to the analyzed sub-task "Planning of Machining Strategy" from figure 2.

REFERENCES

Bainbridge, L (1988). Ironies of Automation. In: *New Technology and Human Error* (Rassmussen, Duncan, Leplat (Ed.)). Wiley, New York.

Bullinger, H.-J (1985). Grundsätze der modernen Dialoggestaltung. In: *Handbuch der modernen Datenverarbeitung - Softwareergonomie*, **22. Jg.**, p. 21-31.

Bullinger, H.-J.; Fähnrich, K.-P.; Thines, M. (1994). Zukünftige Benutzungsschnittstellen an Werkstattinformationssystemen. In: *REFA-Nachrichten 6/94*, p. 5-18.

Cherry, C. (1967). *Kommunikationsforschung.* S. Fischer Verlag, Hamburg.

Dix, A.; Finley, J.; Abowd, G.; Beale, R. (1993). *Human-Coputer Interaction.* Prentice Hall, Englewood Cliffs.

Dzida, W.: Das IFIP-Modell für Benutzerschnittstellen. In: *Office Management - Sonderheft 1/83*, p. 6-8.

Foley, J. D.; van Dam, A.; Feiner, S. K.; Hughes, J. F. (1990). *Computer Graphics: Principles and Practice* (Second Edition). Addison-Wesley, Reading, MA.

Nyce, J.M.; Kahn, P. (1992). A Machine for the Mind. Vannevar Bush's Memex. In: *From Memex to Hypertext: Vannevar Bush and the Mind's Machine.* Academic Press, New York.

Martin, H. (1992). *Erfahrungsgeleitete Arbeit mit CNC-Werkzeugmaschinen und deren technische Unterstützung.* Institut für Arbeitswissenschaft, Kassel.

Rasmussen, J.: Skills, rules, knowledge: signals, signs, and symbols and other distinctions in human performance models. *IEEE Transactions on System, Man and Cybernetics* **SMC-13 (3)**, p.257-267.

Rasmussen, J. (1986). *Information processing and human-machine interaction. An approach to cognitive engineering.* North-Holland, New York 1986.

Wirstad, J. (1988): On knowledge structures for process operators. In: *Tasks, Errors and Mental Models* (Goodstein, Anderson, Olsen (Ed.)). Taylor & Francis, London.

Zimolong, B.; Konradt, U. (1994). Diagnose-Informationssystem. In: *Zeitschrift für wirtschaftliche Fertigung 5/94*, p. 244-246.

RADAR SCREEN INFORMATION ACCESS: AN EXPLORATIVE INVESTIGATION

T. Bierwagen, K. Eyferth and H. Helbing

Berlin University of Technology, Department of Psychology, DFG-Project „Flugsicherung als Mensch-Maschine-System", Dovestr. 1-5, 10587 Berlin, Germany

Abstract: Information flow between machine and man is of basic importance in the design of HMI. This is especially true for complex systems like air traffic control (ATC). Therefore an experiment to control information access on a simulated radar screen is reported. The Method and the design of the experiment are described. A short description of the simulation facility is enclosed. The relative and absolute usage of information is outlined. By frequency of usage, three categories of importance are identified. Strategies on information access are summarised. Task demands are controlled by a NASA TLX test. The results are discussed.

Keywords: Man-Machine Systems; Mental Models; Air-Traffic Control; Cognitive Systems; Human Factors; Simulation; Automation

1. INTRODUCTION

Information flow between machine and the human operator is of basic importance for any investigation on the human operator's mental representation of the system. Concerning the task of a better human-machine-interface design in complex systems, knowledge on this mental representation (the operator's 'model' of the system) is needed. By this a matching between the operator's internal model of the system and the technical model of the system (i. e. the system itself) is to be attained (Mogford, 1991). Any matching process of this manner will affect future states of system automation in a positive way. This ensures not only a safer and more efficient socio-technical system, but also improved working conditions and operator's job satisfaction. To meet this conditions, as a basic step information access of the operator is to be investigated carefully.

Looking on the domain Air Traffic Control (ATC), further automation is to be implemented until beginning of next millennium. So in this extremely complex and heterogeneous system, research on information access is rather promising. The investigation reported was conducted in spring 1994. A computer-based simulation system with a primary radar screen display was used as a simplified working position for air traffic controllers. By using a computer mouse, the subjects were able to access information parameters for each target by choice. Information already accessed was displayed continuously as a *'self-made'* primary target's label. Task was to handle all traffic safely and efficiently.

Results of the experiments will be presented. Hereby the process of information access will be described in detail. Strategies of accessing the information will be exemplified. The authors understand these results as a first step aiming at *keeping the controller inside the loop* in future ATC systems.

2. EXPERIMENT ON RADAR SCREEN INFORMATION ACCESS

The experiment was designed to identify the data needed by controllers for constructing a representation of the traffic situation (a so called *'picture'*, see Bierwagen et al., 1994). It was an explorative study and for that it was not based on hypotheses.

2.1 Method of the Experiment

On the radar display only sector borders and airway information as background and symbols for the air-

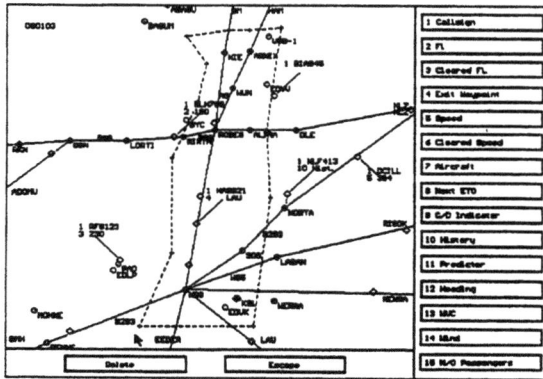

Fig. 1: Primary Radar Screen Display with Button Board

craft targets were presented. Any other information about objects usually given in the secondary radar, was eliminated. On the right hand side of the display, a button board had been installed, which would allow the controller to select that particular information which he needed to exert control on an individual aircraft (see Fig. 1). For this interaction the controller used a mouse. He had to click on the target first and then the softbutton labelled with the information item needed (or vice versa). Each object could carry up to three information items at the same time. For further items, an explicit deletion had to be carried out to exchange items no longer needed by those needed at the moment. This deletion was done by first selecting the information item in the target's label and then the softbutton 'Delete' on the lower button board. The other control button, 'Escape', made reselections possible. Finally, by pressing the right mouse button the label of the object indicated by the mouse cursor could be turned in steps of 90 degrees each. This special feature was necessary to keep the display readable.

By this experimental setting, controllers had to choose actively all information items needed. The parameter field (at the right side of Fig. 1) contained information usually presented on flight progress strips or on the secondary radar display. Additional, the controller could get a connection to a 'ghost pilot', who substituted any pilot being addressed. This connection was established in both forms: as a pseudo radiotelephone link and as computer system link. On the controller's demand, the ghost pilot would alter heading, speed, or altitude of targets in the scenario, thus changing the display process.

One simulation run lasted about 20 minutes. Each subject had to survey two runs. The controller was instructed to handle all traffic presented on the screen safely (not only the traffic within the sector displayed). Targets already on the screen at the onset of simulation could be interpreted only after some of their attributes were chosen actively by the subjects.

Targets entering the screen later simply reported their callsign (and, of course, other information on request) via radiotelephone. Handing over between sectors was eliminated to exclude any verbal exchange of information with neighbouring controllers.

2.2 Design of the Experiment

Independent variables were the rate of conflicts and the amount of traffic. In the low-traffic run 19 targets had to be handled within 20 minutes. Three conflicts were integrated. In the high-traffic situation, 39 targets and five conflicts had to be handled during an equal period. As a variation, another conflict was added right at the end of 50 percent of the scenarios. So half of the low-traffic runs included four conflicts, and half of the high-traffic runs six conflicts. This variation was introduced to control memory performance. The hypothesis was, that after an unsolved conflict at the end of the scenario, attributes of the conflicting aircrafts could be reconstructed better than after scenarios lacking such a conflict. This hypothesis has been tested by a free recall at the end of each simulation run. For all the runs the same sector was used: sector FIR Bremen South Radar 2, complemented by parts of the adjacent sectors.

Dependend variables were the results of the free recall and the requests of information. In order to protocol the calls for attributes of specific targets, all keystrokes were logged in a logfile together with their timestamps. The call of the control parameters ('escape', 'delete', 'label turn') have been included in the logfile as well. Subjective demand characteristics of the tasks were assessed by NASA Task Load Index (TLX) at the end of each run. Communication with ghost pilots was recorded on tape. To avoid sequence effects, the four conditions (two different traffic demands, each of them with or without open conflict at the end) were permuted completely.

2.3 Experimental Simulation System

Due to the demands of performing psychological experiments in the domain of ATC, a new simulation facility was build up. The system called 'EnCoRe-PLUS' (EnCoRe - programmable airspace simulation) is a PC-based simulation of a radar controller working position. ATC-specific data as well as experiment-specific data can be determined in setup-files. The system presents a normal secondary-radar display supplemented by electronical flight progress strips (callsign and route information only). Optionally a primary radar display can be presented, as it was used in this experiment. The image can be varied according to experimental aims. Airspace structure data can be displayed if needed. To perform interactive scenarios with on-line changes in flight data, a ghost-pilot system can be connected to the system. An intercom for verbal communication between controller and pilot including tape recording is available. Every event, especially all kinds of interaction with the system concerning traffic flow or experimental

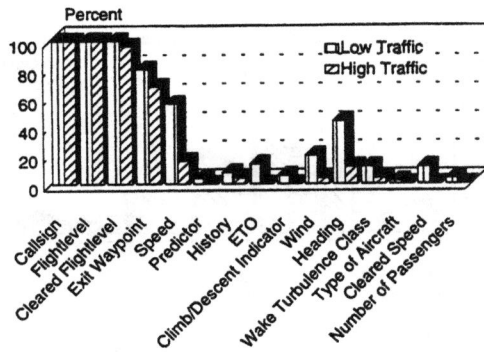

Fig. 2: Percentage of Parameter Requests

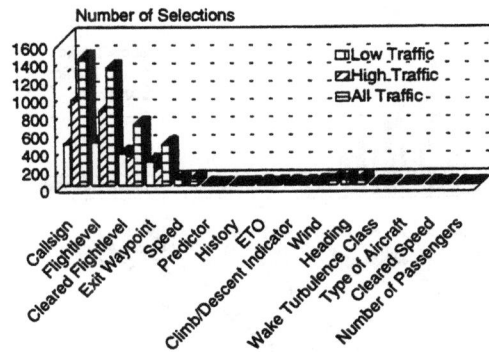

Fig. 3: Absolute Frequency of Parameter Requests

issues, are recorded in a logfile. Computer-aided data analysis is supported. Similar to existing ATC-systems, the primary target is a rhombus.

The experiments were performed in the 'Bremen South Radar 2'-sector of the lower airspace in Germany. This piece of airspace was selected, taking into account usual traffic structure as well as the controllers' familiarity with the airspace. Data preparation started from original data of the most busy days in Germany in 1991 and 1993. Data were modified according to the parameters to be varied.

2.4 Procedure of the Experiment

The general instruction on the system was followed by specific instructions on the experiment. Subjects started with a warming-up scenario with medium traffic load lasting about 10 minutes. During this warming-up, the mouse handling was trained until the subjects reported to know how to carry out all possible actions. Then, the first simulation run started. Immediately after the end of this run, the first - not announced - free recall started. It lasted 5 minutes. The subjects were instructed to name all objects and attributes they could remember. The first part of this experiment ended with a version of NASA-TLX. The second part followed the same sequence as the first one. The only difference was that now the subjects could expect that the free recall followed.

2.5 Results

Information requested. An overview on the different parameters used is given in Fig. 2. This figure shows how many subjects (in percent) called for each specific parameter. The figure shows clearly, that the only parameters used by all subjects are *'callsign'* and *'flightlevel'*. The *'callsign'* identifies the target and is indispensable for communicating with the pilot. However, the callsign is not an attribute that relates the target to other objects within the sector. *'Cleared flightlevel'* holds nearly the same information as *'flightlevel'*, both parameters can mostly substitute each other. According to the frequency of calls the next parameter is *'exit waypoint'*. The four parameters accessed (*'callsign'*, *'flightlevel'*, *'cleared flightlevel'*, *'exit waypoint'*) cover about 93 percent of

all demands for information. We will call these the *'category 1'* of parameters.

All subjects requested information on *'callsign'* and altitude. In addition, frequently an indicator for the target's direction has been used. There were no hints given at altitude and direction in the primary radar. Furthermore, Fig. 2 shows that during the low traffic-scenarios more different items were requested than in the case of high traffic.
A *'category 2'* of parameters consists of *'speed'*, *'predictor'*, *'history'*, *'estimated time over next fix'* (ETO) and *'climb/descent indicator'*. They were used by 10 to 20 percent of the subjects and cover another 6 percent of all requests. All remaining parameters (*'category 3'*) were only marginally used. Altogether they cover merely about 1 percent of all calls.

The absolute frequencies of using each parameter, shown in Fig. 3, proves again the priority of the first four parameters. For example the parameter *'speed'* had been used by about 50 percent of the subjects. But regarding the absolute number of calls, it falls back into the category 2 of less important parameters, covering 6 percent of the absolute usage. This is also true for the items *'predictor'* and *'history'*. *'Wind'* and *'type of aircraft'*, parameters used by several controllers, belong to category 3 of rarely used parameters.

The three categories for parameter frequencies are:

1. *'callsign'*, *'flightlevel'*, *'cleared flightlevel'*, *'exit waypoint'*   **93 percent**
2. *'speed'*, *'predictor'*, *'history'*, *'climb/descend indicator'*, *'ETO'*   **6 percent**
3. *'cleared speed'*, *'type of aircraft'*, *'heading'*, *'WTC'*, *'wind'*, *'number of passengers'*   **1 percent**

The relative frequencies of using each parameter is shown in Fig. 4. Looking on the 'all traffic'-condition, the parameters *'callsign'* and *'flightlevel'* cover 65 percent of all requests, the other parameters of category 1 (*'cleared flightlevel'* and *'exit waypoint'*) a further 28 percent. Here the original categorisation was confirmed. The parameters within category 1 are intercorrelated. *'Callsign'* and *'flightlevel'* correlated significantly (.72, p < .001). The parameters *'flightlevel'* and *'cleared flightlevel'* correlated negatively (-.35, p < .006), which indicates that they

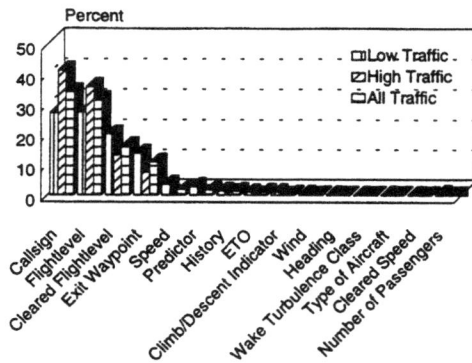

Fig. 4: Relative Frequency of Parameter Requests

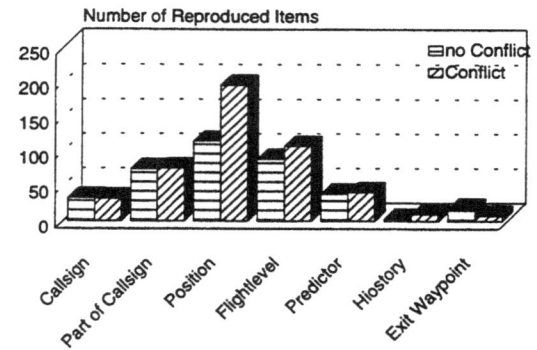

Fig. 5: Number of correctly reproduced Items (Free Recall)

had been used alternatively. In high-traffic conditions, lower relative values are to be found for all parameters except 'callsign' and 'flightlevel'. The variety of parameters in use decreases significantly in high-traffic scenarios. In cases of high demands, a shift in strategies to access and to process information can be presumed. Although the other parameters of category 1 ('cleared flightlevel' and 'exit waypoint') seem in this case to be less important, they still belong to the most frequently used parameters. Some parameters belonging to category 3 disappear under conditions of high traffic.

Information Access Strategies. To assess strategies of information access, each logfile was analysed in detail. Any interaction to the system was taken into account to determine schemes of interaction. These analysis show, that different kinds of information access strategies occur. A parameter-specific access and a target-specific access can be separated. The target-specific access can be devided into more and less strictly target specific procedures. Strictly target-specific interaction stands for a strict rule of information access, e. g. always *callsign* and *flightlevel* of one target, proceeded by the next target. A less target-specific strategy varies in the sequence of the same parameters, e. g. *callsign* and *flightlevel* of one target, proceeded by *flightlevel* and *callsign* of the next one. In contrast, a parameter-specific access represents any access related to one parameter, e. g. *flightlevel* of one target followed by *flightlevel* of the next target.

Looking on all subjects, a shift in information access strategy can be stated. In a preceding phase (called *orientation phase*) most subjects work parameter-specific. After about 3 to 5 minutes - depending on the traffic load - they shift towards a target-specific strategy in a so called *working phase*. This shift is indicated by a longer pause in information requests. The average information access in the *orientation phase* in high traffic scenarios of 20 minutes duration is about 43 percent of all parameters accessed, whereas in low traffic condition (20 minute scenarios as well) it is only about 38 percent. In addition, the *orientation phase* in high traffic scenarios lasted about 4 minutes and 50 seconds, whereas in low traffic conditions it took about 2 minutes and 40 seconds. A clear categorisation between strictly tar-

get-specific and less target-specific access in the *working phase* can not be drawn, although most of the subjects do use a target-specific strategy in this phase.

Results of Free recall. Checking the free recall data, only items which had been reproduced correctly were taken into account. Because all subjects went through one unexpected and one expected test, data were analysed on differences between both groups. As expected, the number of items reproduced correctly was significantly lower for the unexpected run (M = 26.54; s = 12.62) than for the expected one (M = 32.42; s = 14.53). Nevertheless, the further analysis is based on data of both groups.

In Fig. 5, accounts on remembered callsigns are followed by other categories according to the frequency of their reproduction: 'position', 'flightlevel', 'predictor', 'exit waypoint' and 'history'. With the exception of the category 'position' which shows up quite frequently, all categories had been discussed above. They belong to the frequency categories 1 and 2. Parameters belonging to category 3 were never reproduced. The category "position" indicates that visual information could be reproduced quite reliable by entering the object at a sector map.

Reproduction frequencies depend on whether there was an unsolved conflict at the end of the scenario or not. In total, significantly more items are reproduced correctly under the condition of a final conflict (M = 31.63; s = 15.4), as compared to scenarios ending without conflicts (M = 27.33; s = 11.84). As depicted in Fig. 5, clearly more position items and more flightlevel items were reproduced correctly in the case of an open conflict. This focus of the working memory on position and flightlevel of targets in conflict suggests further analysis of the data on the metrics of the representation.

Task demands. To compare the variation of the amount of traffic with the subjective strain of the task, the NASA Task Load Index (TLX) was used. Subjective strain in 'low-traffic' scenarios was judged by the air traffic controllers to be lower (36.8 percent) than under 'high-traffic' (67.1 percent). This difference is highly significant (p < .01). It supports the validity of the variation of traffic density in the scenarios. Even experienced controllers could not indicate parameters of the task load in terms of the number of aircrafts, the number of climbs and descends, or other situational features; that is why these tests were necessary. Actual demands are determined by complex interactions between sector structure and different traffic conditions.

Analysing each of the six dimensions of the test separately brought about significant differences between high-traffic and low-traffic scenarios. *Mental demands, physical demands, temporal demands*, and *effort*, showed significant differences concerning the volume of traffic. No differences were found in the dimensions *performance* and *frustration*. The dimension *performance* was scored slightly higher for low-traffic than for high-traffic scenarios.

3. DISCUSSION

The experiment was an explorative study. It was designed to clarify which data are needed by controllers for constructing a conclusive representation of the traffic situation. This experiment information access was not based on hypotheses. The clear restriction on two up to four parameters, which enabled the controllers to form a representation of the situation and to handle the traffic, was nevertheless quite unexpected. Demands of the scenarios had never been judged as low; under high-traffic conditions they seemed to be even unusually high. Performance data was of no interest at all. For instance, the numbers of not detected and not resolved conflicts were counted to characterise strategies. The experimental reduction of information forbids conclusions from experimental on practical performance.

All data necessary to interpret the primary radar display were presented on request. In the everyday routine, these information are given in the secondary radar display, by the flight progress strips, or by verbal communication. These items are by no means all of the same value for the control task, some are even redundant. Partly they are relevant to specific situations only, or only to approach controllers. In this experiment decisive data were not available on the display. It seems reasonable to suppose that not all items were to be called up with the same probability.

It has nevertheless been unexpected that the majority of the subjects requested only the items *'callsign'* and *'flightlevel'* and - far less frequently - *'exit waypoint'*. In this case, *'callsign'* is not a feature of the target defining its position among other targets, but an identification mark and the address for verbal communication. Controllers reduce the information uptake strictly to those items not presented in the analogous display: to altitude and to destination aspects. In high-traffic scenarios this tendency is even clearer recognisable than under low-traffic conditions.

Basically, the information items requested were all immediately needed for constructing a three-dimensional representation of the present traffic situation. For this purpose, an altitude information had to be added to the analogous radar display. In cases of conflict, precise information on the directions of the involved flights gained importance. In most cases this information had been requested as *'exit waypoint'*, though alternative items had been used (e.g. *'heading'* and *'ETO'*). Obviously, the global analogous representation is supported here by digital information on the conflict area. Under which conditions more precise metric route- or speed- information becomes relevant, is to be examined in further studies.

The results of the reproduction test (free recall) confirm this interpretation. Most often the lateral position parameters of objects have been reproduced correctly, translating analogous information from the radar display. Also the digital data on *'flightlevel'* and the *'callsign'* of targets were frequently kept in memory.

The most distinctive result of this experiment is, that the controllers reduced the tremendous amount of available information radically, and still gained an analogous representation of the three-dimensional traffic situation allowing effective control. Most information presented is irrelevant for this purpose. However, in the case of conflict the demand for information seems to vary: All of a sudden precise parameters gain importance. The notion that in case of a conflict the global analogous representation requires supplementary metric information deserves further investigation.

4. SUMMARY

This investigation of the information access of radar controllers in ATC shows clearly, that only a few parameters are of basic importance. These parameters (*'callsign'*, *'flightlevel'*, *'cleared flightlevel'*, *'exit waypoint'*) cover 93 percent of all information accesses. The higher the traffic load becomes, the more these parameters gain importance. Looking on strategies of information access, two phases can be identified. In an *'orientation phase'* controllers try to put up a representation. Here they work parameter-specific. After 3-5 minutes a shift towards a target-specific strategy indicates the *'working phase'*. A reproduction test at the end of each scenario shows differences: a final conflict in the scenario comes along with significant more items remembered than without such a conflict. Task demands were controlled by means of a NASA TLX test. It supported the validity of the variation of traffic used in the scenarios.

5. ACKNOWLEDGEMENTS

The research is founded by Deutsche Forschungsgemeinschaft (DFG) in 1992 to 1996, titled "Flugsicherung als Mensch-Maschine-System", coded Ey 4/16-1 and Ey 4/16-2. The authors thank Harald Kolrep, Cornelia Niessen, Katharina Seifert, Norbert Wolff and Stefan Husmann. The Deutsche Flugsicherung GmbH (DFS) and Eurocontrol supported the work significantly.

6. REFERENCES

Bainbridge, L. (1987). Ironies of Automation. In: *New Technology and Human Error* (Rasmussen, J., Duncan, K. & Leplat, J., Eds.), pp. 271-283. Wiley, Chichester.

Bierwagen, T. & Helbing, H. (1994). Experiments on Air Traffic Controller's Representation as Approach to MMS-Design. *Proceedings IFAC Conference on Integrated Systems Engineering* (Johannsen, G., Ed.), pp. 459-464. Pergamon, Oxford

Dubois, M. & Gaussin, J. (1993). How to Fit the Man-Machine Interface and Mental Models in the Operators. In: *Verification and Validation of Complex Systems: Human factors Issues* (Wise, J. A., Hopkin, V. D. & Stager, P., Eds.), pp. 381-397, NATO-ASI F 110. Springer, Berlin.

Falzon, P. (1982). Display Structures: Compatibility with the Operator Mental Representation and reasoning Process. *Proc. 2nd Eurpean An. Conf. on Human Decision Making and Manual Control*, pp. 297-305.

Johnson-Laird, P. N. (1983). *Mental Models*. Harvard University Press, Cambridge MASS.

Kolrep, H. (1993). Automation and Representation in Complex Man-Machine-Systems. In: *Verification and Validation of Complex Systems: Human factors Issues* (Wise, J. A., Hopkin, V. D. & Stager, P., Eds.), pp. 375-380, NATO-ASI F 110. Springer, Berlin.

Leroux, M. (1993). The Role of Verification and Validation in the Design Process of Knowledge Based Components of Air Traffic Control Systems. In: *Verification and Validation of Complex Systems: Human factors Issues* (Wise, J. A., Hopkin, V. D. & Stager, P., Eds.), pp. 357-373, NATO-ASI F 110. Springer, Berlin.

Mogford, R.H. (1991). Mental Models in Air Traffic Control. In: *Automation and System Issues in Air Traffic Control* (Wise, V. A., Hopkin, V. D. & Smith, M. L., Eds.), pp. 235-242. Springer, Berlin.

Norman, D. A. (1983). Some Observations on Mental Models. In: *Mental Models* (Gentner, D. & Stevens, A. L., Eds.), pp. 7-14. Earlbaum, Hillsdale NJ.

Rasmussen, J. (1986). *Information Processing and Human-Machine Interaction*. Elsevier, Amsterdam.

Seamster, T. et al. (1993). ATC Cognitive Task Analysis. *The International Journal of Aviation Psychology*, 3, pp. 257-284

Whitfield, D. & Jackson, A. (1982). The Air Traffic Controller's Picture as an Example of Mental Model. In: *Proc. of the IFAC Conf. on Analysis, Design and Evaluation of Man-Machine Systems* (Johannsen, G. & Rijnsdorp, J. E., Eds.), pp. 45-52. Pergamon, London.

Wilson, J. R. & Rutherford, A. (1989). Mental Models: Theory and application in Human Factors. *Human Factors*, 31, pp. 617-634.

TOWARDS A DECISION MAKING MODEL OF RIVER PILOTS

Fulko van Westrenen

Safety Science Group
Department of Technology and Society
Delft University of Technology
The Netherlands

Abstract: This paper presents a navigator model for the approach into a harbour. The model is a three stage decision model, containing tracking, short-term planning and long-term planning behaviour. This model is verified by measuring the mental workload of pilots whose task is to bring a ship to berth. The mental workload is obtained by ECG recording, from which heart rate and heart rate variability (the 0.1 Hz power density) were calculated. It is shown that the workload varies with the difficulty of manoeuvring the ship, and the situation ahead.

Keywords: decision making, mental workload, model, ship control, tasks

1. INTRODUCTION

Technological innovations have made more information become available for the navigator. Good information is particulary important during the approach and departure of a harbour, when the risks of grounding and collision are highest. In most harbours a pilot comes on board to assist the crew with the navigation task. One irony is that the closer the ship comes to land, the less automation is used. Even highly automated ships go to full manual control in harbours, as they did some 50 years ago.

Normally the pilot performs the entire navigation task on his own. If better information is to be provided, we need to know what information is required at what time and place. Very few attempts have been made to do a thorough task analysis. No wonder: navigation is a difficult process to describe. Most information needed is obtained from the pilot sees outside, few orders are given, and performance is hard to quantify. One of the few variables that can be quantified is mental workload. This provides an opportunity to obtain some insight into the decision making process.

When a pilot comes on board he has very little knowledge about the specific features of the ship. A number of characteristics is stated on a manoeuvring sheet, but since this information is only applies at high sea, the pilot does not bother looking at it. Personal observations and interviews have learned that the pilot only knows the basic dimensions of the ship, and he estimates other characteristics by simple rules of the thumb and experience. Yet he will bring the ship to berth with a very high success rate.

2. NAVIGATOR MODEL

Many navigator models have been made over the years. Most of them are based on open-sea manoeuvring. Few of them actually focus on inland navigation, and these often take a control theoretic approach (Papenhuijzen, 1994). The navigator model presented here is a decision model with three levels of control (see figure 1).

On the top level the overall planning of the voyage: long-term planning (LTP). Decisions on this level range from very simple decisions such as a free berth and the availability of tugs, to very complicated decisions concerning an overall plan to sail the ship to its berth. Most planning is done before the pilot

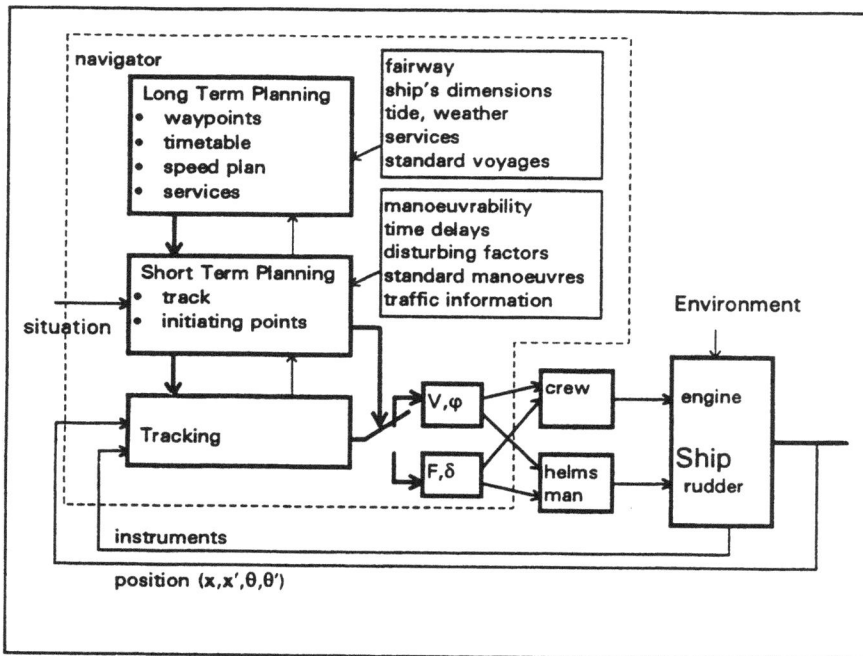

Figure 1: A navigator model

boards the ship, and the pilot will check on board if everything is in order, and change his plan if the situation is not as he expected. Characteristic of long-term planning is that it's procedural and it is done only once.

The middle level is that of short-term planning (STP). When a pilot brings in a ship, he will not plan the track as a whole in advance: he will sail the ship from waypoint to waypoint. Each waypoint has its own objective: a new course, a curve, a change in the current, an intersection. In STP the decision is made about a situation at hand. The controls available available to the pilot are very limited: rudder and engine. Due to the inherent inertia of the ship, decisions often take the form of a point of no return. A similar situation may also be observed during take-off stages of an airplane. This level's decision is that of the track to follow. Short-term planning is based on knowledge about the forthcoming local situation, traffic, standard manoeuvres, and the pilot's personal style. This, together with the manoeuvring characteristics of the ship, will be applied to plan a track. The manoeuvring characteristics and the required accuracy determine what type of control he will apply: speed-course (V, φ) or thrust-rudderangle (F, δ). Ships are slow to react to new settings, depending on the ship's size and type of machine control. Therefore, setpoints must be planned well in advance. Short-term planning is a discrete process, it is based on a set of methods, it is proactive, taking place at specific points in space, depending on the on-coming situation.

The lowest level of control is that of tracking behaviour. The track that has been chosen on the STP-level must be followed as closely as possible. For this the navigator must observe the position and

movements of the ship relative to the environment. When deviations from the desired track occur corrections must be made. Important factors are wind, current, ground-effect, manoeuvring characteristics of the ship, various elements in ship control, and the accuracy with which the position fixing can be done. The bigger the ship, the longer the time-constants will be and the more precise the prediction must be.

Some variables are noted in figure 2. Tracking is a continuous process, it is based on techniques, and it is reactive by nature. The most important characteristics of each level of control are summarised in table 1.

Table 1. The major features of the three levels of control.

level	character	type
long-term planning	procedures	once
short-term planning	methods	discrete
tracking	techniques	continuous

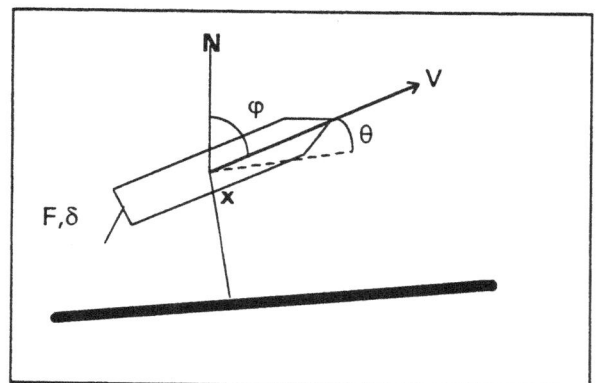

Figure 2. The variables describing the ship's state. N is North, V and φ are speed and heading, F and δ are thrust and rudderangle, x and θ are relative position and relative heading.

3. WORKLOAD

In this paper, the term workload always refers to mental workload. Physical workload resulting from boarding and leaving the ship, or irregular working hours is not considered. Performance shaping factors are also not considered.

218

Not all factors that could add to the workload are present at all times, and some factors present may not be important.

A voyage can be cut down into a number of sections, each characterised by a few elements. These elements determine the track to be chosen. From this track it follows what manoeuvres must be made and how they are to be initiated.

Two elements are needed for tracking: observation and control. After a track is chosen it must be possible to observe the deviation from the desired track. The uncertainty of the position, related to the required accuracy, also adds to the workload. During tracking different types of disturbances act on the ship. This will increase the workload imposed on the navigator. The use of speed-course control is considered easier than thrust-rudder control. This would also linearly add to the workload.. All these workload components are directly related to the situation present.

Short-term planning results in a workload present well before a situation occurs. The workload is linked to a individual situation and occurs at a time related to the manoeuvrability of the ship. The workload depends either on the fairway with its local problems, in which case it is related to a specific location/situation, or it depends on other ships, in which case it is related to the traffic situation.

Long-term planning leads to an increase of workload for a short time when the pilot arrives on board, and play a role later only if the planning has to be changed for reasons unknown at first.

Combining these three levels of workload will result in a workload like in figure 3.

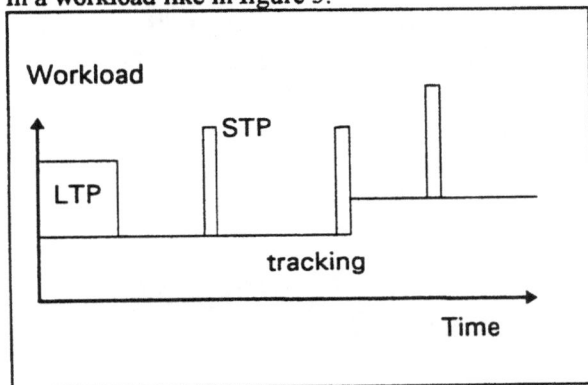

Figure 3: Workload profile of three-stage navigator model. The step in the workload of the tracking-mode results from switching from course-speed to thrust-rudderangle.

4. MEASUREMENT

To verify the model a method must be applied which will provide workload information about the transitions. Generally, there are four ways to measure mental workload: task demand load in relation to performance, primary task measures, secondary task measures, and physiological measures. Much has been written about each of them and about their possibilities and limitations. (see for instance: Hancock & Meshkate, 1988; Sheridan & Stassen, 1979)

Performance measures are very difficult to obtain in navigation tasks. The number of unknown factors is high, and personal preferences lead to different outcomes.

Subjective workload estimates are not very good for measuring transients in workload. They provide an overall rating of the workload. Due to the specific working conditions, one may doubt the possibilities for self-rating by pilots. The TLX-ratings obtained give rise to wild speculation. Our idea is that pilots have learned to ignore their perception of workload.

To apply a secondary task, the primary task must be known exactly. This is impossible due to the very nature of this project. Additionally, a secondary task increases the workload and is intrusive by its very nature, which is considered impermissible.

Despite the fact that the relation between mental workload and various physiological measures is not clear at all, some measures have shown high correlation between workload and the physiological reaction. Blood pressure, hormonal concentrations, and muscle tension are well known examples from among these. Heart rate and heart rate variability (the power in the 0.1 Hz-band) were chosen. These measures are easy to record, sensitive, particulary for central processing and visual perception, and non-intrusive. Although their reliability and validity have often been questioned, there is evidence that these measures are reliable measures of workload (Vincente et al, 1987; Mulder 1983). The major disadvantage are their problematic interpretation, a drawback for all physiological techniques. It was assumed that the effects caused by a little physical exercise would be small compared to the differences caused by mental workload.

Four experienced Rotterdam pilots participated in a series of recordings. Each pilot did at least two inward-bound and two outward-bound voyages. The ships were selected for maximum size from the ships available. Ships for the river were preferred to ships for Europort. All pilots were videotaped, and marker points were plotted on a map for specific locations and traffic.

The ECG recordings were corrected for recording errors, and transformed into power density signals using Carspan 1.99 (Mulder, 1988). The power in the 0.1 Hz-band was calculated using a Direct Fourier Transform based on the Integral Pulse Frequency Modulator model (Rompelman, 1986). The time window of 50(s) was found to be acceptable for both a sufficient spectral resolution and a stationary signal. As a smoothing filter, a modified discounted-least square filter was applied with τ=30(s) (SWOV, 1994).

The goals of the pilot has to be known to interpret the readings. Interviewing during the voyage may be a serious disturbance of the ongoing process, and was rejected for that reason. We compromised by asking the pilot afterwards what had happened during the voyage, what his plans were, and what problems occured when executing these plans. Over the time, some knowledge was obtained about the pilot's task, and this was used to ask specific questions about the events that had taken place. Additionally, some knowledge about standard

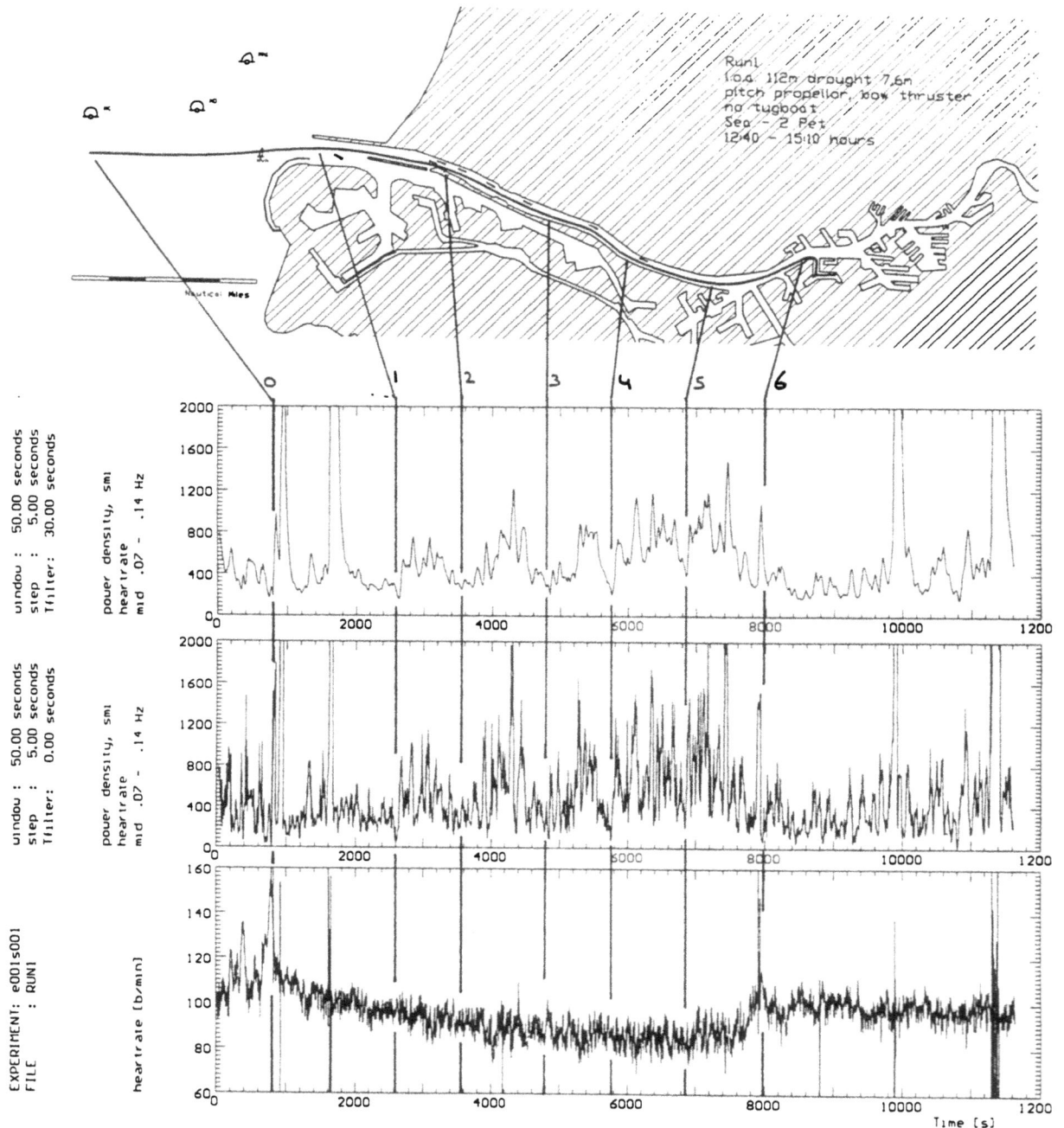

Figure 4: Physiological data from the voyage. From bottom to top: the heart rate, the power density (unfiltered), the power density (filtered), the voyage. Note that the mental workload is reversely related to the power density of the 0.1 Hz-band. The peaks in the power density result from various artefact in the heartrate. The increase in heartrate is clearly visable during boarding at sea (0), and when entering the harbour (6). At marker 2 there is an increased workload due to traffic and construction work (between 2 and 3). Marker 3 is the result of a ship that came to the wrong side of the fairway. Marker 4 is located at a place where difficult currents exist. Marker 5 is placed just before a location of usually intense traffic.

voyages was obtained by interviewing outside the recorded voyages.

5. RESULTS

Because of the many differences between the voyages, no statistical analysis was conducted over pilots, ships, and situations. Voyages were analysed by comparing the voyage with the heart rate and power. One voyage has been chosen as an example. In figure 4 one can see the track and the resulting ECG-data. With the exception of climbing on board the pilot doesn't do any physical excersize. The heart rate of all pilots on board stabilised around 90 beats/min, about 10 beats higher than on the pilot-ship. Climbing on and off board can be seen quite clearly in all heart rate plots. Beforehand inward-bound and outward-bound voyages were distinguished. This difference may be clearly distinguished in the heartrate plots. When the pilot enters the harbour, the heartrate increases significantly, to reach its peak at the start of the turn. This effect could not be found with outward-bound ships. The same pattern could be found during critical manoeuvres and passing bridges. This effect was attributed to arousal, which is directly related to hormonal effects (Gellatly & Meyer, 1992).

When computing the power density, it was found that many artefacts occur in the harbour. This is very troublesome because they act like Dirac-functions, masking all other data in that time segment. It is known that stress-like situations often result in extra-systolic contractions of the first ventricle, recording as an extra heartbeat, and disturbing the Fourier transforming process. A second outcome was that the power density is highly variable, which made graphs difficult to interpret. A Modified Discounted Least Square filter was found to be most useful. This filter is an a-symmetrical exponential time-weighted filter.

The overall profile of the power density gave a good indication of the workload (face value, corroborated by interviews). The peaks in workload as indicated by the power density occurred at pre-defined places, such as before a significant change in the current, or when decisions had to be made about passing a ship coming from another fairway. Orders, passing of other ships and disturbances could often be clearly identified. As is usual with these techniques, a lot of variation is left unaccounted for.

6. CONCLUSIONS

The data seem to confirm our model. All major workload peaks which were found to be related to locations or situations, were confirmed in interviews. For a more detailed model, more detailed analysis is required. With the help of this technique, most dominant factors may be identified.

ECG-recordings seem a valuable tool for the development of a navigator model. The data from interviews and the ECG-recordings gave valuable information about the pilot's task. For a more sophisticated interpretation, a thorough talk-through afterwards is required, making the interpretation much easier. In field experiments such as these, this is not always easy to achieve. There may be very little time between ships, due to high time-pressure, and work is done around the clock. Time for measurements only becomes available with full cooperation of the entire organisation.

The pilot's actions, the goals, the ship's characteristics, it's location and situation must be known to interpret the heartbeat data. For this, a lot of information must be recorded during and after the voyage. Unfortunately, due to many unknown factors, not all changes in power density may be explained.

ACKNOWLEDGEMENTS

We would like to thank the Rotterdam Pilots, the participating pilots, and the captains and crews of the ships for their assistance and hospitality.

REFERENCES

Gellatly, I.R. & J.P. Meyer (1992). The effects of goal difficulty on physiological arausal, cognition, and task performance. *Journal of Applied Psychology*, vol 77, no 5, pp 694-704.

Hancock P.A. & N. Meshkati (eds) (1988). *Human mental workload*. Advances in psychology 52, North Holland, Amsterdam.

Mulder, L.J.M.; H.J. Van Dellen; P. Van Der Meulen & B. Opheikens (1988). Carspan: a spectral analysis program for cardiovascular time series. In: F.J. Maarse; L.J.M. Mulder; W. Sjouw & A. Akkerman (eds) *Computers in psychology: methods, instrumentation & psychodiagnostics.* pp 39-47. Swets & Zeitlinger, Lisse.

Papenhuijzen, B. (1994). *Towards a human operator model of the navigator*. Delft University of Technology, the Netherlands, Thesis.

Sheridan T.B. & H.G. Stassen (1979). Definitions, models and measures of human workload. In: N. Moray (ed), *Mental workload, it's theory and measurement.* Plenum Press, New York, 1979

Rompelman, O. (1986). Tutorial review on processing the cardiac event series: a signal analysis approach. *Automedica*, vol 7, pp 191-212.

SWOV (1994). Control strategies for a highway network; A joint research project of the SWOV, the Technical University Delft and the Institute for Perception TNO, sponsored by the Dutch Ministery of Transport and Watermanagement. Part I, II and III. R-94-34 I/II/III. SWOV, Leidschendam, the Netherlands.

Vincente K.J.; D.C. Thornton & N. Moray (1987). Spectral analysis of sinus arrhythmenia: a measure of mental effort. *Human Factors*, vol 29, no 2, pp 171-182.

THE DEVELOPMENT OF
THE FINAL APPROACH SPACING TOOL (FAST):
A COOPERATIVE CONTROLLER-ENGINEER
DESIGN APPROACH

Katharine K. Lee[1]
Thomas J. Davis[2]

[1]*Sterling Software, Inc., Palo Alto, CA*
[2]*NASA Ames Research Center, Moffett Field, CA*

Abstract: Historically, the development of advanced automation for Air Traffic Control in the United States has excluded the input of the air traffic controller until the end of the development process. In contrast, the development of the Final Approach Spacing Tool (FAST), for the terminal area controller, has incorporated the end-user in early, iterative testing. This paper describes a cooperation between the controller and the developer to create a tool which incorporates the complexity of the air traffic controller's job. This approach to software development has enhanced the usability of FAST and has helped smooth the introduction of FAST into the operational environment.

Keywords: Air traffic control, automation, human factors, interface, simulation.

1. INTRODUCTION

The development of an automation system for assisting terminal area air traffic controllers in efficiently managing and controlling arrival traffic has long been an objective of researchers and engineers. A fundamental issue for researchers in the development of such automation tools is to build functionalities and user interfaces that enhance the air traffic controllers' ability to perform well in their job. Early efforts in the automation of terminal air traffic control was presented by Martin and Willett (1968). Their system provided speed and heading advisories to controllers to help increase spacing efficiency on final approach. Although traffic tests of the system showed an increase in landing rate, controllers found that their workload was increased and rejected the system (Martin and Willett, 1968). An examination of the concept suggests that while some aspects of the design were sound, its acceptance was limited by the technology of the time period, especially the lack of an adequate controller interface. More recently, however, several automation systems have found their way into operational use in Europe due in large part to the introduction of modern computer processing and interfaces, and also because of more careful design approaches (Volckers, 1990, Garcia, 1990). In addition, recent real time simulation studies have confirmed the potential for increasing landing rates with the assistance of active advisories for controllers in the terminal area (Davis *et al.*, 1991, Credeur *et al.*, 1993).

A candidate system for the automated management and control of terminal area traffic, referred to as the Center TRACON Automation System (CTAS), is under development at NASA Ames Research Center in collaboration with the FAA's Terminal Air Traffic Control Automation Program Office. The elements comprising the CTAS are the Traffic Management Advisor (TMA), the Descent Advisor (DA), and the Final Approach Spacing Tool (FAST) (Erzberger *et al.*, 1993). The advisories generated by these tools assist controllers in handling aircraft arrivals starting at about 200 n.mi. from the airport and continuing to the final approach fix. Recently, the elements of the CTAS system have been evaluated in a series of real-time simulations at NASA Ames Research Center and in field testing at facilities serving the Denver, Colorado, and Dallas/Fort Worth, Texas, areas.

This paper describes a cooperative effort between software developers, human factors specialists, and air traffic controllers to develop FAST. The main function of FAST is to provide landing sequence, landing runway assignments, and speed and heading advisories that help controllers manage and control arrival traffic and achieve an accurately spaced flow of that traffic onto the final approach course (Davis et al., 1991, Davis et al., 1994). This paper emphasizes the role of the air traffic controller in FAST development, from providing direction into the controlling strategies incorporated into the FAST algorithms. This paper also describes the use of the Controller Acceptance Rating Scale to track the controllers' perspective of the overall system through the development of FAST.

2. CONTROLLER INPUT INTO FAST DEVELOPMENT

The CTAS development process has differed from more "traditional" approaches to software development by incorporating the expertise of the end-user at the beginning stages of development. Air traffic controllers have been involved in simulations at NASA Ames Research Center, the FAA Technical Center in Atlantic City, New Jersey, and in evaluation activities at the operational facilities into which the initial deployment have been targeted. The extensive involvement of controllers increases their understanding of the software and engineering constraints, and helps them to focus their expertise and provide input that is maximally useful to the software developers (Sanford et al., 1995). In addition, early end-user involvement also helps to integrate human factors issues during design, which can help improve system usability (Small, 1994). Larger questions of the suitability of a system, or how well the system provides for the users' problem-solving requirements (Harwood and Sanford, 1993), can also be addressed with early end-user involvement. The development, and ultimate demise, of the Advanced Automation System (AAS) provided numerous examples of design problems that could have been ameliorated with earlier controller input. Small (1994) describes how controllers involved towards the end of development were inappropriately focused on interface, rather than operational elements of the design. In addition, controller involvement late in the development process created an environment in which the controllers and the developers saw each other on opposite sides. This made reaching compromises much more difficult, as developers did not understand the operational needs of the user, and the controllers viewed development constraints they encountered as arbitrary (Small, 1994).

FAST simulations have been conducted at the Ames Automation Laboratory, where the software was originally developed and tested. Controller participation in these simulations, and other design activities, have helped to shape the requirements, test the software, and provide input into human

factors issues as well as insight into controller strategies to reduce workload and increase throughput efficiency. In addition to a pool of local, retired controllers, controller personnel from the Dallas/Fort Worth (DFW) TRACON where the initial FAST testing and deployment will occur, have been evaluating FAST. DFW facility representatives, including Union members and managers, have been encouraged to participate in the software evaluations. In total, three groups of controllers were incrementally involved in the FAST simulations.

The simulations consisted of traffic scenarios displayed on radar screens with FAST advisories added to the radar display interface. The traffic scenarios were created from recordings of live radar traffic from DFW. Two feeder controller positions (East and West) and three final controller positions (for the three arrival runways) were used for simulating DFW operations. Each arrival sector controller worked from a separate radar display position. The level of traffic during the 1 to 1.5 hour simulations was approximately 80-100 arrival aircraft per hour, reflecting the actual "rush" durations and traffic levels at DFW. Controllers were provided with headsets through which they issued commands to pseudo pilots, located in another part of the laboratory. The controllers were able to accomplish all of the basic entries and inputs into the system that would be expected in the real facility, such as taking handoffs and changing runway assignments.

Developers monitored the simulations in the laboratory; following simulations, debriefings were held to allow the controllers and developers to discuss key decisions made during the simulations, and explain any observed problems. All of the simulation outcomes were recorded for later review, and the debriefing sessions often included a replay of the simulation.

FAST has also been evaluated by controllers using a technique known as "shadowing." Shadowing involved viewing the live operational activities in real time, with FAST operating in the background, superimposing its advisories on an auxiliary computer display. Shadowing enabled controllers and developers to view the effects of the controllers' current procedures and how the traffic outcome was influenced by the presence of FAST advisories. Shadowing activities occurred in a testing environment near DFW, where radar communications were also monitored to provide a more complete picture of the traffic flow and decision making activities.

3. PHASES OF CONTROLLER INVOLVEMENT

Figure 1 shows the development of FAST through varying levels of development milestones and controller involvement. This paper concentrates on the three phases of controller involvement in

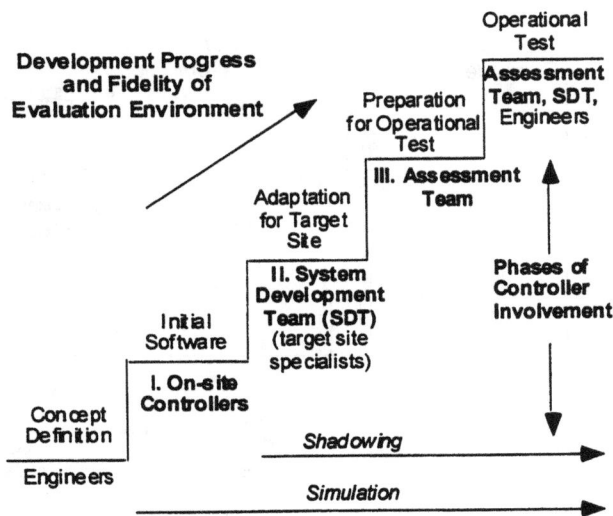

Fig. 1. Controller involvement in the context of FAST development milestones.

simulation, from the initial software development to the preparation for the operational test.

Three different controller teams participated in the three levels of simulation fidelity. Each group of controllers provided increasingly more refined information in their evaluations. The first level, which required the initial assessment of new functionality and new displays, utilized the expertise of a group of local, retired controllers and pilots. The second level focused on expanded functionality and site-specific issues and incorporated the participation of a small cadre of controllers, or System Development Team (SDT), who represented the DFW facility. The third level increased the amount of feedback and input into refining the FAST functionality and determined levels of acceptability for limited field testing, utilizing the expertise of a larger group of controllers from DFW.

3.1 Phase I: On-site Controllers

The first level of FAST development involved creating a system prototype in which new functionality was first tested and graphical user interfaces were generated. Simulations were conducted several days a week with on-site controllers, most of whom were retired from regional facilities in the San Francisco Bay Area. Feedback from this early testing helped direct the iterative development of FAST.

The simulations with the on-site controllers enabled the developers to gauge the software's performance under basic, day-to-day testing conditions. The on-site controllers provided an indication of how air traffic controllers as a whole would react to the displays, controlling strategies, and advisories incorporated into the software. From a human factors standpoint, they provided the most input at the level of FAST's usability, determining

if basic information could be extracted under simulated operations. They also provided some input on suitability issues, evaluating whether or not the information provided by FAST was appropriate for its intended use. In addition, the researchers benefited from on-site controller participation in that they were able to conduct studies under fairly well-defined conditions to determine the impact of FAST on controller workload. These studies showed that FAST reduced the level of self-reported workload by the air traffic controller (Lee, et al., 1995) as well as reducing the number of commands issued to aircraft (Slattery, et al., 1995).

It was clear, however, that the development of FAST functionality soon required more specific input from the DFW TRACON controllers, once basic concepts of ATC were incorporated in the software. The on-site controllers' input into basic air traffic control concerns and attitudes was often accurate, but their understanding of the operations at the DFW TRACON was limited. As a result, the System Development Team was brought into the development process.

3.2 Phase II: System Development Team

The second phase of FAST development incorporated the expertise of a cadre of 3 controllers, one area supervisor, and one training specialist from the DFW TRACON. These controllers formed the SDT. The SDT provided the developers with guidance about the specific procedures of the operational environment and an understanding of operations under the high traffic levels that are common to DFW. The SDT participated in formal software evaluations at NASA Ames and the FAA Technical Center approximately every 6-8 weeks, depending upon milestones in the software design.

Split of FAST Functionality. One of the SDT's pivotal inputs into FAST development was to split FAST functionality into two parts. FAST was originally conceived to provide a suite of advisory information for the TRACON sector controllers. This information would enable the TRACON controller to efficiently sequence and space arrival traffic by providing sequence and runway advisories. In addition, speeds and headings were provided to efficiently meet the sequences and runway assignments. Following an early evaluation, the SDT recommended that FAST be split into "passive" and "active" stages. Passive FAST was the portion of the software that provided only the sequence and runway advisory information and Active FAST was the portion where speed and heading advisories would be added. Together with the developers, the SDT decided that Passive FAST would be tested first.

In the early months of FAST development, the SDT was very cautious towards FAST and their recommendation to test Passive FAST first was directly attributable to this cautious attitude. It is possible that when the SDT first began to evaluate FAST, the utility of the FAST advisories were not fully developed, and hence the benefits were not as obvious as they have become. However, the SDT's acceptance of FAST increased significantly in the three years of their involvement in FAST development, and as a number of their concerns outside of the software's capability (described below) were raised and addressed. This suggests that their initial apprehension towards developing full FAST capability was due to issues in addition to the immaturity of the software.

There were a number of reasons for the SDT's initial hesitancy. First, from the facility's perspective, the involvement of the SDT members could have implied that the facility's controllers were inefficient and that the abilities of the controllers as a whole were being called into question. It was easy to draw the conclusion that the development of ATC automation would prompt the FAA to reduce the controller workforce in response. Second, the SDT was clearly responsible for the outcome of FAST; thus the SDT's reputation as seen by the facility would be at stake. Third, the SDT had concerns about liability. The FAA could decide that FAST should dictate commands to the air traffic controller, yet the controller (and not the software) would be ultimately responsible for the consequences of the FAST advisories. Finally, the SDT was concerned that FAST would automate the interesting aspects of the controller's job and thus reduce job satisfaction.

These concerns were not all obvious in the beginning of the development process. Some of the concerns the controllers discussed directly; other concerns were elicited after many months of development. The developers learned to provide reassurance to the controllers in addition to promoting the potential benefits of FAST. In addition, the developers earned the trust and respect of the SDT through demonstrating a thorough understanding and appreciation of ATC in general, and DFW operations in particular. This knowledge helped the controllers and the developers work together to address the SDT's concerns, resulting in changes to the software, or making compromises when software changes could not be accomplished.

SDT Input into FAST Algorithms. The SDT controllers hoped that FAST could provide a reduction in controller workload. However it quickly became clear that while a basic element of workload was the number of aircraft for which a controller was responsible, other variables, such as the types of aircraft involved, the characteristics of the sector's airspace, and the amount of coordination between positions, had to be considered. Consequently, the SDT showed that some FAST advisories, while intended to reduce controller workload by reducing delay, sometimes

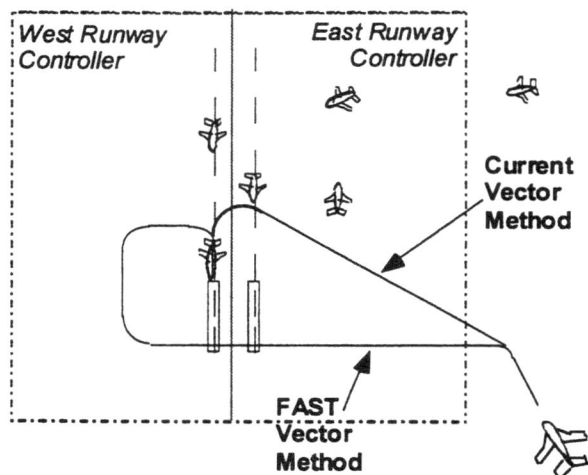

Fig. 2. Possible routes for Southeast cornerpost arrival traffic.

made the traffic situation more complicated from the controller's perspective.

The SDT also pointed out situations where FAST algorithms produced procedural obstacles and controlling strategies that were considered unfamiliar or unfavorable. These issues required the developers and the controllers to work together to determine how the advisories were perceived and what could be done to minimize the impact on coordination and workload. In some cases, the SDT discovered that the unfamiliar strategies FAST suggested were beneficial ones. For example, FAST would propose that aircraft from the Southeast cornerpost fix be vectored over the top of the airport to a West-side runway. The current procedures would have specified a different route to the West-side, or would have directed the aircraft to the East-side runway, as shown in Figure 2. FAST's over-the-top vectoring was not typically favored by the facility, but the SDT found that this routing could provide benefits to their overall operations by decreasing the workload of the East runway controller.

Shadowing observations were a key turning point in the SDT's assessment of the software; they became an evaluation environment in which the benefits of FAST advisories became very evident. Figure 3 depicts a situation observed in many cases at DFW where the arrival rush side of the airport would be overloaded. More experienced controllers would have balanced the runway loading better than that seen in Figure 3, with vectoring strategies similar to those described earlier. By comparing the outcomes of FAST advisories to actual facility operations, the SDT and the developers could see where FAST could provide increased flow and efficiency as well as help determine where FAST needed to adjust for facility procedures.

After two years of SDT evaluations, it became clear that the SDT was providing input that was native to their own style of traffic control that was not necessarily representative of DFW as a whole. As

Airspace Boundary

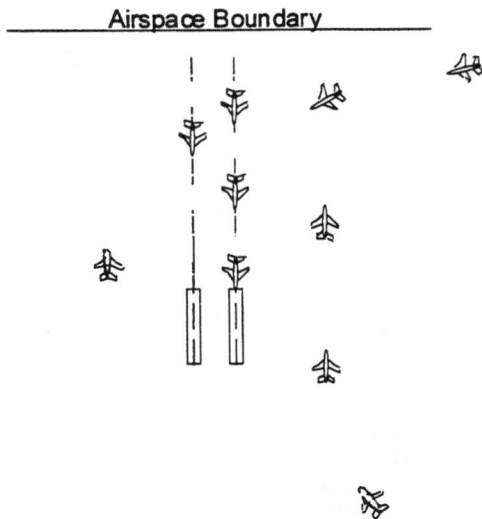

Fig. 3. Shadowing observations demonstrating uneven runway balancing.

more advanced development issues were encountered, further input from the facility would be needed. In addition, FAST development was nearing a phase in which an operational test would be required. In order to conduct such a test, a larger group, or assessment team, of DFW controllers needed to be assembled and made proficient in using FAST.

3.3 Phase III: Assessment Team

The Assessment Team was composed of 6 controllers (3 from each specialty) and two area supervisors from DFW TRACON. The Assessment Team participated in simulations at NASA Ames and the FAA Technical Center and will be participating in the upcoming Limited Operational Demonstration at DFW. The Assessment Team brought a wider range of skill, expectations, and viewpoints to the FAST development process. In their initial introduction to FAST, some of the Assessment Team members were pleasantly surprised and impressed that FAST could increase their throughput capacity without making them feel taxed. Others felt that they operated quite well without the need for added automation. As with the SDT, the developers had to work with the Assessment Team to come to an understanding of where the benefits of FAST's advisories and strategies could be realized. While the Assessment Team also introduced new controlling strategies not previously encountered, they also saw the benefits of changing their controlling strategies in the operational environment based on what they had seen in FAST simulations.

The Assessment Team's main goal was to determine when FAST would be ready for a limited operational test. But because developing ATC automation with active controller input was a new

process within the FAA, there were no established guidelines to help determine system readiness and acceptability. Furthermore, the benchmarks of acceptability from the developer's perspective did not sufficiently address the controller's sense of workload. It was clear that qualitative measures of workload and acceptance had to be assessed along with quantitative measures of reduced delay and increased runway throughput.

3.4 The Controller Acceptance Rating Scale (CARS)

Based on the need for a measure of acceptability, the Controller Acceptance Rating Scale (CARS) was developed by the human factors specialists (see Figure 4). The CARS was an adaptation of the Cooper-Harper Rating scale for pilot assessment of handling qualities of aircraft (Cooper and Harper, 1969). CARS is still being assessed and validated at NASA Ames.

The original Cooper-Harper Scale measured the performance of the pilot and the vehicle working together; this merging of the pilot and the vehicle defined the system being evaluated (Harper and Cooper, 1986). In adapting this scale to the ATC environment, this concept of a system including the user, the software and the hardware was preserved. In addition, changes were made to the scale layout itself: compared to the original Cooper-Harper Scale, the scale was re-ordered so that a rating of "1" was the worst performance level, and "10" was the highest performance level. This change was made to reflect that a lower number was associated with a lower, less acceptable rating, and the higher number to be associated with a higher, more acceptable rating.

CARS was intended to provide the Assessment Team with a means for determining how well they and the FAST software performed together in controlling traffic in simulation. Following a simulation, each of the Assessment Team controllers viewed the scale, produced a rating, and gave an explanation for the rating. In addition, they were asked to provide a confidence indication of their rating, which could incorporate issues of the fidelity of the simulation, the amount of information they had available to make their rating, and other issues that might be external to the software advisories.

The four main rating levels described by CARS were controllability, tolerability, satisfaction, and acceptability. The controllers started at the top of the CARS diagram and first determined if the system was controllable. An uncontrollable system would imply safety violations such as near misses, collisions, or an inability to maintain separation. Such a system would require mandatory improvements.

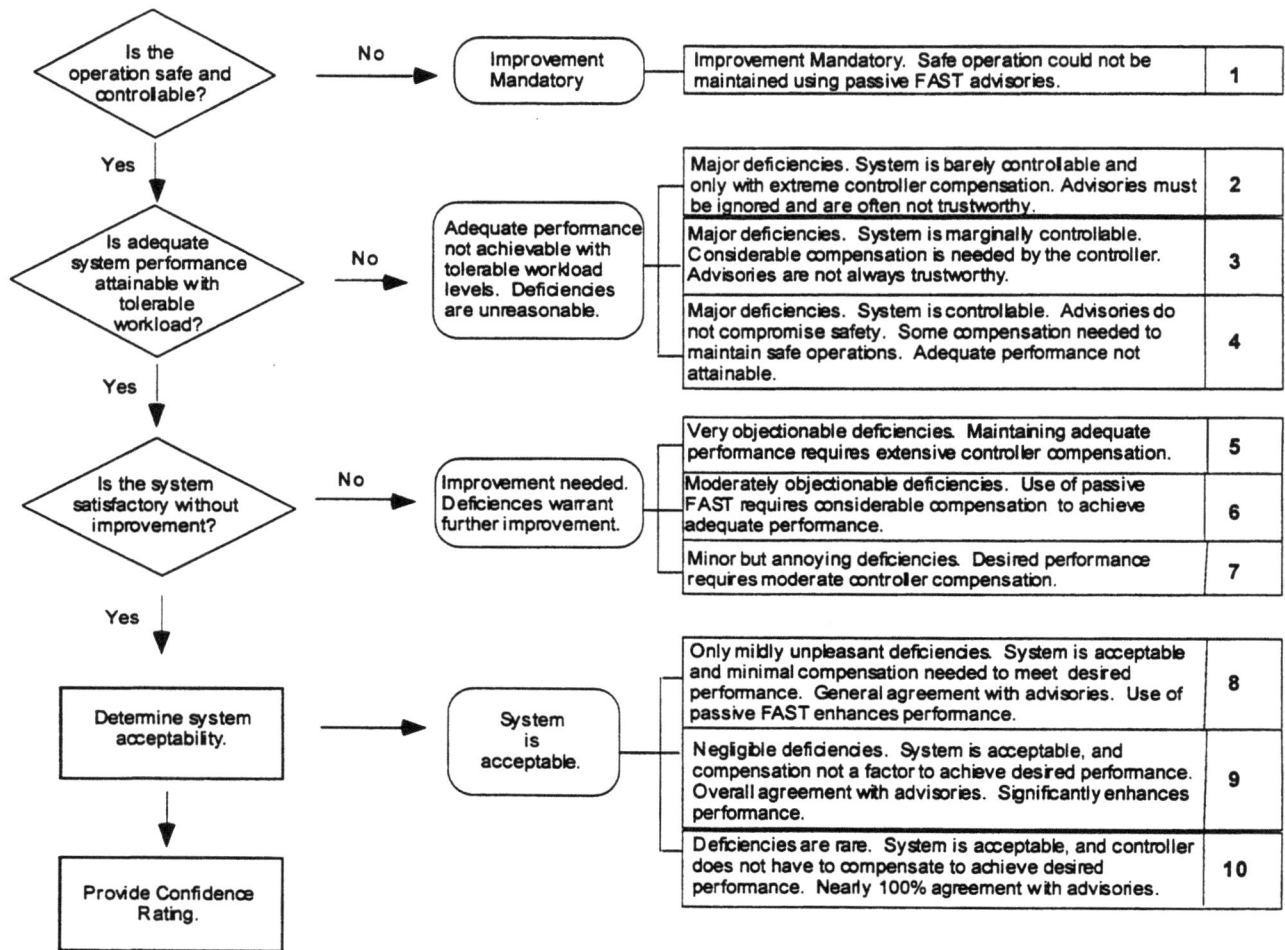

Is the operation safe and controllable? — No → Improvement Mandatory	Improvement Mandatory. Safe operation could not be maintained using passive FAST advisories. **1**
Yes ↓	
Is adequate system performance attainable with tolerable workload? — No → Adequate performance not achievable with tolerable workload levels. Deficiencies are unreasonable.	Major deficiencies. System is barely controllable and only with extreme controller compensation. Advisories must be ignored and are often not trustworthy. **2**
	Major deficiencies. System is marginally controllable. Considerable compensation is needed by the controller. Advisories are not always trustworthy. **3**
	Major deficiencies. System is controllable. Advisories do not compromise safety. Some compensation needed to maintain safe operations. Adequate performance not attainable. **4**
Yes ↓	
Is the system satisfactory without improvement? — No → Improvement needed. Deficiencies warrant further improvement.	Very objectionable deficiencies. Maintaining adequate performance requires extensive controller compensation. **5**
	Moderately objectionable deficiencies. Use of passive FAST requires considerable compensation to achieve adequate performance. **6**
	Minor but annoying deficiencies. Desired performance requires moderate controller compensation. **7**
Yes ↓	
Determine system acceptability. → System is acceptable.	Only mildly unpleasant deficiencies. System is acceptable and minimal compensation needed to meet desired performance. General agreement with advisories. Use of passive FAST enhances performance. **8**
	Negligible deficiencies. System is acceptable, and compensation not a factor to achieve desired performance. Overall agreement with advisories. Significantly enhances performance. **9**
Provide Confidence Rating.	Deficiencies are rare. System is acceptable, and controller does not have to compensate to achieve desired performance. Nearly 100% agreement with advisories. **10**

Fig. 4. Controller Acceptance Rating Scale

If the system was rated as controllable, the controller then assessed the next level: tolerability. If the system was rated as intolerable, this implied major deficiencies and an inability to achieve adequate performance with tolerable workload levels. If the system was rated as tolerable, this implied a reasonable workload with manageable deficiencies and the controller then assessed the next level: satisfaction. An unsatisfactory rating implied that the deficiencies warranted improvement. If the system was considered satisfactory, however, the controller then rated the level of acceptability the system was able to achieve in the simulation. A confidence rating was obtained following the numerical rating. The Assessment Team was quite receptive to CARS.

To date, CARS has been used for one Assessment Team evaluation, in which the controllers participated in 7 simulations with varied runway configurations typical to DFW operations. Each controller who participated in a simulation rated the system based on his own experience working traffic from his particular sector position.

Because CARS data has only been collected from one series of simulations, results are not presented here. In subsequent simulations and operational testing, the developers intend to continue to use CARS in order to track the progress of the software evaluation by the controllers. CARS has indications of being a useful measurement device, and its validation will be useful to other ATC software development processes.

4. THE BENEFITS OF A COOPERATIVE DESIGN APPROACH

Closely linking the air traffic controller with the aeronautical engineer in developing ATC software helps to produce a system that will capture as much of the actual complexity of ATC as is possible prior to final field deployment. The process for incorporating air traffic controllers in the development of FAST allowed the FAST engineers to achieve a greater understanding and appreciation of the overall task of air traffic control, and helped them to create the most effective algorithms for generating efficient sequences and runway assignments. This process also allowed the controllers to become involved in understanding the software and engineering constraints under which the development occurred and helped provide focus in their feedback to the developers. Such a cooperative design approach was mutually beneficial to the controller and the engineer as both parties were able to satisfactorily understand each others' needs throughout development.

Early involvement of the air traffic controller was also extremely valuable for addressing human factors issues. The early and continuous involvement of controllers allowed interface and suitability issues to be tested and implemented throughout development. Issues such as information extraction, the assessment of workload, and especially, the assessment of controller acceptance, were able to be accommodated throughout development when changes could be made easily. Measurements taken by instruments such as the CARS have provided valuable insight into acceptance issues and also provided direction to the controllers on how to evaluate the system in the absence of outside guidelines.

By implementing the process described in this paper, the controllers have become more than just end-users awaiting a product. They are actively a part of the development of FAST and they clearly have demonstrated their enthusiasm for its success. Their active participation has contributed to the FAA's continued awareness and support of CTAS development. In addition, the controllers have become strong advocates of FAST in their facility, working to promote FAST as a beneficial addition to the controller's tools. Once the controllers themselves were satisfied that FAST was going to benefit the facility, and was not intended to replace the controller, they were able to provide reassurance to the facility that the concerns over job satisfaction and workload were being taken into consideration. Clearly, the input of controllers from the deployment site in early stages of development will be critical for upcoming operational testing. The controllers provided the developers with cues towards the overall facility culture, and have helped to prepare the facility for the introduction of new technology.

5. CONCLUDING REMARKS

The FAST development was a collaborative effort between air traffic controllers, developers, and human factors specialists. The development proceeded with increasing levels of expertise and complexity. The concept was first developed by engineers, then tested with retired controllers to gain early data on controller acceptability. Once this initial phase demonstrated benefits both to the controller and the air traffic control system, expert controllers from the initial deployment facility were introduced to gain specific expertise for the FAST system. These expert controllers drove the system development toward a usable system for their facility through real-time simulation and by shadowing live traffic operations with the FAST system. Finally, after a high degree of system fidelity was reached, a controller assessment team for the initial deployment site was brought into the final phases of development in order to prepare for an operational test. The assessment team

controllers contributed to the final tuning of the system through real-time simulation and through the use of a new Controller Acceptance Rating Scale.

The complexity in developing automation aids for ATC is due to more than the complexity of modeling ATC strategies. The successful introduction of automation aids into the air traffic control environment must consider the impact on the air traffic controller in the form of changes to controlling strategies, workload, and job satisfaction. Early controller involvement helps to identify such concerns at a stage when changes to the software and education for the controller and developer can easily take place. In addition, early controller involvement helps to increase acceptance, which is the key measure of the software's success.

If the controllers' input is delayed until the end of the software development process, their sense of involvement is diminished and their operational concerns are likely to be ignored because of the inability to accommodate changes at later stages of development. In addition, preventing controllers from making inputs until the later stages of development assumes that the developers have anticipated all problems and that the facility doesn't need any time to grow accustomed to potentially large changes to the nature of its work.

By using a collaborative development approach, the FAST development team successfully integrated controllers into the development process. The controllers became integral to the development as much for their ATC expertise as for their input into directing the introduction of automation into the field. The support of the controllers helped focus the development of FAST and the controllers themselves have been instrumental in the push towards field testing. This motivation and the direction provided by the controllers have helped the software become a tool that the controller will find useful, beneficial to the overall operations, and which will help them to meet the increasing demands of the ATC environment. The controllers' cooperative participation with the engineers has become key to the successful development of automation for the air traffic control system.

REFERENCES

Cooper, G.E. and R.P. Harper (1969). The use of pilot rating in the evaluation of aircraft handling qualities. *NASA TN D-5153*. Washington, D.C.

Credeur, L., W.R. Capron, G.W. Lohr, D.J. Crawford, D.A. Tang and W.G. Rodgers (1993). Final-Approach Spacing Aids (FASA) Evaluation for Terminal-Area, Time-Based Air Traffic Control, *NASA TP-3399*.

Davis, T.J., H. Erzberger, S.M. Green and W. Nedell (1991). Design and evaluation of an air traffic control final approach spacing tool, *J. of Guidance Control and Dynamics*, **14**, 848-854.

Davis, T.J., K.J. Krzeczowski and C. Bergh (1994). The final approach spacing tool, In: *Proceedings of the 13th IFAC Symposium on Automatic Control in Aerospace-Aerospace Control '94, Palo Alto, California*, pp. 70-76.

Erzberger, H., T.J. Davis and S.M. Green (1993). Design of Center-TRACON automation system. In: *Proceedings of the AGARD Guidance and Control Panel 56th Symposium on Machine Intelligence in Air Traffic Management, Berlin, Germany, 1993*, pp. 11-1-11-12.

Garcia, J. (1990). MAESTRO - A metering and spacing tool. In: *Proceedings of the 1990 American Control Conference, San Diego, California*, pp. 501-507.

Harper, R.P. and G.E. Cooper (1986). Handling qualities and pilot evaluation. *J. of Guidance*, **9**, 515-529.

Harwood, K. and B.D. Sanford (1993). Evaluation in context: ATC automation in the field. In: *Human factors certification of advanced aviation technologies* (J.A. Wise, V.D. Hopkin, and P. Stager, Eds.), pp. 247-262. Embry-Riddle Aeronautical University Press, Daytona Beach FL.

Lee, K.K., W.S. Pawlak, B.D. Sanford and R.A. Slattery (1995). Improved navigational technology and air traffic control: A description of controller coordination and workload. In: *Proceedings of the Eighth International Symposium on Aviation Psychology, Columbus, OH*.

Martin, D.A. and F.M. Willet (1968). *Development and application of a terminal spacing system*. Federal Aviation Administration, Rept. NA-68-25 (RD-68-16), Aug. 1968.

Sanford, B.D., K. Harwood and K.K. Lee (1995). Tailoring advanced technologies for air traffic control: The importance of the development process. In: *Proceedings of the Eighth International Symposium on Aviation Psychology, Columbus, OH*.

Slattery, R.A., K.K. Lee and B.D. Sanford (1995). Effects of ATC automation on precision approaches to closely spaced parallel runways. In: *Proceedings of the 1995 AIAA Guidance, Navigation, and Control Conference, August, 1995*.

Small, D. (1994). *Lessons learned: Human factors in the AAS procurement*. MP 94W000088. MITRE, McLean, Virginia.

Volckers, U. (1990). Arrival planning and sequencing with COMPAS-OP at the Frankfurt ATC-Center. In: *Proceedings of the 1990 American Control Conference, San Diego, California*, pp. 496-501.

HUMAN FACTORS IMPLICATIONS OF DATA LINK RESEARCH FOR ATC/FLIGHT DECK INTEGRATION

K. Kerns

The MITRE Coporation, McLean, VA 22102, U.S.A.

Abstract: Operationally-oriented research on data link was reviewed as a source of evidence on how human characteristics will interact with data link technology to affect the exchange of information between the ground system and the flight deck. The paper discusses the need for ATC/flight deck integration and then introduces key aspects of the integration problem. This is followed by a brief synthesis of the research findings on controller-pilot information transfer via data link. The research findings are analyzed with respect to operational and human factors issues in ATC/flight deck integration. The conclusion of the paper discusses implications of the simulation research for ATC/flight deck integration in the future system environment.

Keywords: Aircraft Operations, Air Traffic Control, Communication Channels, Communications Systems, Human Factors, Information Integration

1. INTRODUCTION

Many believe that a new integrative mechanism, a digital data communications link, can help alleviate some of the problems in the current ATC environment and more effectively couple the groundside and airside resources to support operations in the future environment. This paper examines how a data link could be used in the current and future operational environment to reduce errors and enhance information transfer and to redistribute and balance controller and pilot workload. Operationally-oriented simulation research on data link is reviewed as the primary source of evidence on how human characteristics will interact with this new technology to affect the exchange of information between the ground system and the flight deck.

1.1 Why ATC/Flight Deck Integration is Needed

Today's ATC is based on a centralized, ground-based, human intensive system with the controller having a pivotal role in planning the movement of air traffic and

transmitting instructions to carry out the plan. The role of the pilot receiving ATC services is one of processing advisory information, accepting instructions, and acting upon them (Billings and Cheaney, 1981).

As the NAS has evolved to accommodate more traffic, this role relationship between controller and pilot is contributing to growing inefficiencies in ATC system performance. These inefficiencies may be largely attributable to divergent evolution of the complex airside and groundside human-technology systems in which the controllers and pilot are embedded. The technological capabilities and procedures used by controllers and pilots embedded in the ATC and air carrier organizations have been designed to interoperate within their respective organizational systems, not between them. As an example, controllers work with a two-dimensional, plan view display of traffic that is well suited to radar separation procedures and representation of vector solutions to separation and spacing problems. In contrast, procedures and flight management systems used by airline pilots support vertical profile planning in all flight phases to manage fuel and flight schedule requirements. Consequently, vector instructions from controllers impose a high cognitive demand on pilots attempting to execute and maintain a prescribed vertical flight plan.

Additionally, the operational communications procedures controllers and pilot use to coordinate their activities and conduct operations in different environments have not been developed to solve the overlying air-ground coordination problems. Rather, they have evolved incrementally, adapting to the capabilities and limitations of the voice radio communications system. The evolution of operational communications procedures piecemeal has resulted in a simultaneous over- and under-proceduralization of system operations and an allocation of responsibilities between controllers and pilots that is sometimes unbalanced (Degani and Weiner, 1994). Over-proceduralization reduces system flexibility and opportunities for increased efficiency, while under-proceduralization increases human stress and workload because operational demands and responses are not predictable (Adam et al., 1994).

1.2 Voice Communications

In the current ATC environment, the primary integrative mechanism supporting the exchange of information between controllers and pilots is voice messages carried by radiotelephone. Notwithstanding the care that has gone into designing the language and procedures for controller-pilot communications, the voice communication system does not always operate as intended. Studies of voice communications conclude that both human factors (distraction, forgetting, failure to monitor, nonstandard procedures, and phraseology) and system factors (unavailability of traffic information, frequency congestion, and high workload) contribute to information transfer deficiencies (Billings and Cheaney, 1981). This research also emphasizes that, in the busiest operating environments, the controller's and pilot's duty priorities are most apt to conflict, interfering with effective cross checking of mutual understanding and with timely transfer and acquisition of information.

Research on information transfer problems in the aviation system generally concluded that the human tendency to fill-in information based on expectations when processing voice communications was implicated in many types of communication problems (Grayson and Billings, 1981). The expectation factor contributes to misinterpretations and inaccuracies because pilots and controllers sometimes hear what they expect to hear. This generates what have been called "readback and hearback" errors in which respectively a pilot perceives what he expected to hear in an instruction transmitted by a controller and a controller perceives what expected to hear in the readback transmitted by

the pilot. Other factors that contribute to misinterpretation include (1) phonetic similarities in which the words used in the message lead to confusion in meaning or in the identity of the intended recipient, (2) transposition errors in which the sequence of numerals within the message was inaccurate, and (3) formulation errors in which messages were based on erroneous data or resulted from erroneous judgments. In addition, transcription errors in which messages are received correctly but are recorded erroneously or entered erroneously into airborne or ground systems have been noted in related studies of navigational errors.

Especially in busy environments, the required timing of information transfers when controllers are preoccupied with higher priority tasks accounts for failures to transmit appropriate messages as well as untimely transmissions which are originated too late or too early to be useful to the recipient. For example, controllers are less likely to initiate traffic advisories during high traffic periods, precisely when the pilot's need for the information is greatest.

Analyses of tape recordings of controller-pilot communications highlight the role of nonstandard communication procedures in information transfer problems (Morrow et al., 1993; Cardosi, 1994). Although this research indicates that the actual incidence of communications errors is low, a much higher incidence of procedural deviations in which pilot do not follow standard procedures was observed. Moreover, the research also showed that errors and procedural deviations tended to increase as the complexity of the messages increased, requiring additional transmissions to correct or clarify a message.

1.3 Alternative Means of Information Transfer

Studies of voice communications also point out the urgent need for alternative means of transferring information that is now communicated exclusively by voice (Billings and Cheaney, 1981; Adam et al., 1994). Unlike voice radio, the data link communications medium can transmit coded, digital data to individual addresses. Data link system and transaction status is monitored through a built-in feedback path or protocol that automatically verifies the integrity of the message reaching the addressee and provides information to the sender concerning responses, interruptions, or failures. Once received, message data can be stored for future reference and formatted for easy access by the user. These features of data link can be expected to alleviate problems induced by user interaction with the voice radio system at nearly all stages of the communication process.

Off loading some of the voice radio message traffic onto another communication link will alleviate miscommunication problems caused by congestion. In addition, data link system capabilities such as discrete addressing of messages, preformatted messages and standard protocols, and preservation of information, can minimize radio-based problems like call sign confusion, overlapping transmissions, procedural deviations, and memory lapses (Kerns, 1991; Morrow et al., 1993).

Perhaps the greatest potential for improving information transfer via a data link will come from exploiting device-independent message standards and coding which allow flexible representation of information and direct interface of the data to automation systems. At the most basic level, standardization supports computer-aiding to simplify the human's communication subtasks such as message formulation, communication system monitoring and error checking, and message logging. At a more sophisticated level, standardization supports transfers of data between aircraft and ground automation systems to ensure common databases and consistent solutions, without requiring the human to recode and reenter information.

Counterbalancing the potential improvements in information transfer are significant design challenges. Many of the basic principles and methods of effective voice communication also apply to data link but safety measures for data link must address different albeit analogous design and procedural issues. Visual display and manual control of transmitted information adds load to the human's busy visual information processing channel. Even though research suggests that a visual display may be less prone to misinterpretation than an acoustic display, visual perception is still susceptible to the effects of expectations and errors. The potential kinds of errors in visual perception also are predictable; confusion is no longer between information which sounds like other information when spoken but between alphanumeric characters and graphical codes which look alike or are physically adjacent (Billings and Cheaney, 1981; Morrow et al., 1993; Hopkin, 1994).

A visual communications medium is also more susceptible to failure at different points in the air-ground transfer process. Data link system capabilities can be exploited to standardize the content and procedures used in the ATC/flight deck segment of the information transfer path. However, new procedures will be needed to ensure coordination within controller teams and flight crews when the communications medium is silent and less readily observable by multiple operators.

1.4 Alternative Allocation of Functions

Reducing errors and enhancing information transfer are necessary conditions for effective integration of air and ground system resources, but they are not sufficient. In as much as inefficiencies in the current ATC system stem from the pattern of the workload and the timely availability of the information required to carry out tasks in the operating environment, efficient use of resources may also entail a redistribution and a reallocation of tasks among human and automated system elements to better exploit the resident capabilities. Controller and pilot acceptance of an altered allocation of functions is a critical consideration in ATC/flight deck integration.

Early experiments with cockpit display of traffic information (CDTI) concepts (Kreifeldt, 1980) showed that the preferred mode of air traffic management was a more distributed one in which both controllers and pilots participated. In these experiments, controllers were responsible for establishing an arrival sequence of aircraft and communicating the sequence order to the pilots; the pilots managed their position in the sequence and maintained spacing from other aircraft from that point on. This research showed that distributed management resulted in reduced controller workload and, although pilots reported a higher visual workload with CDTI, they still preferred distributed management. It was also significant that increased pilot workload was more than offset by the additional information provided by the CDTI and the associated perception of increased control of the situation. In contrast, controller acceptance of the alternative allocations of functions diminished with decreasing ground centralization of control.

More recent work on the allocation of functions extends the scope of the integration problem to formally consider the next level of integration, including the role of the air carrier's aeronautical operational control center on the airside and the role of traffic flow management on the groundside. Central to this research is the notion of adapting the allocation of functions in specific flight phases and airspaces to take advantage of existing system capabilities and provide a more consistent level of service across areas that presently have markedly different procedures and technological capabilities (Sorenson et al., 1992).

2. SYNTHESIS OF DATA LINK SIMULATION STUDIES

Although an aeronautical data link system is being developed to support a broad range of digital communications, including air-ground exchanges of weather, surveillance, and navigation information, the majority of the research conducted to date has focused on the use of data link for controller-pilot communications (FAA, 1994). This section synthesizes a number of conclusions concerning the operational impact and use of data link on the basis of a review of simulation studies. A more detailed review and discussion may be found in Kerns (1994).

2.1 Reliability and Efficiency of Communications

The research consistently shows that using data link as an adjunct to voice communications improves the efficiency of pilot-controller communications by reducing the incidence of communications failures and, consequently, the number of attempts required for successful information transfer. In simulations, this effect is primarily attributable to the availability of a clearer, usable representation and a persistent, storable reference of message content; although in actual operations errors caused by noise or blocked transmissions would also be avoided as would some message formulation and transcription/data transfer errors.

2.2 Speed and Timing of Communications

User perceptions and the performance of data link indicate that its superior reliability for controller-pilot communications is generally obtained at the cost of speed in the information transfer. Delay factors associated with message generation and transmission times account for longer total transaction times with data link than with voice. The time required for message interpretation and acknowledgment is comparable for the two media, although accuracy is improved with data link. Simulation results also reveal that, within limits, controllers and pilots can effectively adapt to the added delay by performing other tasks concurrently and adjusting the timing of their communications. Because of such adaptations, execution of ATC instructions seems to take about the same amount of time, regardless of the communications medium. However, the delay factors associated with message generation appear to limit data link's utility in rapidly changing conditions while transmission delays limit its utility for time-critical instructions.

2.3 Workload

A redistribution in controller and pilot workload accompanies data link communications: visual and manual workloads increase while auditory and speech workloads decrease. This redistribution takes on operational significance in specific environments and for specific classes of information. In environments where severe frequency congestion currently exists, the controller's auditory and speech workload can reach an overload state. In such environments, data link helps to achieve more timely performance within acceptable

workload limits by providing an additional channel for message generation and transmission, especially fort repetitive messages issued to each aircraft. Conversely, the pilot's visual and manual resources are already heavily loaded in some environments. In these situations, data link increases pereceived workload and could inappropriately interrupt and disrupt visual scanning and flight management tasks.

2.4 Operational Communications Procedures

Although the data link system inherently supports greater procedural consistency in the ATC/flight deck segment of the transfer, the silent communication process also requires extra measures, such as cockpit and controller team coordination procedures, display layouts, and voice generation technology, that will ensure access and understanding of information by multiple operators.

2.5 Information Access

The research further shows that redundant display formats widen the band of information available and improve the user's access to the particular features and data that are most compatible with the mental representation of the situation and task requirements. Operationally, both analytic and holistic processing of information are combined in many of the user's tasks. The simulation research highlights some specific classes of information, such as weather, traffic, and route, that are most likely to show benefits of more efficient and accurate assimilation when presented in spatially-oriented, graphical formats.

2.6 Utilization of Human Information Processing Resources

Taken as a whole, the research indicates that data link allows controllers and pilots to devote more attention to critical communications functions such as interpreting, evaluating, and formulating messages. It does this by offering them relief from many of the overhead functions, such as repetitive message preparation, transcription of data to preserve information, and entry of data to provide input to other systems. The ability to automate these overhead functions while retaining human involvement in critical functions not only yields greater efficiency in operations by eliminating redundant transcription and data entry tasks; it also has great potential for preventing data entry errors and, when coupled with flexible display formatting, it should reduce the opportunity for errors of interpretation.

2.7 Application of Data Link in the Current Environment

According to the research, successful application of the data link to ATC/flight deck information exchanges depends on the operating environment in which the exchange occurs. Data link is generally more acceptable in less busy operational environments and flight phases, such as predeparture and en route (Kerns, 1991). However, the research further indicates that messages that can be prepared and issued in advance of final approach and landing operations should also be acceptable for data link in terminal airspace.

Apart from considerations of the operating environment, study results also recommend use of data link to enhance or replace party line as a source of weather and traffic information. The failure of controllers to reliably and consistently transfer traffic avoidance information and the difficulty pilots experience in making use of critical party line information reveals that the current voice delivery mechanism is both unreliable and inefficient. The research further shows that in terms of the mental effort and attention required to access and recode information into usable representation of the situation, digital transfers of traffic, route, and weather information promise to improve both controller and pilot situation awareness.

3. IMPLICATIONS FOR THE EVOLUTION OF ATC/FLIGHT DECK INTEGRATION

The strategic direction for the future system is called air traffic management (ATM). Recently, the FAA and the aviation industry reached consensus on a framework that will guide the redesign of the future ATM system. Central to this framework is a shift away from today's ground-based ATC operations toward a more cooperative arrangement in which users have greater freedom to select their flight paths and greater involvement in traffic and airspace management decisions. Closer cooperation will require the direct exchange of digital messages between airside and groundside automation systems to ensure common databases and consistent solutions to route planning and flight scheduling problems. A second design principle that underpins the ATM framework is the mandate to make full use of available airside and groundside resources in delivering air traffic services. Achievement of this goal implies the need to explore alternative allocations of functions, including adaptive assignment of separation assurance to the flight deck. Finally, the changes in controller-pilot role relationships and the flexibility in air-ground function allocation will require greater commonality in the information presented to controllers and pilots as they coordinate their activities.

3.1 Human Role in Air Traffic Management Communications

Data link research on air traffic management communications indicates that controllers and pilots should have final authority to approve the transfer of information to each other and to their automation systems. Consistent with this philosophy of management by approval is the notion of provisional approval wherein the human operator can delegate final approval of flight plan clearances and amendments to the automation system within the bounds of specific operating constraints and parameters. In addition, the application of management by approval should also include a range of approval options that enable the human operator to retain regular and meaningful involvement in the process. For example, human selection of specific messages for automatic transfer would be consistent with this philosophy, as would human selection of operating parameters that govern provisional approval of information transfers. Finally, use of automation to evaluate incoming messages and formulate preliminary approvals or identify potential constraint violations would constitute another variant consistent with this philosophy.

Negotiation and collaborative decision making is an essential element in future ATM communications. Consistent with the previous research are several plausible ways of conducting the negotiation in a data link communications environment. The flexibility of voice communications could be exploited to work out a mutually agreed on decision, which in turn could be converted to a data link message by the pilot or controller, depending on the situation and the available system capabilities. Additionally, data link could be used to make information on situation constraints available for common access by controllers and pilots prior to negotiation or data link could be used in concert with automation for the iterative exchange of candidate flight path modifications. Controllers and pilots would manage the data link process using the various approval options described above.

3.2 Alternative Allocations of Functions

Simulation results on data link applications confirm that while the addition of a digital communications link can address many information transfer problems, this capability alone will not be sufficient to address user acceptance issues. In particular, successful applications of data link in terminal area operations will depend on workload. Clear and unambiguous definition and operator notification of responsibilities will be needed to support shared awareness and expectations of controller and pilot action.

3.3 Convergent Evolution of ATC and Flight Deck Functionality and Information

As user participation increases and air and ground system elements focus more on solving a common problem, the separate air and ground systems will begin to more closely resemble each other (Weiner, 1988). Data link research confirms that graphical presentations of traffic and weather information on the flight deck will improve the pilot's ability to obtain, assimilate, and interpret information and also provide a common point of reference that more closely resembles the controller's situation display. Although the definitions of comprehensive and coherent situation representations for controllers and pilots may never be identical, it is reasonable to assert that there will be greater commonality in terms of the types of information represented and the most efficient formats.

ACKNOWLEDGMENTS

This work was supported by the FAA Aeronautical Data Link Program. Portions of this material were presented previously at the 8th International Symposium on Aviation Psychology, April 1995, in Columbus, Ohio. This paper is adapted from Kerns (1994), which will also appear in D. Garland, J. Wise, and V. D. Hopkin (Eds.), *Aviation Human Factors*, Lawrence Erlbaum Associates Inc.

REFERENCES

Adam, G.L., D.R. Kelley, and J.G. Steinbacher (1994). *Reports by Airline Pilots on Airport Surface Operations: Part 1. Identified Problems and Proposed Solutions for Surface Navigation and Communications.* MTR 94W60, The MITRE Corporation, McLean, VA.

Billings, C.E. and E.S. Cheaney (1981). *Information Transfer Problems in the Aviation System.* NASA Technical Paper 1875. NASA Ames Research Center, Moffett Field, CA.

Cardosi, K. (1994). *An Analysis of Tower (Local) Controller-Pilot Voice Communications.* Report No. DOT/FAA/RD-94/15. U.S. Department of Transportation, Federal Aviation Administration, Washington, D.C.

Degani, A., and E. Weiner (1994). *On the Design of Flight Deck Procedures.* NASA Contractor Report 177642. NASA Ames Research Center, Moffett Field, CA.

Federal Aviation Administration (1994b). *The Aeronautical Data Link System Operational Concept.* U.S. Department of Transportation, Federal Aviation Administration, Washington D.C.

Grayson, R.L., and C.E. Billings (1981). Information Transfer between Air Traffic Control and Aircraft: Communication Problems in Flight Operations. In C.E. Billings and E.S. Cheaney (Eds.), *Information Transfer Problems in the Aviation System.* NASA Technical Paper 1875. NASA Ames Research Center, Moffett Field, CA.

Hopkin, D.V. (1994). *Human Performance Implications of Air Traffic Control Automation.* In M. Mouloua and R. Parasuraman (Eds.) *Human Performance in Automated Systems: Current Research and Trends* (pp. 314–319). Lawrence Erlbaum, Hillsdale, NJ.

Kerns, K. (1991). Data Link Communication between Controllers and Pilots: A Review and Synthesis of the Simulation Literature. *The International Journal of Aviation Psychology, 1* (3), 181–204.

Kerns, K. (1994). *Human Factors in ATC/Flight Deck Integration: Implications of Data Link Simulation Research.* MITRE Paper MP 94W98. The MITRE Corporation, McLean, VA.

Kreifeldt, J.G. (1980). *Cockpit Displayed Traffic Information and Distributed Management in Air Traffic Control. Human Factors, 22* (6), 671–691.

Morrow, D., A. Lee, and M. Rodvold (1993). *Analysis of Problems in Routine Controller-Pilot Communications. The International Journal of Aviation Psychology, 3* (4), 285–302.

Sorenson et al. (1992). *Opportunities for Integrating the Aircraft FMS with the Future Air Traffic Management System. 37th Annual Air Traffic Control Association conference Proceedings* (pp. 534–543). ATCA Inc., Arlington, VA.

Wiener, E.L. (1988). *Cockpit Automation.* In E.L. Weiner and D.C. Nagel (Eds.), *Human Factors in Aviation* (pp. 433–459). Academic Press, Inc., San Diego, CA.

ASSISTED DIRECT-MANIPULATIVE PILOT-FMS INTERACTION
- BRINGING IN PILOTS' AND CONTROLLERS' EXPERT KNOWLEDGE

S. ROMAHN*, R. HECKHAUSEN and G. WERDÜN

*University of Technology, Aachen, Institute of Technical Computer Science
D-52074 Aachen, Germany. E-mail: romahn@rwth-aachen.de

Abstract. Programming the flight management system (FMS) of modern commercial aircraft is observed and reported by pilots to be a demanding and time-consuming process. In order to facilitate pilot-FMS interaction a direct-manipulative user interface has been developed. Furthermore, pilots are assisted when programming the FMS by taking advantage of the knowhow controllers and experienced pilots have collected during their work. This expert knowledge has been transferred to rules and stored in a knowledge base. An inference engine processes the knowledge and guides the pilots on their way to complete a started task or derives operation suggestions.

Key Words. Flight control; navigation; graphic displays; expert systems; rules; human-machine interface; man/machine interaction

1. INTRODUCTION

Automation has become an indispensible feature in the cockpit of modern aircraft. It plays an important role in maintaining a high safety standard, easing pilots' workload and reducing flight costs. But an increasing degree of sophistication of cockpit equipment has meant an increasing complexity. Accordingly, a high quality of user interface is of growing importance, especially since the need to monitor and supervise automated systems has added a new aspect to the pilots' working environment (Billings, 1991).

Almost traditionally the improvement of user interfaces and the examination of human-machine interaction is lagging behind the technical advancement. One particular instance is the flight management system with its user interface, the control and display unit (CDU), which remains almost completely unchanged since its first installation. The difficulties pilots face when operating the flight management system are well known from many collected pilot statements, incident reports and surveys (eg. SAE-Commitee (1991); Sarter and Woods (1991); Wickens (1994)). The criticism addresses basic ergonomics of the user interface layout or the location in the cockpit as well as qualities like expectation confirmity or system overview (Sarter and Woods, 1992).

This situation is further complicated by the nature of the FMS, which like most automation, tends to reduce workload in quiet phases of the flight and to increase complexity of tasks in time-critical phases (Wiener (1988)). This is particularly true of the approach and landing phase of a flight.

The most obvious solution is to improve the user interface so as to facilitate pilot-FMS interaction. This allows faster access to FMS functionality and enhances situation awareness. An even greater benefit may be expected if the need to reprogram the FMS can be shifted from busy into less demanding phases of the flight.

Our work approaches this problem by giving pilots the possibility to make use of the knowhow experienced pilots and controllers have collected. Pilots report that they, after a number of flights on the same route, can foresee what air traffic control (ATC) directives are likely to come about. With this knowhow they program the expected changes into the FMS in advance, eg. during the en-route phase when workload is low. Inexperienced pilots have to await the ATC clearances and must instantly react. For this reason they often do not program these flight plan changes into the FMS but conduct the manoeuvres either manually or use the autopilot. The advantages of FMS control – effectiveness, economy, exact time calculations etc – are lost.

237

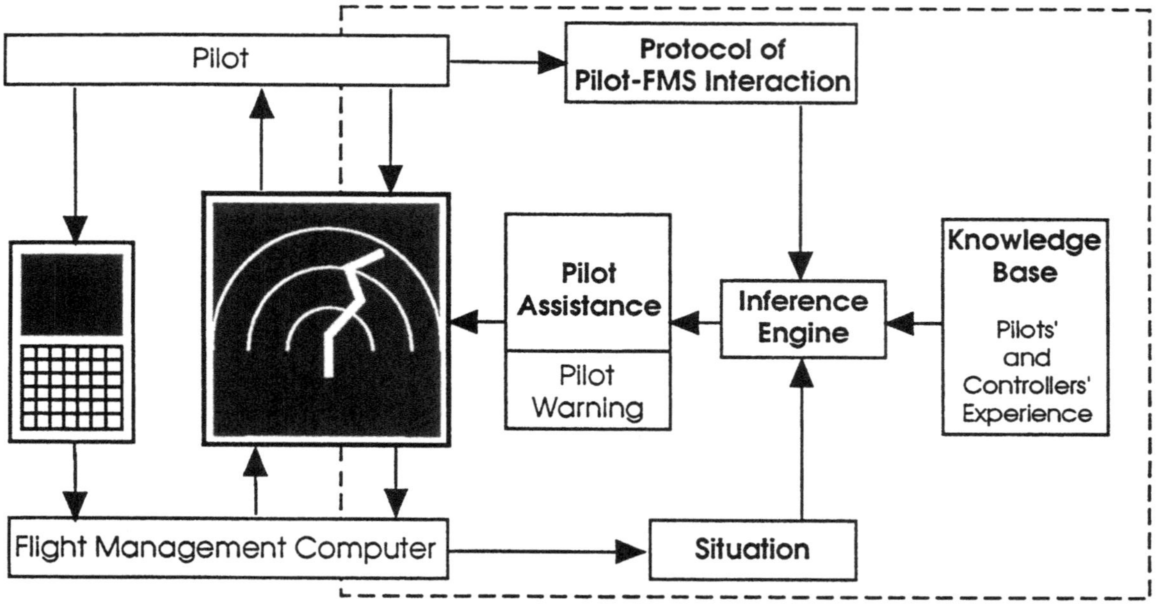

Fig. 1. Architecture of the assisted direct manipulative pilot-FMS interface

2. SYSTEM ARCHITECTURE

This section describes briefly the architecture of our system, as it is depicted in figure 1. It consists of a graphic pilot-FMS interface and a knowledge-based system that is attached to the user interface for pilot support.

The arrows on the left-hand side, outside the dashed box, sketch how the FMS is operated today. Alterations of the flight plan are programmed into the flight management computer through the CDU. Flight data can be reviewed on a navigation display.

Making use of this navigation display, a graphic user interface has been developed which allows direct manipulation of the flight plan. Modifications of the flight plan can be performed easily using a pointer instrument such as a trackball. The system has several tools for entering numerical data. Since all FMS functionality can be accessed via the graphic user interface, an extra input device - such as the CDU - is no longer required. One major advantage of this is that pilots do not have to divide their attention between two different cockpit locations, the glareshield where the navigation display is located, and the pedestal where the CDU is situated. The user interface is described in more detail in section 3.

For further enhancement of pilot-FMS interaction, an assistance system has been attached to the graphic user interface. Implemented as a knowledge-based system it observes pilot-FMS interaction and tries to recognize the pilot's inten-

tion. Two modules take the outcome of the inference and use it for further processing:

Pilot Assistance *SmartInteraction* is the module which assists pilots in reaching their goal by guiding them step-by-step through the operation. Furthermore, when *SmartInteraction* reasons that a flight plan modification is advantageous in a certain situation, it can suggest an operation.

Pilot Warning *SmartTranscript* checks the pilots' operations for incorrect sequences and gives feedback about detected errors for warning.

This paper focusses on the module for pilot assistance, *SmartInteraction*. *SmartTranscript* is mentioned here only in passing. Further details are available in Romahn and Schäfer (1995).

3. THE GRAPHIC PILOT-FMS INTERFACE

This section describes the layout of the FMS graphic user interface and gives an impression about how it is operated.

3.1. *Layout*

Figure 2 depicts the appearance of the graphic user interface of the FMS. The largest part of the screen is covered by the map area in which pilots can review the flight plan and monitor flight progress. Two display modes of the present navigation screen have been implemented, the MAP-mode and the PLAN-mode. In MAP-mode, the screen shows a north-up picture of the flight plan.

If the PLAN-mode is engaged, as in figure 2, pilots are given an aircraft-centered view onto the scene. The aircraft's present track angle defines the screen orientation. Zooming and scrolling the map section is provided in both modes. The area to the left of the map is used to display information about single flight plan objects or map objects on request. The bottom area offers a number of functions to manipulate the flight plan. The status line, the top area of the screen, displays general information about the flight or the FMS condition. The aircraft's callsign (here: LH 5555) is always the first item of the status line, followed by data about active or armed autoflight modes and about accurancy of the navigation data.

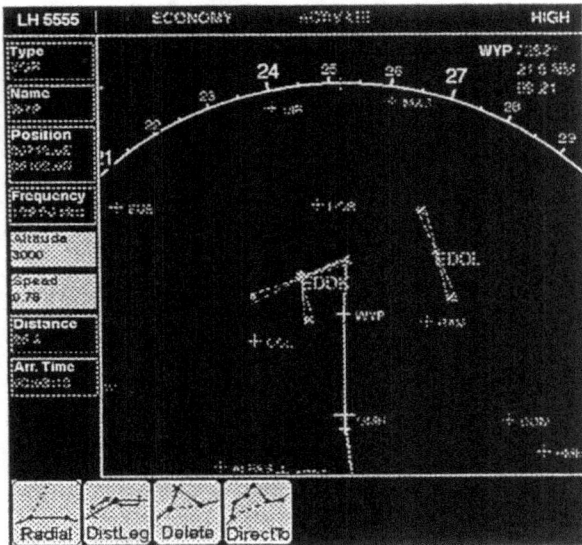

Fig. 2. FMS graphic user interface

3.2. *Retrieving information about map objects*

The plot of the flight plan and the environment is assembled from objects. These objects can represent real items as airports or radio stations as well as abstract items such as defined waypoints or flight legs.

Objects on the screen can have several states which are indicated by different colors:

Grey The *default* state.

White The *selected* state. An object can be selected by moving the cursor onto the object and successively pressing the selection button of the pointer device.

Green The *marked* state, which is entered when the cursor is moved into the vicinity of one object. This helps pilots to select the right one if objects are very close.

Yellow The *proposed* state, which appears when the assistance system suggests this object for selection.

If an object has been selected, several boxes appear inside the information area. The first two boxes always display the type and name of the object. The remaining information boxes vary with different objects. In figure 3 the information boxes of a waypoint (Wipper, WYP) are depicted. The first four boxes contain the data that describe the waypoint, the remaining ones are variable. Data that can be altered by the pilot (such as the altitude or the speed at this waypoint) are marked by a lightgrey background color of the associated boxes instead of the usual black background. Dynamic data, as the current distance to the waypoint or the arrival time, are updated with flight progress.

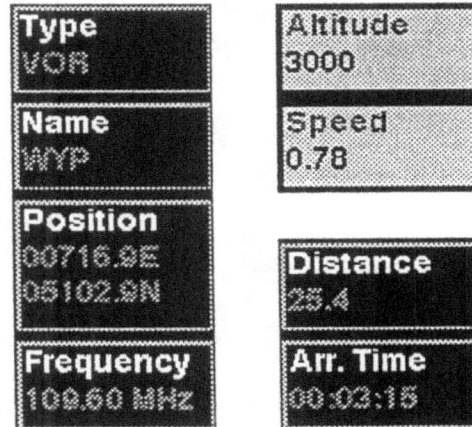

Fig. 3. Information about map objects

3.3. *Entering data*

When pilots intend to enter numerical data, eg. to apply an altitude constraint, they can access the appropriate tool by selecting the corresponding data box. Figure 4 depicts the *altitude/flight level*-tool.

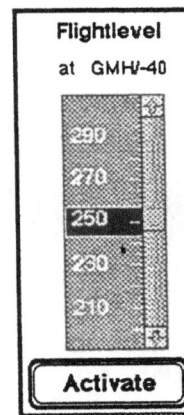

Fig. 4. Tool for entering altitude constraints

The altitude can be altered with a slider bar or in defined steps for precise adjustment. All numerical tools usually open with the current value preselected. The current value is displayed in green in contrast to the white color of selectable values with only one exeption: if the assistance system suggests a value, then the tool turns up with the suggested value preselected. The suggestion is diplayed in yellow whereas the current value remains green.

3.4. *Manipulating the flight plan*

Figure 5 shows two of the function buttons that can be used to manipulate the flight plan. The function buttons appear in the bottom area of the screen whenever an object is selected.

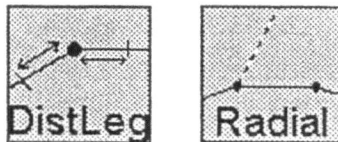

Fig. 5. Sample functions for flight plan modification

The allocation of functions to objects allows a situation adaptive presentation of the functions, so not all function buttons have to be present at all times.

Due to the limited space, not all functions can be explained here. The *Distance-on-leg*-function and the *Radial*-function shall give an impression.

The *Distance-on-leg*-function creates a defined waypoint on the flight-leg to or from the selected waypoint, NOR in figure 6. The distance to the selected waypoint is shown at the cursor position and can be adjusted using the cursor, which in this case can only be moved along the flight legs. Negative values indicate that the defined waypoint will be created prior to the selected one.

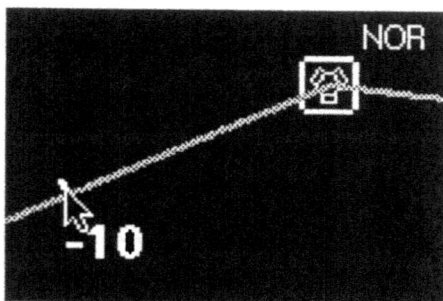

Fig. 6. Creating a defined waypoint

The *Radial*-function opens a tool to create a line on the screen starting from the selected waypoint, OSN in the example shown in figure 7. The line follows the cursor movement, the current angle is again displayed at the cursor position. Usually the *Radial*-tool opens at the 0 degree (north-)position unless the assistance system suggests a radial. The two marks (thick lines of red color on the screen) on the compass-rose indicate the extension of the flight-legs to and from the selected waypoint.

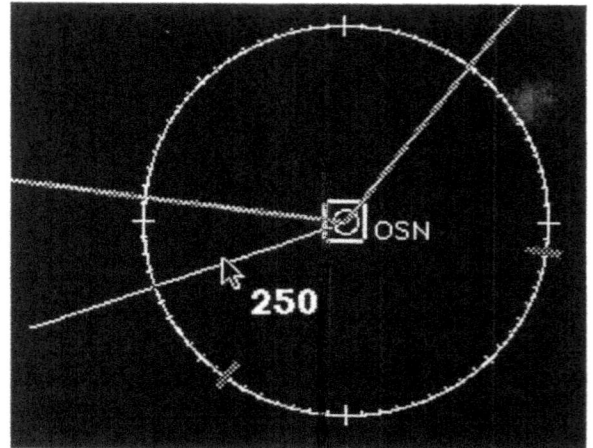

Fig. 7. Creating A Radial

4. PILOT ASSISTANCE

SmartInteraction is the module of the system that assists the pilot who is operating the flight management system. It shall enable the pilot to perform every task with the minimum effort and within the shortest possible time. Furthermore, whenever possible, it shall release the pilot from the need to operate the FMS in those phases of the flight in which workload is very high, and allows the pilot to perform the task in an earlier quieter phase. This section introduces the fundamentals of *SmartInteraction* and its implementation.

When the pilot operates the FMS, every step is temporarily recorded. *SmartInteraction* takes this protocol and constantly attemps to identify the pilot's intention. If it succeeds in finding a possible goal, it looks up the procedure that is to be executed to reach this goal. The pilot is then guided step by step towards his or her goal. If the pilot deviates from the suggested path and violates permissable tolerances, guidance is instantly terminated, and plan recognition is restarted. If the assumed intention matches the actual goal, then the overall time for task completetion is reduced, not only because of an accelerated interaction but also because pilots execute the optimal procedure.

In additon, *SmartInteraction* can identify situations in which a certain FMS operation is appropriate or helpful. It then suggests this operation to the pilot, and, if the suggestion is accepted, again guides the pilot through the execution of the associated procedure.

Our intention is to offer a step-by-step-guidance to the pilots instead of offering complete procedures for choice. One reason for this is, that almost all procedures are not static but have parameters that may or must be defined. Even more important to us is to ensure that pilots are always aware about what the automatic is about to do. This can only be guaranteed if pilots enter every step of a complex flight plan modification themselves. Further positive effects are, that pilots can abort FMS operation after each step, and that pilots will not lose their ability to program the FMS.

SmartInteraction has been implemented as a knowledge based system. Separating the domain knowledge from the inference engine allows an easy maintenance of knowledge. Furthermore, pilots and controllers usually formulate their experiences in if-then clauses, so the rule-based architecture of the system is particularly suitable for the process of knowledge acquisition.

4.1. *Knowledge base and knowledge acquisition*

As a first component the knowledge base of *SmartInteraction* contains the description of possible FMS functions and associated procedures. This knowledge has been taken from operating manuals and is stored as *rules*. These rules form the main body of the knowledge base and are always available to the system.

With respect to different origin and destination airports as well as varying flight routes, further knowledge bases are loaded by the system before the flight. These keep the knowledge about how the flight will usually be controlled by the ATC. Typical transition points and altitudes between ATC centers for instance are included, or the description of shortcuts of flight routes, that are usually offered or cleared to pilots if the density of traffic is low. This knowledge could be gained from interviews with controllers and from the analysis of recorded flight trajectories.

A third kind of knowledge base that is also loaded pre-flight contains the knowhow pilots have collected concerning FMS usage. It decribes at which point of the flight experienced pilots use to perform helpful FMS operations.

4.2. *Inference and engagement with the user interface*

After the inference engine has determined a pilot's goal or has reasoned what might be a helpful operation, it then informs the pilot about the next approriate step. As described below, *SmartInteraction* has several ways of engaging the user interface available. The yellow color is reserved for this engagement, which means that whatever is displayed in yellow on the screen is an outcome of the inference and a suggestion of *SmartInteraction*.

Instruct/Inform *SmartInteraction* can suggest an action by printing the corresponding instruction onto the screen. In the example in figure 8 it was reasoned that selecting the function *Create Radial*, the left-most of the function-buttons, is the next step appropriate.

Fig. 8. *Instructing* the pilot

It is also possible to use this means to inform the pilot, for instance about an ATC frequency to be expected.

Propose An object of the map or a function can be suggested for selection by highlighting (ie. changing its color to yellow) the referenced item.

Preselect Some functions or tools to enter numerical data are opened with an initial parameter value. *SmartInteraction* can open these tools with a proposed value preselected.

Position The topmost level of engagement is *SmartInteraction*'s ability to position the cursor anywhere (esp. onto an object) on the screen itself.

Of course, not every way of engagement is equally suitable for each suggested action. Moreover it has to be carefully researched what means of engagement pilots accept.

4.3. *Applications*

As one possible application, the knowledge bases for a domestic flight to a German airport, Cologne, were created and implemented. Controllers and pilots reported that it is very common to shorten the standard arrival routes to Cologne and to vector the airplane towards the runway so as to save time and fuel. Since these ATC clearances are given usually at a very late point, pilots tend to reject reprogramming the FMS in favor of performing the manoeuvres either manually or using the autopilot. The problem with this is, that in this case the programmed and the actual arrival route can differ drastically in terms of the track miles to the threshold. Since the track miles are the basic data for the FMS to compute the descent profile, pilots have to quit the vertical autoflight mode and perform – and calculate – the descent on their own. Experienced pilots help themselves by entering expected flight plan changes into the FMS in advance, ie. in the en-route phase of the flight.

5. SUMMARY

This paper has introduced a graphic user interface of the FMS. Furthermore a knowledge based system, *SmartInteraction*, has been described, that assists pilots in operating the FMS. Pilots can be guided through the execution of FMS operation procedures, and pilots can be given advice about helpful FMS operation based on the knowledge collected by expert pilots and controllers. There are three kinds of knowledge bases which can be loosely decribed as:

- 'handbook'-knowledge, which ensures a proper execution of procedures
- 'controller'-knowledge, which allows the plan recognition and
- 'expert pilot'-knowledge, which enables the system to suggest FMS operations.

REFERENCES

Billings, C.E. (1991). Human-centered automation: A concept and guidelines. Technical memorandum. NASA Ames Research Center. Moffet Field.

Romahn, S. and D. Schäfer (1995). Automated classification of pilot errors in flight management operations. In: *6th IFAC Symposium on Analysis, Design and Evaluation of Man-Machine Systems*. Boston.

SAE-Commitee (1991). FMS taskforce.

Sarter, N.D. and D.D. Woods (1991). *Pilot Interaction with Cockpit Automation I: Operational Experiences with the Flight Management System*. Department of Industrial and Systems Engeneering. The Ohio State University.

Sarter, N.D. and D.D. Woods (1992). *Pilot Interaction with Cockpit Automation II: An Experimental Study of Pilots' Model and Awareness of the Flight Management System*. Department of Industrial and Systems Engeneering. The Ohio State University.

Wickens, C.D. (1994). Designing for situation awareness and trust in automation. In: *IFAC Integrated systems engineering*. Baden-Baden. pp. 77–82.

Wiener, E.L. (1988). Cockpit automation. In: *Human Factors in Aviation* (E.L. Wiener and D.C. Nagel, Eds.). pp. 433–461. Academic Press, Inc.. San Diego.

Wiener, E.L. and D.C. Nagel (1988). *Human Factors in Aviation*. Academic Press, Inc. San Diego.

TECHNOLOGY IS THE ART OF REALISING CRAFTSMANSHIP

M J Platts

Manufacturing Engineering
University of Cambridge

Abstract: Developing people through the process of teaching technology is as old as civilisation itself and central to it. In Europe there is a very long tradition of technical education being synonymous with human education, in the broadest sense. However, technology in America arose in a different cultural context and still carries with it a different set of social assumptions.

Keywords: Cultural aspects of automation, Education, Human factors, Social impact of automation.

1. THE HISTORY OF TECHNICAL TEACHING

The origins of modern technology in Europe are the translations from Arabic of Euclid's Geometry and the Zij of Khwarizmi (which was about astronomy, but which included sine tables and introduced the use of Arabic numerals and mathematics into Europe) by Adelard of Bath in 1126, at the end of the dark ages (Cochrane, 1994). It is not simply that these translations both encouraged and enabled the building of the great cathedrals. The understanding so gained formed the core of the method of teaching used by the masons, which was still the method of teaching employed by Henry Maudsley in his famous workshops circa 1800, wherein were trained most of the next generation of mastercraftsman engineers, Whitworth, Nasmyth, Clement, Roberts, Seaward, Muir and Lewis, who were so much the heart of the industrial revolution in Britain.

This teaching wove together what was at one level a straightforward technical training - in mathematics, geometry, measurement - with explicit teaching about personal maturation, self observation, self control and self development, using the technical tools of the stonemason's trade as analogies for different aspects of the psyche which also had to be understood and brought under control to develop a well rounded human being (MacNulty, 1991). It is not surprising that masons who spent their whole lifetime building a house for God should at the same time take care to build of themselves fit people to enter it. This continued a long tradition of an interweaving of the building of the tabernacle and the development of the people who were to become Israel, which is documented in Exodus (Halevi, 1980). A similar interleaving is demonstrated in the monastic recivilising of Europe (in which the 'technology' was reading and writing) from Ireland in the 5th, 6th and 7th centuries (Cahill, 1995), and the craft tradition of the Islamic Qurmations (Robinson, 1994) and the Islamic engineers who built the Dome of the Rock in Jerusalem, whose teaching (and names even) became the core of Christian masonic teaching, via twelfth century Spain (Shah, 1971).

This catalogue - and it could have included oriental examples as well (Platts, 1994) - simply serves to establish the long tradition of viewing the training of a craftsman by a mastercraftsman as *human* training in the deepest sense, and the conduct of a mastercraftsman as therefore carrying a profound responsibility. The skill is not simply to do with a *craft* of manual dexterity, it is also to do with an *art* of seeing what is the right thing to do. Every mastercraftsman's understanding is thus a living treasure, as the Japanese would designate it, and no mastercraftsman's work is complete until he has trained the next generation of mastercraftsmen to understand and pass on what he understands.

2. ABSOLUTE MEASUREMENT

The cathedrals were built in an age before a standard measurement system was established, so at the beginning of every construction project a dimensional standard was set, often literally in stone, and titled 'The Great Measure', from which all subsequent dimensions of the cathedral were derived by geometry and proportion (Heyman, 1992). In cathedral construction the proportions themselves were considered to express the relationships between different stages of human understanding of life - of spiritual development - and the use of the tools of making (the gavel, the chisel and the rule), the tools of proportion (the plumb bob, the level, the square) and the tools of design (the marker, the line, dividers) became training tools in analogy in developing the soul - the moral capacity - of people who knew at that time that they were training to be the craftsmen and mastercraftsmen of *civilisation*.

In an exactly similar way, in his workshops in 1800, Henry Maudsley had a bench micrometer which could measure to a thousandth of an inch, which he called 'The Lord Chancellor' (Jaikumar, 1988) (In Britain, the Lord Chancellor is absolute head of the judiciary, in fact ranked higher than the prime minister (Gladstone, 1982)) against which his men had to measure their work, and there were true planes and squares throughout the workshop for them to work to.

3. CONSTRUCTING CIVILISATION

It helps to look at the origins of the industrial revolution, to put Henry Maudsley in context. In 1800 he is a hundred years on from Abraham Darby's development of cast iron in Coalbrookdale (Raistrick, 1989). Abraham Darby was a Quaker. Quakers were not capitalists. They did not subscribe to the idea of limited liability. They are the nearest modern equivalents (if they can be called modern, starting in the 1650's) to the Franciscan and Dominican monks of the 13th century, effectively taking vows of non-selfcentredness, meticulous honesty and absolutely honourable custodianship of anything which would grow humanity. Their industriousness was very carefully focused. By 1762 the Midland Association of Iron Masters was in being, inaugurated under the auspices of the Darbys, and with the vast majority of its members also Quakers. By 1779 the Darbys had created the first iron bridge, another Quaker, William Reynolds, was masterminding the canal system which opened up East Shropshire to manufacturing trade, and Thomas Telford was working with him (Trinder, 1991). Telford went on to inaugurate and be the first president of the Institution of Civil Engineers. The first professional engineering body in the world, called civil engineers because they were not military engineers, the Institution's Royal Charter of 1821 defines engineering as "the art of directing the great

sources of power in nature for the use and convenience of man". Note that engineering is defined as an *art*. It *uses* science, but *is* an art - and *embodies* an ethic. By 1821 also, another Quaker, Edward Pease, finally got approval to build the first public railway in the world, between Stockton and Darlington. It followed nearly a century of steady development of the elements of railway technology in Coalbrookdale, and was called "The Quaker Line" (Emden, 1939).

What is visible here is an ethic as well as a technology. Henry Maudsley was absolutely rooted in a tradition already a century old which later came to be called the Protestant Work Ethic. That title precisely captures the notion of integration of human and technical purpose in the very *process* of industry, as does the title *civil engineer*. These Quakers, and others of the same mind, knew that each and every one of them, in their own modest way, was a custodian of civilisation, and they could *only* sustain, grow and pass on that civilisation (civilisation is not an abstract entity, it is a human attribute - it is the ability to be civil) through what they *did*. They were aware that their working practices produced *people*. And it is for that long string of the master engineers of the next generation who grew under his mantle, that Maudsley is chiefly remembered. Indeed he is chiefly remembered via what they themselves wrote of him.

Nasmyth observed (1883)

> "the importance of having Standard Planes caused him to have many of them placed on the benches beside his workmen, by means of which they might at once conveniently test their work .... This art of producing absolutely plane surfaces is, I believe, a very old mechanical 'dodge'. But, as employed by Maudsley's men, it greatly contributed to the improvement of the work turned out. It was used .... whenever absolute true plane surfaces were essential to the attainment of the best results, not only in the machinery turned out, but in educating the taste of his men towards first-class workmanship".

Whitworth also observed (1856)

> "the vast importance of attending to the great elements in constructive mechanics, - namely a true plane and the power of measurement. The latter cannot be attained without the former, which is, therefore, of primary importance.... All excellence in workmanship depends on it".

Maudsley not only taught his craftsmen to be numerate, he genuinely imparted to them the significance of the numeracy itself - the whole, significant realisation that somewhere, somehow, buried in the heart of the process, the flame of Life itself flickers and warms and illuminates - and can be found, and can be cherished, and can be passed on:- *must* be found, *must* be cherished, *must* be passed on.

If the story could stop there all would be well. But, sadly it also has to include Eli Whitney.

4. OPPRESSION

The sorrow of America is Britain's fault. By 1775 an over-oppressive British government (which had by this time spent over a century trying to repress the meditative, peacegiving Quakers, and was still failing spectacularly), over-believing in the force of arms, finally drove its American colonies to despairing revolution, overdriven by a distant, insensitive government in which they had no representation, demanding more than they could give, while at the same time denying adequate support. The fire of that emotion is still visible in America today.

Sadly, that line of tradition of teaching, handed down from mastercraftsman to mastercraftsman had not been taken to America and the young America was now cut of from it - and because of continuing hostility from Britain (and France) would remain cut off from it for decades to come. Ingenuity there was a-plenty, and Whitney had it, but desperation, mixed with anger and desire for attention, is not a good fuel. The end is too important, and overrides consideration of the means.

Whitney, born in 1765, and a child in 1776 only a few miles away from the fighting, was himself emotionally deprived, his mother dying when he was 12 and his sister commenting that their stepmother, who appeared two years later, gave them no real affection or warm attention (Green, 1956). By his own efforts and ingenuity, from an early age, Eli Whitney was not only productive (more-or-less in a blacksmithing sense) but made money, and was self taught and sufficiently self-driving to achieve a place at Yale. However, while Maudsley was teaching all his craftsmen to use a micrometer in their every day work, Whitney was only able to look at a micrometer, exhibited as a wonder of modern science, in Yale's museum. While he may have read Newton's 'Principia', Whitney did not study what would today be called engineering. He was in fact hoping to become a lawyer. His studies and his ingenuity were quite separate, except that his legal training did perhaps feed into his argumentativeness about patents, and about contracts, later.

Still in search of a living, he was on his way to a tutorial appointment with a family in the south, when he was quietly captivated by the open warmth and charm of a highly respected lady, Catherine Greene, in her late thirties, the widow of General Nathaniel Greene. He never made it to the family in the south, but spent some months sojourning on her estate in 1792-3. At that time agriculture in the south was in great difficulty. They needed a commercial crop but didn't have one. Cotton was being thought of as a crop, but separating the cotton seeds from the cotton, to make it usable, was a hugely time consuming task. It is said that an offhand remark by Catherine Greene that "Oh, Mr Whitney can do anything" triggered his work. What is certainly true however is that during his stay there he produced the cotton gin (cotton engine), a hand cranked carding machine which combed the seeds out of the cotton.

5. SLAVERY

While Whitney's ingenuity was admirable, the price was too high and too desirable, both for the south in general and for him personally. For the south, the cotton gin cemented in place three generations of slavery and an attitude which a civil war did not stop. For Whitney, with the financial encouragement of Phineas Miller, Catherine's plantation manager, the desire to become both rich and powerful overcame reason and, dreaming of monopoly, the terms and cost of using the gin were set too demandingly high for the southern farmers. Legal and technical battles for the rest of his life gave him no peace and no wealth. He was even deprived of the hand of Catherine, by his financier mentor Phineas Miller.

Still with a huge ingenuity and huge self-confidence, he looked about for something else to do and alighted on the fact that America was desperately wanting to make rifles, because there was still the threat of British or French invasion.

Whitney was a man of great personal ingenuity but seriously distrustful of people. Given this, and extending it by him being entirely surrounded by innumerate people, and devoid of the history of teaching of craft skills, it is hardly surprising that he spent his entire career focusing on producing ingenious machines, guiding jigs and so on, which would ensure the accuracy of the produced artefact, without requiring any awareness, or skill, or even interest, from the hired hand who did the work.

This is the beginning of the American system of manufacture. But it is not only that, it is the beginning of the American system of capitalism, where what is important is *mine*. It is in the machines that I put together and I own it and I control it and not only is it not yours to take away, I am not even going to tell you how to do it yourself. You can come here and do what I say, and I will pay you for that, but *you* are *expendable*.

Note at this point that in the relationship of a mastercraftsman to a craftsman, the craftsman is indentured to the mastercraftsman for five years, during which time the mastercraftsman takes him into his household and is responsible not only for his welfare but for the whole of his moral and spiritual training, and that is documented in the indenture certificate and signed by both parties. It is a document about completeness of relationship for life, not in any way a documentation of limitation of relationship or responsibility.

But the logic of slavery naturally progresses into hire-fire employment policies. And Whitney's focus on the machines as being the totality of the technology, extends into the view of Frederick Taylor, and Ford, who sought to control the movements of their employees so tightly they became pseudo-machines. Here, *people* are viewed as a *problem*.

6. TECHNO-LOGOS

These different meanings of technology (*literally*, from the Greek word components, *the knowledge of how to make things*) do not come from random influences, they come as part and parcel of two totally different world views. One of these is self-insecure and thus self-centred, placing me in between you and what you need, so that I have your attention, by control. In contrast to Whitney's desire for control, many of those Quakers didn't patent anything, and even when they did, they held the patents lightly. In the line of tradition of mastercraftsmen it is understood that technology - the *craft* of knowing *how* to make things and the *art* of knowing *what* to make - is not only a public good, it is a *central* public good. It is one of the cornerstones of civilisation itself, and for individuals just as much as for communities, it is one of the keystones of self respect. You do not pull it away and you never deny access to it. Nor, when you receive it, do you deny where it came from, or forget how it came.

You honour it.

Seek a blessing on it by exercising it rightly.

And pass it on.

REFERENCES

Cahill T. (1995) *How the Irish Saved Civilisation.* Hodder & Stoughton, London.

Cochrane, L. (1994) *Adelard of Bath - the First English Scientist.* British Museum Press, London.

Emden, P. H (1939) *Quakers in Commerce.* Samson Low, Marston, London.

Gladstone, F. (1982) *Charity, Law and Social Justice.* Bedford Square Press, London.

Green C. McL. (1956) *Eli Whitney and the Birth of American Technology* Little, Brown. Boston.

Halevi, Z'Ev Ben Shimon. (1980) *Kabbalah and Exodus.* Rider, London.

Heyman, J. (1992) How to design a cathedral: some fragments of the history of structural engineering. In: *Proc. Instn. Civ. Engrs., Civ. Engng.* **92.** Feb., 24-29.

Jaikumar, R. (1988) From Filing and Fitting to Flexible Manufacturing: a study in the evolution of Process Control. *Harvard Business School Working Paper 88-045*

MacNulty, W. K. (1991) *Freemasonry - A Journey through Ritual and Symbol.* Thames and Hudson, London

Nasmyth, J. (1883) *Autobiography of James Nasmyth.* London.

Platts, M. J. (1994) Confucius on Leadership. In: *Journal of Strategic Change*, **3**, 249-260

Raistrick, A. (1989) *Dynasty of Iron Founders.* Sessions Book Trust, York.

Robinson, J. J. (1994) *Dungeon, Fire and Sword - The Knights Templar in the Crusades.* Michael O'Mara Books, London.

Shah, I. (1971) *The Sufis.* Anchor, New York.

Trinder, B. (1991) *The Darbys of Coalbrookdale.* Phillimore, Chichester.

Whitworth, J. (1856) Presidential Address. *Institution of Mechanical Engineers* , London.

THE HUMAN BRAIN: A RESERVOIR OF DIVERSE FLEXIBLE STRENGTH OR CHAOTIC RAGING VIOLENCE?

By Katherine Benziger, PhD * and Sue Holmes**

* KBA, PO Box 116, Rockwall, TEXAS 75087. 214 771 3991
**Clifton Hill Cottage, Clifton Hill, BRISTOL BS8 1BN. 117 9096425

Abstract: The human brain is a complex, elegantly wired machine which is designed to help people live, negotiate reality and ultimately to thrive. How well the brain does this is a function of two very different but inter–related processes: the internal communication in which people "listen" and respond respectfully, or not, to the brain's evaluation and signals concerning what is being experienced or done; and external pressures exerted by the environment (eg climate, and by the dominant social, economic and work patterns) which differentially use or reward specific capabilities. Thus, although the human brain is designed to respond to many, many "problems" or "situations" appropriately, utilising those capabilities that match the situation – in practice current reality often leads over time to diminished flexibility.

Key words: Human brain, social impact and cultural aspects of automation.

1. THE STRUCTURE OF THE HUMAN BRAIN

For the purposes of this discussion, it is useful to view the human brain as comprised of 3 categories of neuronal structures, each with specific responsibilities and functions:

1.1 The Primal Brain

The older or deeper structures, sometimes referred to as the primal brain, are comprised of the reticular activation system which mediates arousal level; the reptilian core which manages autonomic processes and crisis response patterns; and the limbic structures which mediate emotion and facilitate the formation and accessing of memory. The processes performed by these structures are ones that are "done for us" as it were. As such, they do not involve conscious thinking.

1.2 The New Brain and Forebrain

The next set of neuronal structures form what has been referred to as the new brain and forebrain – or cortex. Significantly, this cortex is itself divided into 4 highly specialised chunks: in the back, the left and right sections of the posterior cortical convexity; in the front, the left and right frontal lobes. Each of these four areas has its own specialised processing mode that uses the information it perceives to accomplish tasks that contribute to one of four Generalised Life Tasks: Establishing and Maintaining Productive Foundations; Establishing and Maintaining Peaceful Foundations; Adapting; and Directing (eg evaluating current reality and determining the best response). More specifically:

The left side of the posterior cortical convexity sees mostly bounded shapes or masses that it: (1) labels with the words it hears and uses; and (2) grasps or handles in order to produce a product or service. Moreover, this area of the cortex specialises in SEQUENCING. As a result it excels at performing routine physical tasks that are the basis of the PRODUCTION of man's food, shelter and clothing – especially in a highly automated society. This is the area which uses divisions and lines to process information in either/or, black/white thinking. It breaks things down, step-by-step, bit-by-bit. This mode can appear objectively detached, or gain control through its use of rules, orders, structures and systems.

The right side of the posterior cortical convexity is also very immediate and concrete. However, it perceives very different elements of its environment nonetheless. Specifically, it perceives the presence or absence of HARMONIC relationships – auditory, visual and tactile or kinesthetic – in its environment. Moreover, its inbuilt specialised processing enables it to act to establish harmony and connection where it is missing. As a result, this area excels at building good will, trust, loyalty – the basis for peace, co-operation and collaboration.

In comparison to both posterior or basal areas, the frontal lobes of the forebrain are abstract and conceptual. Yet, again what they each perceive and how they process what they perceive differs dramatically.

The right frontal lobe for example perceives abstract PATTERNS and RELATIONSHIPS. Where the first mode sees largely a bounded shape (eg a person's head), the second sees the concrete relationships within spaces (eg the face, its expressions, the eyes and their expressions), this third mode perceives the abstraction or caricature of the face. In practice, the most useful abstract patterns this area perceives are what are called trends, whether in statistical data or long term macro economic activities. These trends, meaningless to the "producer" and "bonder", signal change and trigger this region to use its internal processes – the imagination – to invent a successful response to the change it has noticed: a new product, a new service, a new strategy. As a result this third specialised area is superbly suited to help people adapt to change.

The fourth area, the left frontal lobe, is structured to perceive function and functional relationships; what supports what; with what degree of tolerance; what contributes what; what stimulates what; what blocks what. Moreover, given its gift for logical analysis, it can calculate, evaluate, diagnose and prescribe very effectively. As a result, this region excels at directing, prioritising and strategising how to accomplish its own or a group's goals.

Again, all four cortical areas CAN be used consciously – they are the primary tools for "thinking and deciding". And taken together, they perceive both the detailed and

the long view of reality in a holistic manner that if LEVERAGED can lead to success.

1.3 Corpus Callosum

Finally the third set of neuronal fibres comprise what is called the corpus callosum – a bundle of 200,000 to 300,000 neurons which serves as a hard-wired communication conduit between the right and left posterior cortical convexities and the left and right frontal lobes. An interesting fact is that the female corpus callosum is almost 1/3 larger and fires significantly faster than the male corpus callosum. This fact of life accounts for women's habit of simultaneous multiple processing or their ability to track several things, tasks or ideas at one time. Possibly more important are two other facts:
1. that there are no "bridges" linking the brain's diagonal areas (ie the posterior left and frontal right; the posterior right and frontal left); and
2. that to get from back to front appears to require a person to talk or take action.

Viewed as a single system the human brain appears to be perfectly structured to succeed – in life and in business: it is able to produce a product dependably; build good will; creatively adapt as needed; and evaluate its current reality to decide what combination of functions will constellate "the best response". Truly, the human brain is a reservoir of diverse flexible strength.

2. THE DYNAMICS OF THE HUMAN BRAIN

Two very different sets of inner forces guide people to ensure success: (1) the emotional response system and (2) the brain's innate preference for one of the cortex's four modes. When people pay attention to and act on the feedback received from each of these systems, they find themselves naturally happy and naturally functioning as a "team player". When they ignore either or both of these guidance systems, they invite trouble.

– As the "seat" of the emotional response to life – the "yes" or "no" – portions of this primal brain serve as people's guide, constantly informing them about the pain (discomfort, danger, threat) or pleasure (comfort, enthusiasm, joy, ecstasy) a given environment or task creates within.

If and as people allow this feedback to guide them, follow its advice, they find themselves in healthy places, doing things that are personally rewarding and meaningful or seeking to change or transform their reality so that it will be rewarding and meaningful. Moreover, the more people are in situations which are validating, which honour and use their gifts, the more enthusiastic they become about life and others. It is the life experience known as "my cup runneth over" – similar in some way to how one feels when in love or feels oneself to be loved – from abundance or validation comes positive self-esteem and joy, and people naturally reach out and give to others.

If on the other hand people cut themselves off from the brain (mind-body's) inner communication, if they ignore or override it, they are more likely to find their enthusiasm waning, and frustration, fatigue and fear growing. What is more, it is probable that people's interest in true collaboration and the ability to function in a team are significantly diminished as anger and/or fear cut them off from others.

– Additionally, although everyone can learn to use any and all of the four cortical modes, it is almost always true that one of the four modes is naturally easier for an individual to use. This natural preference for one of the four cortical modes seems to be the result of that one mode enjoying a naturally lower level of electrical resistance in comparison to the other modes. Thus, each person is more naturally interested in, motivated by and adept at using the preferred mode; as well as less interested in, less motivated by and less adept at tasks or activities utilising the other three modes. In other words, people are all naturally "biased" in what they find energising and meaningful. In other words, like the tools and machines people build and use, each person is specialised in what they do efficiently and hence well.

Significantly, from a system's perspective, this bias within each person, not only contributes to their individuality, it also establishes people as social beings who need each other. Moreover, when they use their two guidance systems in a context that recognises, values and uses the contributions of all modes, people are naturally energetic about life and work, as well as accepting, helpful and compassionate to the people who surround them.

3. HABITUATION OF THE BRAIN AND CULTURAL DIFFERENCES IN RESPONSE TO CLIMATE

In practice, however, the ability to access and use the full range of the brain's diversified strength is often limited by historic and contextual factors, such as climate. In the North central areas of Europe, for example, where there is land to grow crops, but there is also a hard winter, using the posterior left to plant and harvest in a dependable manner is critical, as is using the frontal left to evaluate the environment to know when each task must be done, to ensure sufficient food for everyone during the winter.

As a result, it can be argued the Germanic or Allamanic cultures have over time rewarded those who develop and use both left modes – thereby shaping a culture, known for its focus on and valuing of precision, dependability and order.

By contrast, the Mediterranean Cultures, which have the land to grow crops as well as a year round growing season, have no such left tilt. In fact, one might argue that over time their environment has led to the lack of appreciation for and deficiency in more left brained skills or competences, such that most Mediterranean Cultures appear to lead with one or both right modes.

4. HABITUATION OF THE BRAIN ALSO OCCURS IN RESPONSE TO A SOCIETAL COMMITMENT TO A GIVEN SET OF ECONOMIC PRACTICES

In countries which embrace technology, automation and assembly-line manufacturing, for example, a clear pattern of educating and rewarding the brain's left capabilities evolves. In fact, to be more precise, in response to over 500 years of increased technological development, culminating in 100+ years of heavy automation, the economies of Europe and the United States have shaped their respective cultures to produce a small percentage of 'real leaders' with frontal left skills and training and (as was required by their industrial infrastructure) a workforce comprising 80% of the population trained almost exclusively in posterior left skills.

Thus, the social impact of automation on the human being has been to:

– build a large foundation of PRODUCTION workers highly skilled at performing routine procedures, but only minimally skilled at critical thinking, ingenuity and collaborative harmonising;

– create a man-made environment with few opportunities or rewards for either right mode, in which fewer and fewer shapes, colours, relationships and communication patterns are easily perceived by either right mode.

On other words, automation has cut people off from 1/2 of their diverse flexible resources – perceptive capabilities and processing competencies needed to:

• achieve true practical peace, both locally and globally; and

• be creative enough to solve the global economic and environmental crises humanity's own actions have brought on.

The good news is that habituation is just that: a learned set of responses, not a genetic recasting of the human brain. The full range of diverse resources is still available.

The bad news is that the result of this habituation, in combination with having learned to disconnect from or

249

ignore both inner guidance systems which have been telling people this left bias was not healthy, has been a rise in the level of stress, fatigue and illness, especially in people with gifts or preferences in one of the right modes. Indeed, evidence suggests that the cost of adaptation for people forced by environmental factors to develop and use competencies in a mode other than their natural lead is shockingly high – enough to significantly compromise the natural 20:80 oxygen utilisation balance between the brain and body, setting the stage for illness. Moreover, surrounded by objects and environmental elements that they do not easily perceive – life becomes meaningless.

These twin horrors – exhaustion and meaninglessness – can and often do lead to increased failure of the system as it is, as well as outbreaks of anger and violence. In such contexts, the brain's powerful capabilities, its diverse flexible strength and natural enthusiasm for life and other people, deteriorate into chaotic raging violence, wars and death.

Fortunately, knowledge of how the brain works and how people have created this present reality can also help them reshape it, make it whole and balanced, and set it right.

5. LISTENING TO THE EMOTIONAL RESPONSE SYSTEM

Changing the way people do things, by listening to and respecting both their emotional responses to life and their innate need to use their gifts (or preferred mode), can transform work environments into highly energetic, exciting and rewarding places to be. It can also open doors for people to progress to full maturity, by growing and blossoming as thriving, loving individuals.

The starting point is to make sure that people's patterns of response listen to and value the emotions as signals. People need to welcome rather than deny their feelings. Emotions must be included in the thinking and decision-making processes. People are compassionate beings who care about themselves, others and the future of the world, as well as being computer-like processors of information.

On one level this "noticing and valuing" of the two inner guidance systems will be easy – for indeed, once people know what to listen for, once they understand that the tension and fatigue they are feeling or the difficulty they are experiencing in concentrating, is telling them they are doing the wrong job for their brain, telling them to change to doing something which energises and charges them, people will be surprised they lived so long ignoring the signals. So easy are they to read.

On another level, this including of emotions may be difficult for a time, in that the human brain stores up unexpressed emotion and experiences of chronic invalidation and non-use of one's preferences as not merely threatening (eg making one angry or frightened) but as overwhelmingly hopeless (eg making one depressed).

To be healthy, the energy of the emotional response has to flow. People have to express their grief, fear and anger as well as their compassion. All four emotions are essential to well-being. Each is an appropriate response to particular life experiences. Anger tells a person they are meeting resistance and activates them to "defend themselves or their family", to move through that resistance. As it is a signal preparing them to stand up to the resistance, anger adds tension to the body by preparing the muscles to act. Compassion tells a person they are experiencing their own fullness as well as seeing the humanity in another person, especially one who may not be living in the same fullness. It also activates people to reach out, to connect in love to uplift the other person. As such, compassion is healing or relaxing for the giver as well as the receiver.

By contrast, fear and grief cause people to withdraw and tend to separate them one from another. Yet each is valid. Each has a message. Fear tells a person they are meeting resistance that may be too big for them to take on directly. As such, it signals the need to check out information and possibly consider ways of getting out of the situation without damage. If a person manages to re-evaluate a frightening situation and determines it is not a danger, they will feel relief and joy. If on re-evaluating, one confirms its danger but one manages to get out of the situation unharmed, people will also feel these uplifting energising emotions. When one listens to fear and acts appropriately and successfully, one returns to the "comfort zone".

Grief, an even more overwhelming emotion, is also an important signal. It tells a person that the situation is hopeless and that they need to change the situation if possible. When people are sad and depressed, often their emotions may be telling them they are in a job or situation that shames or devalues their gifts. Here, they can act, change jobs for example, to select a job which will validate and use their gifts. When grief is from the death of a friend or family member, changing the situation is not possible – but grieving is – and with grieving comes release of the pain of loss.

Emotional responses can be very wide. Anger can be mild frustration or rage. Fear can be hesitancy or horror. Grief can be just a last look back as one says goodbye to a home one has lived in for years, or the inconsolable loss of a child's death. And compassion can range from a feeling of familiarity to being truly at one with another person.

Significantly, people learn in their families as children which emotions are "acceptable" as well as the proper or "right" way to express them. The result is that many people have learned not to listen to and not to express all or some emotions.

As demonstrated by the work of the renowned 20th century German biologist, Konrad Lorenz, many mammals express anger, fear and grief. Apes are known to nurture their young and grieve if one dies. Elephants have been seen to go out of their way to encircle the grave of a dead herd member as if in mourning. Dogs whose human or canine companion dies often lay grieving for days or weeks. Moreover, dogs, horses and dolphins have been known to act to help people in danger.

But humans are the only animals that can reflect on and talk about their emotions. Thus they can notice they are sad or frightened, find out the reason and act to change their lives. They can withdraw to consider their options and return with a new plan of action. They can also invite another person to share their views and subsequently change their own as a result of seeing another person's point of view.

If the emotions are not noted, shared, listened to and responded to in an individual's or group's decision-making process, those emotions will work against change and against decisions. Things will not go according to plan, no matter how carefully thought out the plan. Or an individual's body will go on strike, like workers who have not been consulted about changes their company wants to make.

When openness to the emotions is present, tension or anxiety can be released in tears, which includes the tears of laughter and joy as well as sadness. And, as with the situations mentioned earlier, when these are released and their messages heard, people return to their natural state of interest and enthusiasm, to the business of getting on with life and the living of it.

As part of listening to their need and society's need to utilise their preferences, people are very fortunate in the troubles they experience around them. The pressure to develop more and better new ideas and products at work and in the world, as well as the pressure to find ways to manage moral and public relations problems before they turn into expensive law suits, walk outs or violence, is so great that they are all ready to make a change for the better. All that is needed is to educate everyone that they simply need to make use of the innate, God-given gifts all employees have. This will dramatically improve their ability to respond to the interpersonal problems by including in their decision-making the advice of those with natural basal right gifts for harmonising, building trust and getting along with others. Similarly, they will significantly improve their ability to respond creatively by making use of the natural frontal right gifts for inventiveness.

6. CONCLUSION

Thus today, people face an exciting challenge, the challenge to change the way they use their brains so that they think as they have been designed to think, and do what they have been designed to do, using their own innate sensitivity and feelings to guide them as they act.

Existing corporate and government systems, that have been built under a different set of assumptions that did not want or need the contributions of either right mode or the emotional response, will need to be effectively educated so that they get behind the changes rather than resist them. However, in the end they will be won over. They will be transformed. For, not only are these changes natural and evolutionary, honouring God's and Nature's plan for each man, woman and child on earth to live a meaningful, contributing life full of joy and love – such changes will also transform corporate and government organisations so that they are more productive as a result of their employees' collaborating more readily and effectively; having more flexible and agile thinking; making better decisions, and developing more and better creative solutions to the problems people are facing all over the world today.

REFERENCES

Benziger, K (1989). *Art of Using Your Whole Brain*, KBA Publishing

Benziger, K (1990). *Building Positive Self-Esteem*, KBA Publishing

Benziger, K (1993). *Maximising Individual & Team Effectiveness*, KBA Publishing

Benziger, K (1993). *Overcoming Depression*, KBA Publishing

Chopra, D (1993) *Ageless Body, Timeless Mind*, Harmony Books

Haler, R (1988). *Cortical Glucose Metabolic Rate Correlates of Abstract Reasoning and Attention*, Intelligence, June 1988

Holmes, S. *Unclothed*, to be published

Justice, B (1987). *Who Gets Sick?* Jeremy P Tarcher Inc

Konner, M (1982). *The Tangled Wing: Biological Constraints on the Human Spirit*, Holt, Rinehart & Winston

Odent, M (1986). *Primal Brain*, Century Hutchinson

Pribram, K (1991) *Brain and Perception*, Lawrence Eribaum Associates Publishers

THE SCULPTOR AS AN ARTIST AND ENGINEER

H. Bigelmayr* (Transl. D. Brandt**)

* Feichthofstr. 100, D - 81247 Munich, Germany

*** University of Technology (RWTH), Department of Informatics in Mechanical Engineering
(HDZ/IMA), D-52068 Aachen, Germany*

Abstract--In this presentation, the author reports on his experiences as a sculptor to create and build a large wooden sculpture for public view in Munich, Germany. During the process of modelling and building the sculpture, the artist turned engineer solving genuine engineering problems. His work included designing a new building construction to obtain wheather shielding for his work area; designing new cranes to move about the parts of the sculpture for assembly etc. This paper is based on an interview with the artist.

Key-Words: Human-Centred design, Human factors, Modelling

In 1990, the Bavarian State Research Institute of Nuclear Radiation, Health and the Environment, Munich, advertised the commission to create a large sculpture to commemorate the opening of its new laboratory compound.

For this competition, I submitted the model of a corn stalk broken off at a certain height and its main part lying on the ground. The stalk itself was broken off at a height of about 10 cm. The other part formed a kind of arch which extended over a distance of about 20 cm. One of its leaves was bound around the stalk. The model was, thus, about natural size. My concept was to enlarge this model 100 times its size. It was to be built of the trunks of German oak trees so as to last several hundred years in the open air. I had modelled the sculpture first by carving it very delicately out of light wood (ash) and then by casting it in bronce for submission at the competition. The model was already designed to correspond in detail to the architecture and the setting of the laboratory buildings from the point of view of people entering the compound by car or on foot and continuing towards the buildings which would lead them around the sculpture. My vision, thus, was to integrate the following elements:
- The first view of the arch was to re-model the line of the horizon far beyond the sculpture. This horizon is characterized by the very obvious and strange shape of a man-made mountain. It contains the bulk of the rubble which was transported out of the city of Munich after its air-raid destruction 1944/45 in order to make it possible to rebuild the city.
- The overall shape of the sculpture was to clearly resemble life-forms as an obvious contrast to the geometrical-rectangular or circular shapes which characterize the laboratory buildings.
- The vertical parts of the stem and the leaves were to correspond to the two large chimneys of the (never used) nuclear research reactor which dominate the laboratory buildings.
- The sculpture was to stand on a small mound in order to be easily visible against the buildings. Trees were to be planted behind the sculpture to form a background from one aspect. Thus, the sculpture would be more visible than if it were merely seen against rows of windows. The sculpture was to be dark in colour against the green of the trees and the white of the buildings.

I made these pre-decisions while talking to the architects designing the new laboratory buildings and while these buildings were being erected. Thus, I was trying to identify with the architects and the builders as well as with the users of the laboratories: how they see the buildings and their environment; how they experience their different professional tasks and challenges taking place on this compound.

I was very surprised when I was preliminarily informed that I had won the competition. I had not expected my strange non-geometrical design to appeal to the taste of today's architects and scientists. Now I suddenly had to develop an engineering strategy to build this broken stalk with an overall length of about 30 m, its main height being nearly 7 m. Up to then, I had already created several large sculptures symbolizing different life forms: large leaves or stalks, also an ear of grain to walk around or under them. I had carved them from light wood, e.g. maple or ash trees, to be placed indoors: in a large stairwell of a government building, in an exhibition hall etc. Now I had to find about 10 large oak trees, cut them to size and link them in a way to last almost for ever in the open! Therefore I started to look for such trees even before I had the official contract. I travelled almost the whole of Germany in order to find such oak trees. Nobody, however, was willing to cut down such beautiful trees for my sculpture - quite understandably. Then the famous hurricane „Wiebke" hit Germany and threw down a whole forrest of ancient oak trees near Munich! I had exactly 1 week given to me by the forrester to disentangle the huge mess of old trees lying on the ground, and to get hold of those trunks best suited for my sculpture, because the week after, the whole forrest was to be sold in bulk to one of the biggest German saw mills and I would no longer have been allowed to remove even one trunk. At that time I did not even have a place to put these big trees, neither did I yet have the official contract. I had to risk everything if I wanted to have a chance to complete the commission.

An old uncle of mine had a farm not far from this (former) oak forest. It had been in the family for several generations. There was no direct heir when he was forced because of his ill health to leave it to go into a retirement home. When he heard my story he was at once willing to sell the farm to me for a very reasonable price - but still a fortune for me. It was about half the money I was expecting out of the commission. On Monday, I had seen the forrester; on Tuesday I had met my uncle; on Thursday I signed the contract to take over the farm, on Saturday, Sunday we got hold of the 10 huge trunks each weighing many tons, and we transported them to the old farm. Several weeks later I finally got the signed contract for the commission which allowed me to officially start on the sculpture.

I decided to use the old stable of the farm as my new workshop. I cleared the stable of all the old farm gear and of the remains of 100 years farming history. I had to pave the whole stable and to demolish one of the walls in order to integrate the pigsty for additional

length of the work area. Thus, the new sculpture would just fit in if mounted diagonally. The main problem, however, turned out to be the height of the sculpture. The horizontal ceiling of the stable was far too low. Hence, I took out the whole ceiling. In order to keep the stable from collapsing, I designed a completely new interior building construction - against all official building regulations. On each side of the stable, I introduced 3 large supports which reached from the foundations of the two side walls straight up into the roof to carry the main beam of the roof. Thus I turned the stable in its construction into a tent like structure. Only after I had finished everything I did dare to show the new design to the regional Board of Builders and Architects - and they were both shocked and excited. Since then, the building has withstood several very severe storms without showing any sign of weakness.

In the next phase of work, I had to design two special cranes for the tasks involved. One crane was to become a fourlegged gantry crane moving on four wheels. It was to pick up the trunks to enable me to move them about manually. No forklift would have been strong enough to handle these trunks within the space available on the farmyard and through the stable door. I designed this crane to have the 4 legs adjustable in length, so that it would fit in height through the frame of the stable door. It was build by a small local metal company according to my design.

The other crane was to be standing firmly secured on the floor of the stable its arm stretching across the full inner width of the stable. It was to hold and move about the full weight of the trunks to be worked on and to be linked together for the final design of the sculpture. I had a lot of arguments with the engineers of our local metal company about how to build this crane because I was fighting for every cm of height or horizontal extension to fit into the tent-like structure of the stable. After all, the trunks making up the main arch of the sculpture had to be moved upwards to a height of more than 3 m. In its final shape, the arm of this crane moved freely with just about 1 cm to spare before touching one of the main „tent" supports. The foot of the crane was buried several meters deep in concrete.

After all this preparatory engineering work I was able to start on the sculpture itself. I was originally trained as a wood sculptor creating human-size sculptures and developing further the art of creating nature-like shapes in light wood. Additionally I had developed thorough competence and experience in metal casting. Therefore I firstly made another larger model cast in bronce, as the main source of data for the construction of the sculpture. During this process I learned from viewing tests that I even had to increase the overall height of the main arch in order to correspond to my vision of the final design - which made my task even more difficult.

Now I was able to choose the parts of the 10 different trunks which came closest in shape and strength to the different parts of the sculpture - very much as the old shipbuilders did when they cut down exactly those trees in the forest which came closest to their drawings of the different parts of the ship's hull.

From my previous experience in light wood, I knew that I had to hollow out the trunks' marrow in order to make it last without cracking radially all the way through the centre of the trunk. It is easily done in ash wood - it turned out to be a major task to do it in old oak trees. With my hand-held chainsaw, I made a slit along the main length of the trunk. Through this slit, I made deep cuts into the trunk changing both angle and depth of cutting until only a few centimeters of the outer 'skin' of the trunk was left. Thus the slit just widens or narrows minutely according to whether the trunk becomes humid and wet or the wood dries out. I have never heart of anybody else using this technique - except the carvers of the famous wooden sculptures of the Middle Ages and the Renaissance, who hollowed-out their sculptures from the back and closed the opening afterwards with a matching piece of wood.

In the final design, these slits in the trunks would point towards the ground to prevent damage by rain. The joints of the trunks, however, had also to be protected against the intrusion of water. Hence, I designed a type of joint which is completely hidden from the exterior view except for one seam: One of the two trunks to be joined was cut open from underneath so as to take the bulk of the material of the other trunk. Thus the skin of the one trunk protects all the inner seams of the joint against rain. The joints were secured by both tightly - fitting wooden wedges and steel nuts and bolts. The arch was set on bolts mounted in concrete. The upright part of the stem which extends nearly 10 meters vertically, was to be sunk into a concrete bed to a depth of nearly 3 m.

In order to test the final assembly I had to prepare my stable exactly as the mound would be prepared to take up the sculpture. It was a major task of earth-moving and construction of foundations, adjusting and re-adjusting the many different parts of the sculpture and the foundations. During this time I became seriously ill and had to postpone the final assembly for nearly a year. I continued my work, however, between phases of treatment and despite my weakness - the sculpture had become a kind of obsession.

Today the sculpture stands on its mound exactly as I envisaged it. It has developed its dark „tan" exactly as I hoped, due to the impact of the weather on the old oak wood. Most viewers are impressed by its „living" quality, resembling both forms of plant life and the strange shapes of huge insects - set in an environment which has been thoroughly shaped by man.

Fig. 1: The Broken Corn Stalk - The Limits of Growth. Oak wood, length 30 m, height 7 m, Munich 1992

AUTHOR INDEX

www.ingramcontent.com/pod-product-compliance
Lightning Source LLC
Chambersburg PA
CBHW082305210326
41598CB00028B/4445